Multidisciplinary Approaches to Ethics in the Digital Era

Meliha Nurdan Taskiran
Istanbul Medipol University, Turkey

Fatih Pinarbaşi
Istanbul Medipol University, Turkey

IGI Global
PUBLISHER of TIMELY KNOWLEDGE

A volume in the Advances in Information Security, Privacy, and Ethics (AISPE) Book Series

Published in the United States of America by
IGI Global
Information Science Reference (an imprint of IGI Global)
701 E. Chocolate Avenue
Hershey PA, USA 17033
Tel: 717-533-8845
Fax: 717-533-8661
E-mail: cust@igi-global.com
Web site: http://www.igi-global.com

Library of Congress Cataloging-in-Publication Data

Names: Taskiran, Meliha Nurdan, 1960- editor. | Pinarbași, Fatih, 1991- editor.
Title: Multidisciplinary approaches to ethics in the digital era / Meliha Nurdan Taskiran and Fatih Pinarbaşi, editors.
Description: Hershey, PA : Information Science Reference, [2020] | Includes bibliographical references and index. | Summary: "This book examines the concept of ethics in the digital environment through the framework of digitalization"-- Provided by publisher.
Identifiers: LCCN 2020002820 (print) | LCCN 2020002821 (ebook) | ISBN 9781799841173 (hardcover) | ISBN 9781799853350 (paperback) | ISBN 9781799841180 (ebook)
Subjects: LCSH: Mass media--Moral and ethical aspects. | Digital media--Moral and ethical aspects.
Classification: LCC P94 .M86 2020 (print) | LCC P94 (ebook) | DDC 175--dc23
LC record available at https://lccn.loc.gov/2020002820
LC ebook record available at https://lccn.loc.gov/2020002821

This book is published in the IGI Global book series Advances in Information Security, Privacy, and Ethics (AISPE) (ISSN: 1948-9730; eISSN: 1948-9749)

British Cataloguing in Publication Data
A Cataloguing in Publication record for this book is available from the British Library.

Advances in Information Security, Privacy, and Ethics (AISPE) Book Series

Manish Gupta
State University of New York, USA

ISSN:1948-9730
EISSN:1948-9749

Mission

As digital technologies become more pervasive in everyday life and the Internet is utilized in ever increasing ways by both private and public entities, concern over digital threats becomes more prevalent.

The **Advances in Information Security, Privacy, & Ethics (AISPE) Book Series** provides cutting-edge research on the protection and misuse of information and technology across various industries and settings. Comprised of scholarly research on topics such as identity management, cryptography, system security, authentication, and data protection, this book series is ideal for reference by IT professionals, academicians, and upper-level students.

Coverage

- Internet Governance
- Telecommunications Regulations
- Security Information Management
- Data Storage of Minors
- CIA Triad of Information Security
- Privacy Issues of Social Networking
- Device Fingerprinting
- Global Privacy Concerns
- Security Classifications
- Tracking Cookies

IGI Global is currently accepting manuscripts for publication within this series. To submit a proposal for a volume in this series, please contact our Acquisition Editors at Acquisitions@igi-global.com or visit: http://www.igi-global.com/publish/.

The Advances in Information Security, Privacy, and Ethics (AISPE) Book Series (ISSN 1948-9730) is published by IGI Global, 701 E. Chocolate Avenue, Hershey, PA 17033-1240, USA, www.igi-global.com. This series is composed of titles available for purchase individually; each title is edited to be contextually exclusive from any other title within the series. For pricing and ordering information please visit http://www.igi-global.com/book-series/advances-information-security-privacy-ethics/37157. Postmaster: Send all address changes to above address.

Titles in this Series

For a list of additional titles in this series, please visit:
http://www.igi-global.com/book-series/advances-information-security-privacy-ethics/37157

Handbook of Research on Digital Transformation and Challenges to Data Security ad Privacy
Pedro Fernandes Anunciação (Polytechnic Institute of Setúbal, Portugal) Cláudio Roberto Magalhães Pessoa (Escola de Engenharia de Minas Gerais, Brazil) and George Leal Jamil (Informações em Rede Consultoria e Treinamento, Brazil)
Information Science Reference • © 2021 • 529pp • H/C (ISBN: 9781799842019) • US $285.00

Limitations and Future Applications of Quantum Cryptography
Neeraj Kumar (Babasaheb Bhimrao Ambedkar University, Lucknow, India) Alka Agrawal (Babasaheb Bhimrao Ambedkar University, Lucknow, India) Brijesh K. Chaurasia (Indian Institute of Information Technology, India) and Raees Ahmad Khan (Indian Institute of Information Technology, India)
Information Science Reference • © 2021 • 305pp • H/C (ISBN: 9781799866770) • US $225.00

Advancements in Security and Privacy Initiatives for Multimedia Images
Ashwani Kumar (Vardhaman College of Engineering, India) and Seelam Sai Satyanarayana Reddy (Vardhaman College of Engineering, India)
Information Science Reference • © 2021 • 278pp • H/C (ISBN: 9781799827955) • US $215.00

Blockchain Applications in IoT Security
Harshita Patel (Vellore Institute of Technology, India) and Ghanshyam Singh Thakur (Maulana Azad National Institute of Technology, India)
Information Science Reference • © 2021 • 275pp • H/C (ISBN: 9781799824145) • US $215.00

Real-Time and Retrospective Analyses of Cyber Security
David Anthony Bird (British Computer Society, UK)
Information Science Reference • © 2021 • 267pp • H/C (ISBN: 9781799839798) • US $195.00

Transdisciplinary Perspectives on Risk Management and Cyber Intelligence
Luisa Dall'Acqua (University of Bologna, Italy & LS TCO, Italy) and Irene Maria Gironacci (Swinburne University of Technology, Australia)
Information Science Reference • © 2021 • 273pp • H/C (ISBN: 9781799843399) • US $195.00

Handbook of Research on Cyber Crime and Information Privacy
Maria Manuela Cruz-Cunha (Polytechnic Institute of Cávado and Ave, Portugal) and Nuno Ricardo Mateus-Coelho (Polytechnic Institute of Management and Technology, Portugal)
Information Science Reference • © 2021 • 753pp • H/C (ISBN: 9781799857280) • US $425.00

IGI Global
PUBLISHER of TIMELY KNOWLEDGE

701 East Chocolate Avenue, Hershey, PA 17033, USA
Tel: 717-533-8845 x100 • Fax: 717-533-8661
E-Mail: cust@igi-global.com • www.igi-global.com

Editorial Advisory Board

Table of Contents

Section 2
Ethics in the Media and Marketing World: Post-Truth Era

Section 3
Business Ethics as a Changing Phenomenon in Its Historical Depth

Detailed Table of Contents

The digital transformation caused significant changes in marketing activities. It has become effortless to collect, process, use, and transmit personal information of consumers. However, marketers' obtaining consumers' personal data in such an easy way has raised privacy ethics. This study examines privacy ethics, which has become an important subject in marketing due to digital transformation. For this, in the first part, the emergence of the ethics issue in marketing and the theories developed for making ethical decisions are mentioned. The second part focuses on the digital transformation of marketing and the matter of privacy ethics. It is also said who is responsible for protecting the privacy and how measures can be taken. This study also includes solutions, results, and suggestions for future research.

With 3.8 billion users, social media created ethical problems as well. The Cambridge Analytica scandal has been a serious issue with data security lately. Contents not in compliance with general moral rules is another important violation of ethics. TikTok application, the fastest-rising short video-sharing website, is examined. As a result of the literature review, it is observed that TikTok application also had ethical violations issues like lack of private data safety, not sufficient precautionary system barriers for the young generation, and addiction risk. Furthermore, considering that the young people can interact with malicious users through the fake accounts and the risk of sharing their exceptional videos to have more viewers, it is suggested that the sanctions should be arranged as a deterrent in violation of the rules. Parents and young users must be educated about the risks and ethical violations of social media.

Social media users have increased to 3.5 billion including children users. Social media affects children as well. Although children's rights are protected by United Nations Convention, and advertising addressing children is not ethical, the embedded, integrated advertising in social media is more difficult for children due to their lack of advertising literacy because it is much easier to differentiate advertising in traditional methods compared to digital advertising via social media, where social media integrated advertising is used. This study aims to clarify social media and ethics concepts and how social media advertisements affect children in the digital era.

Ethics is about the behaviors that people enact on a social scale and that have consequences affecting

Foreword

The effort to reach the "summum bonum" is the responsibility that discreet people have felt on their shoulders for centuries, like the Atlas carrying the sky. Although the world is not a perfect place, it would not have been better than if there weren't anyone looking for ways to achieve absolute goodness. Maybe we owe being human to it. The book you are holding is the product of such anxiety.

Multidisciplinary Approaches to Ethics in the Digital Era was created with the great effort of two valuable academicians, Taskiran and Pinarbasi, and significant and sensitive academicians' contributions provide a useful window on ethics. The book consists of three sections: "Contemporary Ethical Approaches for Post-Moral Issues", "Ethics in the Media and Marketing World of the Post-Truth Era", and "Business Ethics as a Changing Phenomenon in its Historical Depth". The first section allocates to concepts such as "political life", "environment", "animals", "cybersecurity", and "health". The second section investigates the changing media and marketing. The third section focuses on business ethics, historical background, organizational culture, and the tourism industry.

The complications in ethics are both challenging and interesting. Professionals are working on them to provide new solutions in the face of current issues. Therefore, the "Multidisciplinary Approaches to Ethics in the Digital Era" is a valuable source that provides professionals, scholars, and students with knowledge of ethical developments. This book is a good start in that current direction.

Recep Yilmaz
Ondokuz Mayis University, Turkey

Recep Yilmaz, *Ph.D., has been working as associate professor at Ondokuz Mayis University, Public Relations and Publicity Department since August 15, 2018. He authored ten books (Handbook of Research on Narrative Interactions; Handbook of Research on Narrative Advertising; Handbook of Research on Transmedia Storytelling and Narrative Strategies; Narrative Advertising Models and Conceptualization in the Social Media Age; Traditional and Digital Advertising in 150 Questions; Handbook Research on Effective Advertising Strategies in the Social Media Age; Advertising in 99 Questions; Keywords in Advertising; Scenarios: How is it written? How are they writing?; Leadership and its Contemporary Dimensions), and he published numerous papers in several journals, conference proceedings and chapters in edited books. His major is narrative theory and advertising. He also studies on theoretical communication.*

Preface

As we gaze over the history of pure ethics or ethics as moral philosophical theories, we find that it began with ancient Greek philosophers such as Socrates, Sophists, Socratic Ecoles, Plato, Aristotle, Epicurus, Stoics. Investigations on Ethical Philosophy from the Homeric Age to the Twentieth Century, signify that ethics, as a content, has undergone absolute changes and expansions since the day it emerged, that researches made prove the idea.

It will be fruitful to look at ethics before introducing the chapters briefly; that is, readers will be more easily grasp the essentiality of today's ethical problems.

The origin of the word "ethics" goes back to the ancient Greek word "ethos" which comes from the word "Ethika". The word Ethika here means "issues related to "ethe" which is the plural form of the word "ethos". "Ethe", which is the plural form of "ethos", means "the place of a living being," the place where it always goes, where it takes shelter "if it is said in its oldest sense.

The root "Ethica" derives from "Ethos" which means temperament, character when it is "singular". The plural form of "Ethos" is "Ethe" which means space, place of refuge; traditions.

The fact that the word 'ethics', which has this meaning in its origin, is the name of a field of knowledge, a discipline of philosophy in Western languages, is based on the definition of "character" in the word Ethos used singularly. The word "ethics", which means "character" or "temperament" in terms of its origin, actually refers to a situation related to the person, a side that is unique to him. However, the plural of Ethos, ethe, also means "tradition, custom", it also refers to "the way of life of a certain group, a community". However, the word ethos refers to the character, character, nature, and ethical character of a person.

Stoics, on the other hand, regarded Ethos as the source of behaviour. Plato has called ethos "habit" in the sense of what is influential in forming the whole of human character.

Similarly, Aristotle handled and examined the types of Ethos as human characters or habits. The Latin for Greek' ethos' is 'mos', and the plural of 'mos' is 'mores'.In Latin, this word includes both custom and character meaning.

However, in ancient Greek, it had a meaning expressing the lifestyle of a community. Nevertheless, philosophers kept the word Ethos separate and independent from its sense of custom in ancient times. Then instead of the Ancient Greek word ethikos, the word moralis started to be used in Latin. Cicero was the first to use the word "moralis".

However, there is a difference between these two terms that should be considered conceptually. Although the word ethos in its plural sense is open to be understood in such a way as to express "the way of life of a group and a community", in terms of its origin, the primary meaning of the word ethics is related to the person, and this is the meaning that expresses it entirely and correctly when it comes to ethics today.

The ancient philosophers, who established and developed ethics as an entire field of philosophy, relied on a framework of understanding based on this meaning of the word.

For example, Socrates (469-399 BC), who is accepted as the philosopher who laid the foundations of ethics, said that "an unquestioned life is not worth living".

What is at stake here are the state of the person himself and his relationship with himself. A person's importance in reviewing his / her relationships with other people, the world and life, learning about himself/herself, and living with self-knowledge is emphasised.

To put it more clearly, these are how the person lives, what he does, his decisions, what he gives importance and priority in life; These are questions that take place in everyday life, such as what their value or meaning is, and which we are all related in some way. In this respect, the point where ethics arises as a field of research is the self-knowledge of the human being or the questions about what a "right" and "good" life is for human beings, and how to live a "right" and "good" life.

These are questions that almost every person can encounter at some point in life, one way or another. Being human and living means making such questioning. Although Socrates is the philosopher who laid the foundations of ethics as a field of knowledge, it is seen that the philosophers before Socrates also emphasised ethical problems and expressed essential ideas about what "right" and "good" life was. The primary and crucial thing here is the person himself and his life's value, that is, his righteous life. In this questioning, the philosopher's question is not about what "morality" is, but about correctly organising one's connection with life.

Ethics is the discipline examining qualitative situations that make an action a right action and a sub-field of philosophy. Ethics is reinforced as the essential attitude of virtuality; but it is sometimes used to emphasise the traditions, character, and morality. Scope of Morality covers what people have to follow in society; behaviour, rules, goodwill, and good manners whereas Ethics covers the study of human actions within moral rules and the set of behaviours that a professional group must respect.

Today, we live in an age where yesterday's invention developed at the speed of light, and today's story is outdated. Shortly, every innovation and content are consumed very quickly. In this process, virtual environments, which emerged as high-tech products, started to highlight many different professions. The legal gaps and deficiencies in this field were new, supporting professional groups' freedom that became apparent in virtual environments.

Apart from classical professional ethics, ethical problems arising from the intense use and sharing of social media environments require urgent measures to be taken in this area.

We have prepared this book to discuss the extent to which moral values are reflected in professions and practices in an environment where visuality and technical possibilities have progressed to this degree and the ethical situations that these new environments have created.

This book includes 17 chapters of which are directly related to Ethics in the Digital Age. All chapters discuss Ethics from a perspective, such as medicine, politics, Islamic business, business life, social media and children, environmental ethics, animal rights, etc.

The topics are covered from the perspective of different disciplines; therefore, chapters are compiled under three sections:

Section 1: Contemporary Ethical Approaches for Post-Moral Issues
Section 2: Ethics in the Media and Marketing World – Post-Truth Era

Section 3: Business Ethics as a Changing Phenomenon in Its Historical Depth

In the first section, there are six chapters, as follows:

Chapter 1, "Ethics in Political Life," explores the role of ethics in political life. It is in two parts. In part I, the relation between ethics and politics and the moral standards on which the politicians and policies framed by them should be judged is considered. It is also analysed why certain ethical norms well accepted in a society may not be appropriate for politicians due to their obligation to the public office or the nature of politics. In part II, the problem of political corruption and areas in which infraction of moral standards in political life occurs across the globe in different countries is discussed. It is also analysed whether objective norms are feasible on which politicians can be judged for their ethical values when they deviate from moral standards applicable to ordinary citizens. While focusing on this, it is argued that greater public vigilance, more vital international institutions and an ethically strong community can improve moral norms in political life. The chapter will provide the readers with a vast provision concerning w's of Political Ethics and its digital future.

Chapter 2, "Environmental Ethics: When Human Beings and Nature Are Not Two," introduces some core issues in environmental ethics (EE) from a multidisciplinary perspective. This approach's primary purpose is to make explicit the richness and complexity of the subject, where arguments coming from natural sciences, politics, philosophy, economics, or psychology often intertwine. Moreover, although it is risky, the diversity of views tries to engage readers from different fields and non-specialists. After the introduction, the chapter focuses on the characteristics of the different opinions on non-anthropocentric EE. Next, five of the main ethical frameworks are presented and followed by some specific points of view: capital in the subject (intergenerational ethics, precautionary principle, deep ecology, environmental justice, indigenous peoples and feminism). An exploration of consumption and population follows. Finally, some critical points on sustainability and human development precede the conclusion of the chapter. The author investigates and enlightens the social aspects of the environment in which we live.

Chapter 3 is "The Rise of Ethical Concern for Animals as a Major Social Phenomenon." The second half of the 20th century represented a significant rise in new ethical concerns, including, in the 1970s, the moral status of animals. Until then, animal rights had never been on the agenda of political actors. Pets have been one of our family members since earlier times. The ones at home on our laps are the lucky ones if we have to compare with the ones at laboratories, who have to expose many medical experiments and suffer for the good of science. It is just the question of the animals' rights issue that the author investigates and highlights. As a specialist, the author takes up ethics from animals' rights point of view and suggests legal solutions before it is too late. The author claims that Veterinary medicine already knew many modalities to alleviate experimental animals pain, but there had been no fulfilment in this way.

This author and others wrote American law that required effective analgesics, which the US Congress mandated in 1985. The research community eventually recognised that failure to control pain in animal research subjects involved not only immoral pain to the animals but also caused invalid research results. The increasing number of people owning companion animals also influenced their view of farm animals' moral status. The larger number of companion animal owners results from the alienation of human social relations, with animals' replacing the emotional value of human companionship. The article reminds us that we must protect and help them survive; that we have to consider the human and the other living things around us on earth. It provokes many questions and raises awareness of animal ethics for the readers concerned.

Chapter 4 is "Ethics of Security: From Personal Safety to Cyber Security." The purpose of this chapter is to identify and analyse the ethical problems of security, with some particular references to cyber and digital threats. The concepts of security and safety are defined based on existing literature. The chapter addresses the key results and research gaps in the field, i.e., security issues in different areas, and future challenges, both theoretical and empirical. Moreover, the discussion is linked to an analysis of the relationship between utilitarian ethics and deontological ethics, which brings a new perspective to the debate on security ethics in general and cybersecurity. It is a remarkable and stimulating work revealing the fine points we need to know as digital natives or digital immigrants living in the digital age.

Chapter 5, "Healthcare Digitalized: Patient-Counsellor Interaction in the Digitalized Era," deals with medical ethics constructed on a patient-counsellor relationship. With the advent of bioethics in the 1970s, patients' right to participate in decision making led to proclaim autonomy as the primary principle in clinical medicine and biomedical research, practically exercised as informed consent; and yet, the issue remains contended and poorly regulated. The author claims that healthcare digitalisation disassembles people into clouds of data and causes individual decision-making interfered with and replaced by dominant algorithms. The ethics investigated through Biomedical aspects and its development into medicalisation, marketisation contractual client/provider relationship, and neglect of personal care for the ill and frail. The chapter provides us with multidisciplinary substrata of psychosocial, existential elements of health/disease, and ethically unsatisfactory medical services replaced. In other words, it reveals the other side of the coin. This beautiful work sheds light on the points that we have always wondered about but cannot access. It is a great chance to have this here!

Chapter 6, "Health Professionals' Attitudes to Obese Patients and Ethical Problems Experienced," is another article concerning ethical problems in medical issues. As we all know, obesity is one of the health problems that threaten humanity considerably. People have the right to receive equal and just health care. However, sometimes obese individuals may suffer from health professionals' negative attitudes in providing protective health services and inpatient treatments. The chapter aims to draw attention to ethical conflicts between obese individuals and health professionals in their health care and raise awareness of these problems. If we consider obesity a disease of the digital age, we can realise how precious being tolerant the patient is. The author successfully raises our attention to the importance of sensitivity and communication with doctors in the obesity treatment process.

There are five chapters concerning ethics related to media, social media, marketing, and advertisements in the second section.

Chapter 7 is "Media Ethics: Evaluation of Television News in the Context of Media and Ethics Relationship." The authors emphasise that as technology advances and develops, new sanctions and regulations are needed, especially in audiovisual print media. Besides the laws, the audience underlines the necessity of supporting media literacy education from an early age and lifelong education processes as necessary for the digital age. The chapter is a kind of media guide with valuable information to unconscious groups of people who instantly spread the news on social media. The authors try to raise awareness for those who believe in everything they read or watch in the media.

Chapter 8 is "Advertisement Ethics in the Context of the Media Industry."

In this research, the authors discuss the mass media from the point of view of economic interest versus Corporate Social Responsibility. The authors prepare a high-quality sociological comparison of the three most crucial Slovenian print journals from the point of view of articles that contain covert advertisements. The paper indicates that it is more desirable for the media industries to profit than a socially responsible company. The results of this research provide insight into covert advertising in

Slovenian daily newspapers and show several unique features that characterise an objective picture of a daily newspaper.

Chapter 9 is "Marketing Ethics: Digital Transformation and Privacy Ethics." The digital transformation caused significant changes in marketing activities. It has become effortless to collect, process, use and transmit personal information of consumers. However, marketers' obtaining consumers' data in such an easy way has raised privacy ethics. This study examines privacy ethics, which has become an essential subject in marketing due to digital transformation. For this, in the first part, the emergence of the ethics issue in marketing and the theories developed for making ethical decisions are mentioned. The second part focuses on the digital transformation of marketing and the matter of privacy ethics. The study also includes solutions, results, and suggestions for future research.

Chapter 10 is "Social Media Ethics and Children in the Digital Era: Social Media Risks and Precautions." Social media users have increased to 3,5 billion, including children users. Social media affects children, as well. Although children's rights are protected by United Nations Convention and advertising addressing children is not ethical, the embedded, integrated advertising into social media is more difficult for children due to their lack of advertising literacy because it is much easier to differentiate advertising in traditional methods compared to digital advertising via social media, where social media integrated advertising is used. This study aims to clarify social media and ethics concepts.

Chapter 11, "Social Media Short Video- Sharing Tik Tok Application and Ethics: Data Privacy and Addiction Issues," reveals social media violations with its multi faces. With 3.8 billion users, social media has always brought ethical problems since its existence, and despite deterrent sanctions, technology continues to have gaps. The author claims that the contents are not in compliance with general moral rules which causes another essential violation of ethics—taken as a subject matter, TikTok application, the fastest-rising short video-sharing website investigated and found that TikTok application has been lacking private data safety, not sufficient precautionary system barriers for the young generation, and addiction risk. The author suggests that the sanctions should be arranged as a deterrent in violation of the rules. Parents must guide young users about the dangers and ethical violations of social media. As one of the millions of users, it is a must-read chapter informing us how to keep our rights and privacy away from the malicious people.

The third and the last section includes ethics from multidisciplinary respects: business ethics, human resources, and organizational culture and islamic tourism.

Chapter 12 is "The Historical Importance of the Concept of Business Ethics and Evaluation in terms of Employees in Human Resources." Throughout the history, philosophy has been after the search of 'summum bonum' (ultimate goodness) to direct people's behaviours in a better way and provide them with a better and happier life. As a part of Philosophy, ethics still discusses the man's behaviour within society and environments. The chapter focuses on the importance of the image in human life and its impact that has never changed or lost its actuality up to the present time. The author depicts the historical background of ethics and conducts business ethics and human resources evaluations from an ethical perspective successfully. Readers in the business world will enjoy reading him.

Chapter 13 is "Business Ethics: The Changing Face of Working Life." We all know that Ethics is about the behaviours people enact on a social scale that affect others and the thinking processes that shape them. The author claims business life is a part of social life, and it must develop principles compatible with value judgments in society. She points out that increasing professional corruption is where values are complex within the community, utilitarianism overcomes everything over time, and an ethical process is required. Digital age conditions require strong ethical sanctions. The chapter describes the need for new ethical regulations in the new business life. The author also emphasises that only an outstanding leader will achieve success by solving this ethical issue and making suggestions on ideal business conditions.

Chapter 14, "Evaluation and Importance of Business Ethics in Terms of Organizational Culture," discusses the interest in business ethics has been on the increase. The impact of economic development and globalisation are among the reasons. The author claims that the economy and the adaptation to the global market have brought unethical practices to the schedule. It is considered that models that improve the decision-making mechanisms of managers against corrupt practices are deemed necessary. Many studies have been cited as references to explain business ethics' individual and organisational factors in the study. It aims to contribute to the literature with the results and results of these models' academic research. The content is entirely dependent on economics and its associated contexts concerning ethical issues. Mostly business people and other related ones will enjoy their reading.

Chapter 15 is "Meeting Customer Expectations In Islamic Tourism: Effects of Islamic Business Ethics." Islam is a supreme religion that 23% of the world population believes. Islam has excellent creeds and virtues concerning social and domestic life. However, the author claims that the change in consumer behaviour and increasing consumer demands in many new markets have negatively affected the Muslim world. The ever-increasing Muslim population has become more affluent and more demanding with the rise of modernisation, especially after 1980's. The author describes contemporary Muslims' increased demands for better recreation and vocation facilities, necessitating establishing a new sector referred to as "Islamic tourism". Although adopted warmly, it brought about new anxieties regarding ethical issues along with itself. The study investigates non-ethical issues regarding Islamic creeds in this new sector, which is considered an alternative tourism approach. The chapter will also embrace those who have questions in mind and those interested in Islam, Islamic life, Islamic beliefs as a lifestyle.

Chapter 16, "Contrast in Ethics, Morality, Justice and Behavior: Some Reflections on Business and Islamic Ethics," is another chapter dealing with Islam within the scope of Ethics. The authors aim to analyse and apply the understanding of dichotomy of Contrast in Ethics, Morality, Justice and Behavior. Since ethics matters a lot in discharging Corporate Social Responsibility in business and trade, the research examines the application of these ethics, morality and behaviour from an Islamic perspective. Authors claim that "one must strengthen contentment, remembrance of death, the firm belief in Allah's mercy, generosity" while refraining from common contemporary business problems such as window dressing of financial results, fraud, deception, bribery to get contracts and among others. Alike with the previous chapter, comprehensive interdisciplinary research will enlighten the readers from Islamic constitutents perspective. The readers concerned are lucky because the researchers are Muslim writers and readers may have the answers they wonder.

Chapter 17, "Universal Principles of Ethics and the Future of Tourism and Hospitality Ethics," examines and discusses ethics in the tourism sector. Mainly the author investigates ethical principles determined by the World Tourism Organization that all stakeholders in the tourism sector should adopt and share. The problems caused by ethical violations are indicated, and suggestions for solutions are presented. The author makes some crucial suggestions to stop ethical issues in the tourism sector, such as a shared infrastructure based on domestic behaviours for all touristic organisations. The general scope of the chapter fosters interesting and inspiring proposals on the issue.

The book we prepared for you addresses a large scale of readers who mostly will find it erudite and develop themselves with the new scopes in the Ethics and related disciplines as well. I hope you enjoy your reading.

Acknowledgment

I want to start by thanking the dear contributors to this book. As they quickly complete every detail requested, even under pandemic conditions. Thanks to everyone on our publishing team, who contributed to our book's publication and produced successful solutions during the pandemic process, Jan Travers, the ever-patient director of IGI, and Maria Rohde.

We would like to thank our field referees, and our special thanks go to Ozlem Arda and Recep Yilmaz, our dear colleagues, who support our busy moments with their exceptional assistance.

We also would like to send sincere thanks to our colleagues Mehmet Hulusi Ekren and Enver Sait Kurtaran for their unique assistance in the promotion stage of our project.

We are grateful to our families for their unique tolerance and affection. Consequently, this book is a product of having the right team workers at the right time and right places, without the experiences and supports from our peers and team, this book would not exist. We thank you all so much, indeed!

Section 1

Contemporary Ethical Approaches for Post-Moral Issues

Chapter 1
Ethics in Political Life

Bal Krishna Chaturvedi
Earlier Planning Commission, Government of India, New Delhi, India

ABSTRACT

This chapter explores the role of ethics in political life. It is in two parts. In Part 1, the relation between ethics and politics and the moral standards on which the politicians and policies framed by them should be judged is considered. It is also analysed why certain ethical norms well accepted in a society may not be appropriate for politicians due to their obligation to the public office or the nature of politics. In Part 2, the problem of political corruption and areas in which infraction of moral standards in political life occurs across the globe in different countries is discussed. It is also analysed whether objective norms are feasible on which politicians can be judged for their ethical values when they deviate from moral standards applicable to ordinary citizens. While focussing on this, it is argued that greater public vigilance, more vital international institutions, and an ethically strong community can improve moral norms in political life.

INTRODUCTION

Politics is understood to mean “winning and using of power to govern society ” (Collins English Dictionary, 1996). It includes any activity concerned with acquisition of power. Politicians devote themselves to governance through institutions evolved for this purpose. Ethics on the other hand is concerned with moral values and norms based on which individuals or the society decide various issues and people live their life. The process of governing a state involves security and public order. It implies framing laws and rules for social and economic development which are in accordance with the ethical values of the society. As the business of the state is to look after the interest of its citizens and conduct international relations so as to best promote the country’s interest, ethical principles are often ignored in political life. The author plans to describe in this chapter the ethical concerns which influence life of a politician, the extent to which these differ from the moral values applicable to ordinary citizens, the conflict faced by politicians between requirements of governing a State and adherence to certain moral standards during the process of governance and the problems which may arise when normative ethical principles are applied during governance. It also describes how politicians have lust for power and misuse the state

DOI: 10.4018/978-1-7998-4117-3.ch001

power to enrich themselves at the cost of common man. While realising the limitations of use of moral principles in governance for reasons of State, it is suggested that stricter societal vigilance can vastly improve adherence to ethical norms by politicians.

The objective of this chapter is to explore role of ethical values in political life, nature of deviation from such values by politicians during governance due to reasons of state, and possibility of developing a more moral system of governance.

BACKGROUND

For more than two millennium societies have evolved various institutional mechanisms to provide for an orderly conduct of their affairs. Ethical principles in political life of the society had a key role in governance during this period in political systems. According to Kanglee (1972) the famous Indian sage Kautiliya, in his treatise on governance written about 350-400 BC, advised the king that in the happiness of the subjects lay the happiness of the king and what was dear to the king was not beneficial for him but what was dear to the subjects was. (p 47).In a cross country study of India, China, the Greeks and Christianity, while reviewing the political advices given by different advisors and thinkers Girardin (2012) mentions about another Indian sage Manu who advised that within his realm the king should act in accordance with rules with similar thoughts echoed in Taoism, by Confucius and later Men Zi in China. Confucius thought that exemplary virtue of king was being just, clever, sincere, patient and detached. Men Zi later advised that quality of law matters and effective rule requires not only humanity but just laws (pp 23-25). In Europe after the adoption of Christianity by the Roman kings, church had a major say in political life. Later the people presented a charter to the king in England and asserted their rights. All these civilizations laid emphasis on ruler to follow ethical norms while dealing with their subjects. In the 16th century Niccole Machiavelli an Italian political thinker strongly disagreed with this approach and propounded a theory off extreme political realism. He argued in his famous work "The Prince" that in matters of state end justifies means. He advised the king as below

For that reason, let a prince have the credit of conquering and holding his state, the means will always be considered honest, and he will always be praised by everybody, because the vulgar are always taken by what a thing seems to be and by what comes of it: and in the world there are only the vulgar,… (para 6, chapter XVIII)

He thought that tyranny, deceit or dishonest means were justified in the interest of the state. He did not see much need for moral values in handling issues concerning state

While many societies were governed by Kings for long and had advice for adherence to ethical values in their system of governance as mentioned above, there were also democratic traditions in different parts of the world since nearly 2500 years back. These systems being based on public reasoning or balloting provided a strong foundation for promoting ethical values in political life'. Greece at that time had a strong tradition of democracy. According to Adkins (1984) Greek philosopher Aristotle considered politics being for the practical good. He thought of it as an association of people where they come together for sake of an excellent life. He also felt that human beings whether by themselves or acting together as a group have common goal and hence there is no distinction between best man and best constitution (pp,29-30).

Amartya Sen (2009) has described the prevalence of democratic traditions in different parts of the word since the early Greeks set up their system. Arguing that democracy is interpreted in terms of ballots and elections or government by discussion, he talks of open public discussions in democracy in Athens and the beginning of balloting for selecting representatives in Greece. Largely under the Greek influence, he mentions balloting was used in many Asian countries. Specifically he mentions the city of Shushan in South West Iran which had an elected council, a popular assembly and magistrates who were elected by the assembly. Similarly in ancient India there was widespread practice of municipal democracy. Some of the open general meetings for discussion on various issues were held in Buddhist tradition and the first of these councils was immediately after Buddha's death. In Japan the Buddhist prince Shotoku gave a seven point charter in early seventh century which said that decisions on important matters should be taken by discussion amongst many (p 339-331).

With development of nation states, many countries accepted democratic institutions while framing their constitutions. A lot of them were influenced by the strong tradition of the French revolution, the US constitution and British tradition which was initially signalled by Magna Carta. In the post war period democratic ideals have expanded rapidly and many societies have adopted this form of government. Countries were also signatories to UN charter which emphasized human rights. In the post war period the concept of a welfare state accepted by many democracies led to emphasis on social justice, more equitable distribution of incomes and access to basic social goods including health and education for all. Government policies had to be evolved to meet these objectives. With expansion of media and especially social media, there was increasingly greater public scrutiny of actions of government. This involved judging the politicians not only in terms of their actions in government but the ethical norms followed by them. They, therefore, had greater pressure to adhere to moral norms in the political life.

The relationship between politics and ethics reflects the philosophical beliefs, nature of political institutions and moral values of the society in different nation states. Man has a responsibility in a state as a man and is governed by certain ethical norms sanctioned by the society. He follows those or some religious commandments dictated by his religious beliefs. He also has his responsibilities and duties as a citizen. He likes to learn music, has his ideas on a good education system, and likes close contact with international community and many ideas on economic equality, abortion, justice and other social issues. He believes in certain political ideals as liberty, equality, freedom to freely engage in public discourse on social issues and equality of all religion. He does not want a dictatorship or abuse of rule of law. It is these values he would like to be part of his nation's politics. Many of above are part of philosophical studies. It can be argued that ethical norms in politics are evolved in the philosophical studies which deal with these and similar other concepts. This approach is supported by Kantian or Rawlsian beliefs and their approach to conception of political institutions. Wells (2015) has commented upon this aspect and argued that the moral law of Kant implied that the constitution of the State, its governing institutions and legislature must align itself with dictates of this law and also required citizens to obey the laws made by legislature and other rules in pursuance of the above. The Rawlsian approach for creation of a flourishing well-ordered and stable state was creation of moral norms so that basic structure of such a society is governed and regulated by principles of social justice which were enunciated by him. In such a state citizens can consent to be governed by it even where conceptions of good life differ as there is overlapping consensus. In both approaches ethical norms in politics were integral part of the way the state institutions were constituted (pp, vi-ix). MacLauchlan (2006) mentions that in sharp contrast to above are views of Hannah Arendt, a political thinker. She believes politics to have its own skill set and not devoid of values. She considers a political action as ethical not in the meaning of morality of society but judged in terms of its 'greatness'.While she does not define greatness specifically, its meaning is discussed in her writings. Drawing a parallel, the greatness is judged by her like Socrates who did not want to live if his conscience did not permit. Forgiveness is a quality she supports as it frees political action from a cycle of revenge. She talks of Winston Churchill's greatness as "nobility, dignity, steadfastness, and a kind of laughing courage". She clearly sees political action to be judged in terms of its own morality –greatness of an act, conscience and principles. (p 20)

There are also thinkers like Max Weber who consider politics as having very little relationship with ethics (Frazer, 2006). He was a believer in value pluralism, and considered politics and ethics as of separate value domain. Reinforcing this view he considers politics with little ethical content and remarks 'certainly, politics is not an ethical business. But there does nevertheless exist a certain minimum of shame and obligation to behave decently which cannot be violated with impunity, even in politics' (p 26). He has, however, held certain qualities necessary in a politician –passion, responsibility and judgement. He considers it necessary that politician must have the courage-wisdom implies courage as well- to face the consequences of his political action, both intended and some others which were not intended or are the side effects (pp 31-32) It is indeed a powerful and important ethical characteristic since often the bad consequences are passed on to the bureaucracy by politicians when uncomfortable questions are raised about their decisions.

Several questions arise as one looks at the above relationship and actual functioning of politicians. First, the nature of relationship between ethics and politics and whether the requirements of political office necessitate certain ethical norms for a politician which are different from ordinary ethical concerns of the society and the citizens and the nature of such deviations.. Second, the personal qualities of politicians and standards of higher ethical norms expected of them as they have an accountability to the people for their actions. Third, ethical concerns in framing public policy. Fourth, in international relations nations have concerns and objectives of a different nature. Nature of ethical norms which politicians can follow while framing and implementing policies in this area. Fifth, whether the ethical concerns are being followed by politicians globally, the level of corruption amongst politicians and their misuse of authority to capture and keep power or managing to continue with it. Sixth, measures required for a cleaner and more ethical political life. In the next two parts of this chapter these are discussed in some detail.

PART I

There is a strong belief that politicians are unprincipled persons who often go to any extent to retain or acquire political power. Nikita Khrushchev, General Secretary of former USSR communist party had once remarked that politicians in various countries behaved in the same unprincipled way and they may promise to build a bridge even where there is no river. George Bernard Shaw, the famous English playwright did not think much of intelligence level of many politicians and had observed that when a person knows nothing and thinks he knows it all, it clearly indicates that he is suitable for a political career. The poor image of politicians is due to a variety of factors, particularly large scale corruption in many countries of which they are beneficiary and their unprincipled acts to get and keep power. This trait is exhibited in most economies including USA, China, India, Russia, UK, Italy and Japan. In 2013 the Chinese Railway Minister Liu Zhijun received death sentence for receiving bribes and misusing his office. Despite the view which may indicate politicians as persons who have little use for morality in their field of activity, considering the complex nature of governance and politician's involvement in it, there is a well-argued case for taking a more comprehensive look at the relationship between ethics and politics. Thompson (2018) has argued that political ethics is a separate branch of study and dealing with practice of making moral judgments about political action and study of that. He divides the study in to two parts: one, ethics of process which concerns public officials and the institutions in which they act and two, ethics of public policy dealing with policies and laws made by the government. (para 2) While dealing with these issues several other critical areas emerge. There is need to look at whether norms of conduct can be identified, which politicians must follow, to have better governance and a more moral political life. A key area of ethical questions namely international relations, which poses difficult questions even for those politicians who are otherwise strongly inclined to follow ethical norms, needs to be looked at some depth. These issues are discussed below.

Ethics of Process

The process of governance of a state is very critical in all societies. Generally nations provide for a constitution which outlines how the political executive, which represents the will of the people, is to be elected. They suggest laws which are discussed in the parliament or such other bodies which has elected representatives of the people and after its approval and a further assent by the head of the state, these are implemented. The day to day governance is done in accordance with the laws and rules or regulations made in accordance with that. As these are framed after public debates and by people's representatives, they generally accord with ethical norms of a common citizen. It must, however, be clear that all laws are not always in accordance with ethical norms. Many laws made in Nazi Germany were clearly immoral. With expansion of non-state actors like the civil society, wider media presence and huge expansion in inter connectivity amongst people, the principles and rules concerning governance have a stronger moral base.

Adhering to ethical norms during the process of governance, even in countries with strong democratic tradition has problematic dimensions. One of the key issue according to Thompson (2018) is the extent to which, due to responsibility of public office, politicians should have moral norms different from those of ordinary citizens. For example politician may decide to sacrifice innocent citizens arguing that it is for the larger good of the people and the state. So where are the lines drawn? The politician, since he professes to act on behalf of ordinary citizens, has responsibilities for them. The ethical dilemma, when alternative course of actions presented for decision to him are both bad, is often part of politician's life. He lies, practices deceit and tortures in the interest of the common man. All this makes him commit immoral acts and create the problem of 'dirty hands'. (Para 5)

According to Walzer (1973), who has analysed the problem extensively, the origin of the phrase is from a play 'Dirty Hands' by Jean Paul Sartre performed in 1948 in theatre Antoine in Paris. In the play the communist leader Hoerderer says that "I have dirty hands right up to the elbows. I've plunged them in filth and blood. Do you think you can govern innocently?" (p161) Walzer, while discussing various dimensions of the problem, has argued that the problem arises when the politician may do the absolutely right thing in 'utilitarian terms' and yet be guilty of moral infringement .On the other hand if he takes the absolutely moral stand, he is guilty of not measuring up to the official responsibilities and duties apart from an action which is not correct in utilitarian terms. To illustrate his point, he gives example of a good politician who during his election campaign has to promise to some of his constituents a contract for construction of a school if he wins. His constituents want to vote for him because he is a good man and would not indulge in immoral acts if elected. He needs to negotiate with ward bosses to win the election .For that to be clinched, they require assurance that they would get contract for construction of school to them in the coming years. He would not like to make such deals as he is a good man. But the constituents want him to win as he is a good man. If he denies the deal, he is likely to lose. His constituents want him to win. So he cannot afford not to strike the deal. So he dirties his hand and strikes the deal and wins. In another example he mentions how a well-meaning good leader, wanting to give freedom from colonial rule to a country, decides to order torture of a rebel as its party, working for freedom, has planted bombs in residential colonies and information from the rebel can save lives. He knows torture is morally wrong. It is not even clear whether the rebel leader was himself involved in the bomb plot. That makes it even more morally incorrect. Yet he dirties his hand .If he did not do that, he could be accused of a serious moral law infringement if the bomb were to explode and kill hundreds of people. He must act as per his responsibilities as a politician.

The problem of dirty hands also raises another critical issue: how should the politician be looked at by the rest of the social class. Walzer considers three options. First is the approach of a Machiavellian politician. He may be a good man but has to learn bad things to succeed in politics. He is judged by the result. If he adopts bad means and succeeds, power and glory are his. He is considered a hugely successful politician. In this view it is the result which finally justifies a bad action. But he does not suffer any punishment for the wrongs that he did. The second approach is of Max Weber in his essay *Politics is a Vocation.* He argues that God did not choose the vocation of politics for a man. A politician choses it himself. Hence when he does bad to do good, his soul should finally suffer for his bad actions. The third approach is taken using an analogy from Albert Camus essay *The Just Assassins*. It is argued that the punishment for the bad action of a politician should be given but be commensurate with the act and good that he caused. Walzer supports the last approach (pp161-180)

It is well accepted that the process of governance is complex and quite often gives rise to situations where strong initiative has to be taken to meet emergent situations. Such situations can be either a before a politician is elected to a public office or during the course of his functioning while governing the state. For election to an office politicians need to raise funds. This process in any country involves cutting deals with powerful and rich groups. Also, he needs support of different segments of society. Without their support politician may not get enough votes to get elected to the public office. He has to promise favours to their important leaders to get them on his side. So even before entering public office, even a good moral man has his hands dirtied. While in office several situations arise where a politician needs to take immoral decisions. The problem in the recent years has been compounded by rapid expansion of terrorist activity across the globe. Some of the regional groups wanting an independent state or change in the political governance system have also resorted to violent acts in which women and children have died. So occasions when politician may have to authorise torture or other unethical acts like snooping have multiplied. Politicians, to meet these challenges, have no order invading of privacy of its citizens or using force on them. So the dirtying of their hands is a genuine problem. What needs to be seen is whether the good that comes out of it for citizens would really require the immoral act or was there a better course of action.

It has to be, however, recognized that a large body of men coming into politics have poor moral standards. Walzer too recognizes this in the above referred to article. They may have committed many immoral acts before coming in to politics. They thus have their hands already dirty before entering political office. So while compulsions of political life force them to commit more immoral acts, it is hardly an ethical problem for many such politicians. They may have no compunctions doing an unethical act particularly if it brings them glory and fame. So it may be fair to refer to it as a problem of *more dirty hands* in case of most politicians.

In many situations there may be a choice between saving one life or sacrificing one and saving many. Consider the case of a terrorist hiding in a building. His two friends have blown up a cinema hall in which a very large number of persons have been killed. His friends have blown themselves as suicide bomber. But material recovered from them indicates further plans. The politician orders capture or killing of the remaining terrorist. He is killed. Now his killing has saved many lives. Is the politician guilty of an immoral act? Considering that his action has saved so many lives, he cannot be blamed and his act should be judged accordingly. It can be argued that If the action has resulted in savings of lives and the alternatives were worse, then one should look at the problem of dirty hand, not as a problem but an essential part of the profession. Just as a soldier has to kill in war or policemen chasing a group of bandits have to fire in return, a politician has to issue orders to prevent a larger harm as part of his duties. There is however merit in the argument that the action of politician should bring good which is commensurate with its bad consequences. In other words he should do his job well. It is not possible to defend an act where a few lives are saved and say a hundred men get killed. There is merit in Walzer's view that the bad consequences must be judged against the good that he caused to assess moral worth of his action.

The process of governance throws up other dilemma too. Consider a case where a politician has to decide on an issue where voters in his constituency may hold an opinion opposite to that of his or a wide range of differing opinions and his government looks at the problem differently. He has a genuine problem in such cases. It is more so when different powerful groups which had backed him in election to the constituency hold different views. But in all such cases he has to take a view consistent with the requirement of his office. The problem of ethical concerns is very genuine in such cases.

The Ethical Norms

Responsibilities of politician require that in several areas his moral standards are of a higher order than the ordinary citizen. An extensive study of the issue was done in UK by the Nolan committee in1995. It recommended that during the process of governance, the politicians must exhibit certain moral norms which may be more stringent than may be applied to common citizens. They must act with selflessness or purely in public interest, avoid placing themselves in obligation, act objectively and make decisions fairly, impartially and on objective evidence. They should while governing must be accountable to the people, act in a transparent manner, act honestly and truthfully and show leadership qualities while supporting and motiving others to follow the above norms. While several of these are good ethical norms even for normal citizens, the focus on accountability, transparency, not placing themselves in obligation are not required by a common man living in a democratic country. While ordinary citizens have right of privacy and the right not to publicly disclose their income or financial assets, politicians cannot have this luxury as they deal with huge sums of public money and people should have faith in their probity. The politicians cannot claim privacy as many parts of their life has to be in public domain for scrutiny. They are required to declare their financial assets, in many countries. Since a number of persons with criminal records were participating in elections, in India there is provision mandating declaration by all candidates about details of their criminal history so that electors can make an informed choice. An ordinary citizen is not required to do so. They must also avoid conflict of interest. Though he may act fairly as suggested by the Nolan Committee, people should also perceive it in that manner .Hence if he has interest in any issue of financial or any other nature, he must recuse himself from such decisions.

The committee on standards in public life report prepared by Phillip (2014) has while endorsing the Nolan Committee recommendations emphasized the point that political ethics is quite like the moral standards to be followed by doctors, lawyers or other similar professions where to retain public trust in the responsibilities of their office, a moral code different from ordinary citizens has to be devised. He gives examples of talks going on with a country and argues that details about that if given in public may jeopardize the result. So while politicians in order to effectively discharge responsibilities of public office, which they hold, may abide by the precepts of the above committee on ethical standards, this will always be subject to the requirement of the public office held by them. This may often pose problems in deciding whether a politician is deviating from norms of morality due to such requirements or is just using it as a cover to explain his infraction.

International Relations and Ethics

While nations focus on preserving their territorial integrity, issues in international diplomacy often raise moral dilemma before the politicians. He has to choose often between being ethical or protecting national interests which his office requires him to do. In such situations, he may at times be required to hide facts and tell lies. Or permit his victorious army to occupy the losing nation's land or even merging part of it which may be in interest of his nation or taking a moral stand and withdraw. According to Korab- Karpowicz and Julian a theory of political realism was propounded by Thucydides in about 400BC in Greece and later Hobbes and Machiavelli in the 16th century according to which the states acted in pursuit of their own security. Interest of the state was considered paramount and states acted to further their interest and pursuit of power. In the international arena there was later a contrasting theory of liberalism which believed in cooperation among states. Later a more moderate form of realism theory was developed which believed in possibility of moral judgment in international relations. Unlike Machiavelli, who believed in justifying anything for reasons of state, the new tradition developed by Morgenthau and Niebuhr while accepting the primacy of national interest, assigned very high value to prudence and thus accepted possibility of ethical values in international relations. Prudence meant ability to assess political reality and judge worth and correctness of a successful political action after considering various alternatives and their political consequences. Neo realism theories which were later developed attempted to put this theory on a more scientific basis. (2018,para 1-3),

.A very interesting analysis of some of these issues has been given by Thucydides while describing the Peloponnesian war. This war took place between Athens and Sparta in 404-431 BC. The description is at times in terms of a dialogue. Several issues of ethics arise as it is described. Athens, a leading military power attacks Melos. The "Melian Dialogue" captures the essence of the argument. The Athenians argue that Melos has no chance of winning a fight against them and have a choice of surrender or face complete destruction. There is no place for justice in these relations, the Athenian envoy argues. Thucydides has described it as below

For ourselves, we shall not trouble you with specious pretences—either of how we have a right to our empire because we overthrew the Mede, or are now attacking you because of wrong that you have done us—and make a long speech which would not be believed; and in return we hope that you, instead of thinking to influence us by saying that you did not join the Lacedaemonians, although their colonists, or that you have done us no wrong, will aim at what is feasible, holding in view the real sentiments of us both; since you know as well as we do that right, as the world goes, is only in question between equals in power, while the strong do what they can and the weak suffer what they must.(History of Peloponnesian War, Trans 1910, Book 5 Chapter 89)

The principle of might is right is clearly reflected in the above thoughts. Machiavelli and Hobbes too belonged to this tradition and believed that for reasons of state, all actions were justified. Moral values did not seem to figure as important in international relations for them.

Primafacie wars are clearly immoral as result in untold miseries and destruction. Lazar (2020) has discussed the ethical dimensions of a just war theory and mentions how the traditional view supports waging of war if certain conditions are met. It.emphasizes presence of just cause, being fought by legitimate authority, right intention, reasonable prospect of success, proportionality and last resort as requirements before a war is started. It specifically mentions that non-combatants be not attacked and only military objectives should be involved. It also mentions proportionality and necessity as two other ingredients. In addition there are post war norms too. A contrary view is that of pacifists who argue that war is wrong and support other interventions. While deciding on these questions a politician has often a dilemma on the human tragedy involved and benefits which may accrue to the state.(para 12) .

In international platform, there is no super state which can ensure that different states abide by certain norms of morality and political; economic and military power decides interest of which state has to prevail. But three important developments have changed the equation. First, the development of nuclear weapons, which leads to assured destruction of warring nations, has put a restraint on their propensity and desire to acquire greater power. Second, the setting up of UN which has Security Council as its important arm and development of various international conventions to which most nations are signatories, particularly the Geneva Convention, has acted as a check on nations for acting irresponsibly and not following ethical norms in international relations. Third, the development of information technology and globalization has increased economic interdependence of nations. This is forcing them to abide by ethical norms which are part of bilateral treaties or international convention. It has to be, however, realized that these developments have only improved possibility of ethical norms in international relations. The views of the classical realists still have merit. Example of Russia annexing Crimea which was part of Ukraine is a recent example. US decision to increase tariff on imports from China and Chinese retaliation are examples of nations pursuing their interests irrespective of its impact on global economic system.

Ethics of Public Policy

An area of ethical concern in political life is the objective of a public policy in framing of which politicians have a key role.. This raises a host of ethical problems. In countries like China, India, USA, Australia and Canada, there are regions with different levels of economic development. Availability of resources in these regions varies. In order to provide a certain quality of life to citizens, it is necessary to give additional financial resources to not so well off regions. This naturally must be funded by better off areas. While deciding on this policy a moral dilemma arises: to what extent should there be inter region resource transfer and why should the economically prosperous regions fund the development of poor regions at all. At times different societal groups have conflicting interests and to decide which one should prevail is a difficult choice. Countries have societies with diverse social and economic needs. In large countries with different geographical features in the regions of the country, economic developments is often uneven. The policy which should be framed for welfare and growth, raises very often difficult questions. Should the current generation curb consumption of energy so that coming generations have a good life ? Why should the state resources be not given preferentially to fund the lot of the poor population? Similar dilemma arise when politicians wish to tax the rich to provide for development of the poor. Deciding on these policies invariably is a difficult choice.

There are other questions in public policy which raise moral dilemma in developing economies .Many of these countries are having limited facilities for hospitals, schools roads, bridges or clean drinking water facilities. Imagine a politician who has fought from a constituency and promised them development. So when a new school is to be opened he wants it in his constituency. But there are other areas which are worse off and they have a better claim. In fact several areas may have almost equal claim. How does he decide? All the regions need the school to provide education to children. No child is more important than the other. This is a difficult question of ethical policy.

The problems becomes especially difficult when one considers questions of health. If a new hospital is to be opened or doctors provided, choosing an area for setting the hospital is problematic. There are several criteria which are equally important. Does he decide on the basis of levels of nutrition or number of mother's dying during child birth or babies dying at birth or backward areas needing health support? Since human lives are involved here, the choice is very difficult to make. Such questions raise ethical dilemma for the politician who is to decide on this issue finally.

Economic inequality has worsened during the last three decades across the globe. The world inequality report (2018) has shown that it has worsened in all major economies including USA, China, India, Russia and Brazil. In 2016 top 10% earners had income share of 37% in Europe, 41% in China, 46% in Russia, 47% in US- Canada and around 55% in Sub Sahara Africa, Brazil and India. It was about 61% in Middle East which showed highest level of income inequality (p9). In this situation taxing the rich to reduce inequality may seem proper. But the rich could argue against it mentioning how the wealth earned by them is legitimate, earned by their entrepreneurship and they should not be asked to subsidize the poor. These ethical questions would arise before politicians in almost all countries when they frame policies

Thompson (2018) has discussed the issue in great detail and mentioned various dimensions of the problem in the article referred to earlier. The politician while deciding on public policy has to see what is just and fair. Different group of people see this differently on the same issue. Many people who have cosmopolitan approach and not bound by nation states see the need for more equitable distribution of global natural resources or free immigration policy. This may be strongly opposed by many others. Another important social question is need for affirmative action for supporting women and other socially or economically disadvantaged groups. There are others strongly opposed to such preferences in jobs or other areas and support non-discriminatory approach. Similarly, animal right activists or those arguing for the right of the unborn child and against abortion have powerful support and arguments in their favour and an equally powerful lobby opposed to it .There are a host of other areas which are very contentious. These include

1. The question of capital punishment as against not taking life of a person accused of crime
2. The reformative approach to criminal justice as against punishing criminals with tough measures
3. Prohibition as against permitting alcohol consumption freely
4. Government support of religious institutions as against non-interference in religious matters.
5. Civil Disobedience, conscientious objection to joining armed forces or fighting in war or other matters and secession as against strong measures to preserve the state
6. Some new developments in science like nano technology, Genetic testing and embryo research

While deciding on many of these issues the dilemma before politicians is whether there would be support for his decision in the people, the nature of popular beliefs and diversity of population and its approach to these issues.(pp 4-5)

While deciding on these questions, problems arise in making a choice in favour of an option. It is difficult to decide the basis on which certain actions may be considered ethical. Take the case of setting up a hospital where the choice is between two areas both of which have poor health facilities .While one could use cost benefit analysis to make choices but it cannot detract from the needs of both the regions. A similar problem arises when one wishes to decide whether capital punishment should be abolished or those accused of murder or rape or other serious crimes be awarded this sentence in certain cases. It seems patently unfair if a man murders ten persons on cold blood and he is not given death sentence. However, one can argue that since state cannot give life, it has no right to take away life. But the same argument applies to the murderer .He too has no right to take life and yet does so. It has been argued that since the objective of giving punishment is to reduce crime, while punishing guilty this has to be kept in mind. Global experience shows that death sentence has not proved a deterrent to crime.

It has to be realized that while deciding on these issues the politician has no norm on which he should measure the ethical value of a decision or vote for an option. Gert (2017) has described morality as being looked at as either descriptively or normatively. It is referred to as descriptively when it talks of codes of conduct prevalent in society or as a religious command accepted in various groups and by individuals in their conduct. An alternative view is to judge morality in the normative sense as the code that all rational persons, within certain conditions, would follow (para 45). Russel (1954) gives a different view on the ethical values. He argues that reason has no place in choosing it and we decide on our objective or end, as he call it, based on our values .Whether one considers a thing as good or bad depends on it.(p8)

The descriptive approach of morality has serious limitations. While policies based on this may be well accepted in society, it precludes all reforms.in the society which a new law can bring which is not consistent with social beliefs or practices of the society. Consider the case of abortion. In Christianity under the Catholic Church this practice is not permitted as it amounts to taking the life of the child which is in mother's womb. In Protestants and many other societies it is considered appropriate in certain conditions, particularly when mother's health so requires. Imagine a politician in a catholic society who receives a request from a mother and supported by the hospital that her life is in danger unless abortion of the child is permitted. The politician has a dilemma: which life to save, that of the mother or the unborn child. While he may consider norms of society in which he is living as ethical, the normative view of moral code may be different. In fact, this problem may arise in many areas. India had the problem of Sati in 18th century, a social practice in which a widow was expected to sit on the funeral pyre of her husband and be burnt along with his dead body. Reforms were made in the system and this practice was banned. It would not have been possible if the politicians had stuck to the code of social behaviour prevalent in those days in India.

PART II

While politicians are expected to adhere to ethical norms and in certain areas exhibit even higher moral standards than societal moral standards for ordinary citizens, across the globe,they have failed miserably in keeping up these values in public life. While the requirement of their office may at times explain several infractions, it does not explain their conduct in many areas. The problems in many developing economies are of corruption and grabbing of power by undermining democratic institutions using unethical means.

Corruption and Politicians

Corruption amongst politicians in the societies across nations has been a major cause of concern. It implies legitimate rights of an individual being usurped by a corrupt man .It harms the poor to the maximum extent as they have limited power to fight against a corrupt system. There is a strong belief that politicians exploit this to the maximum extent. In the modern world a corrupt deal can touch many nations. In 1950s a US aircraft manufacturing company Lockheed was found to have paid bribes of $ 22 million to various politicians and official to bag the contract for supply of aircraft. Later investigations revealed that West German Minister for Defence Minister, as well as Prime Minister and President of Italy were also involved. During the course of enquiries it was found that a number of other politicians from many other countries were involved. It was alleged that bribes of $3 million were paid to Japanese to buy L-1011, instead of DC 10, by All Nippon Airways (ANA) and the Prime Minister Kakei Tanaka was later tried and convicted for four years imprisonment for the offence. It was also alleged that Prince Bernhard of Netherlands received a bribe of $1.1 million for purchase of Lockheed -104 instead of Mirage -5.After an enquiry a constitutional crisis ensued as the queen threatened to resign (Chaturvedi and Chandra, p 104). Finally the prince was not prosecuted as a compromise. Across the world, Governors of States in US, ministers in China and in India, Presidents and Prime Ministers in Indonesia and Malaysia have been accused of corruption apart from a number of other nations. A few years back Zambian President was acquitted of graft charges though it had been found that he had spent five times of his income in a purchase from a shop in Switzerland. .Recently, PM Netanyahu of Israel was accused of corruption charges. In India there have been a number of enquiries of charges of corruption and a former Chief Minister of a state is in prison after conviction in a corruption case.

Love for Power

In all democracies, for fighting elections, politicians have to raise money. While some of it comes from membership of parties, collecting the rest involves unethical practices. To win elections politicians adopt many unethical practices. In India they at times resort to fake news, advertisement masquerading as news, putting up dummy candidates to confuse the voters and giving gifts of radio, television sets, alcohol and various other lucrative items. To stop voters who are likely to oppose you, criminals are often used to stop them from exercising their right to vote in the elections. Milan Vaishnav, a fellow at Carnegie foundation based on his book "The Crime Pays" has argued that as elections are getting costlier in India, political parties are bringing in their fold individuals who are wealthy even with dubious reputation. Even voters have shown great faith in politicians with criminal record as they can help them. The distinction between politician and criminals is disappearing.(Reuters, Jan 2017)

The conditions in some of the other democracies is quite similar. During the last presidential election in Indonesia there were complaints by the Presidential candidate of existence of millions of fictitious names on the voters list, exploitation of state apparatus and ballot papers with votes already cast.(The Strait Times May 14, 2019). The 2016 election in Zambia were criticized on the grounds of being unfair, police acting in support of ruling party denying opposition right to hold rallies, use of state events and air force helicopters for ruling party campaign, and the state television and radio providing nearly 45% coverage to ruling party candidate .The election results were rejected by the opposition candidate who was later arrested for treason charges and released much later. In Brazil former president Lula was convicted of corruption charges and handed a ten year sentence. Later there were allegations that this was meant to prevent popular left wing candidate from fighting election by justice Minister. After the elections and coming in power of Mahathir Mohammad , in Malaysia the former Prime Minister Najib Rajak was charged with embezzlement of $ 4.5 Billion funds recently and a series of raids revealed that he had misused a fund meant for development.

CONCLUSION

There is clear evidence that despite ethical values and norms suggested and a code of conduct of ministers in some countries, there is very little observance of norms of good conduct. Across the globe politicians have behaved in the same manner and showed little regard for ethical values with very few exceptions. Nolan Committee had suggested extensive norms of ethical conduct and these could form the basis for an ethical code for politicians which countries could suitably modify according to their needs. There is a clear need for transparent funding mechanism for fighting elections. In particular it is essential that the financial status of politicians in public office is in public domain. It is also essential that other healthy norms of behaviour and conduct are put in public domain in addition to Nolan Committee. Of these the most important is the impartial conduct of elections. There is also need for an oversight committee or a watch dog like the Ombudsman to oversee adherence to these. All these measures can reduce corruption and improve adherence to strong ethical values norms.

It has to be, however, realized that by the nature of politics, which deals with getting power and retaining it, and the requirements of their public office, the politicians cannot be judged by the normal standards of ethical values. They may take decisions in interest of people which may be considered unethical as discussed above while discussing the problem of 'dirty hands'. They may fight 'just wars', as against the Nazis, millions of their countrymen may die in it, and may not be accused of immoral acts. While judging them is a difficult job, one could support their adherence to Nolan Committee norms of conduct and assess them possibly on the anvil of 'greatness' as done by Hannah Arendt

Globally, a critical area is of International relations where ethical values applicable to ordinary citizens across most societies fail. While countries can be encouraged through UN bodies, bilateral treaties and multilateral agreements to cooperate, perception of needs of nation state would always lead to unethical conduct. Diplomats may have to tell lies and give false impressions of situations to protect their nation's interest. Even in domestic politics, the lure for power and office will cause many to deviate from upright path of good conduct. Societies can minimize the extent of deviation from ethical norms by strong public oversight and measures suggested above. But unethical conduct amongst politicians will always be a fact of life.

REFERENCES

Adkins, A. W. H. (1984, February). The connection between Aristotle's ethics and politics. *Political Theory*, *12*(1), 29–49. doi:10.1177/0090591784012001003

Chaturvedi, B. K., & Chandra, S. (2015). *Corruption and economic growth*. Academic Foundation.

Frazer, E. (2006). Max Weber on ethics and politics. *Politics and Ethics Review*, *2*(1), 19-37.

Gert, B., & Gert, J. (2017). The Definition of morality. In *The Stanford Encyclopaedia of Philosophy*. https://plato.stanford.edu/archives/fall2017/entries/morality-definition/

Girardin, B. (2012). *Ethics in politics*. Retrieved from https://www.globethics.net/documents/4289936/13403252/FocusSeries_05_EthicsinPolitics_Benoit_text.pdf/bf28841e-216a-41ff-bca6-4ef9d4022d86

Indonesia elections: Prabowo alleges fraud, says he will reject official vote tally. (2019, May 16). *The Straits Times*. Retrieved from https://www.straitstimes.com/asia/se-asia/indonesia-elections-prabowo-alleges-fraud-says-he-will-reject-official-vote-tally

Kautilya. (1972). *The Kautilya Arthashastra*. New Delhi: MLBD Publishers.

Korab-Karpowicz, J. W. (2018). Political Realism in International Relations. In *The Stanford Encyclopaedia of Philosophy*. <https://plato.stanford.edu/archives/sum2018/entries/realism-intl-relations/>

Lazar, S. (2020).War. In *The Stanford Encyclopaedia of Philosophy*. https://plato.stanford.edu/archives/spr2020/entries/war/

MachiavelliN. (2006). *The Prince*. Retrieved from http://www.gutenberg.org/files/1232/1232-h/1232-h.htm

MacLachlan, A. (2006). An ethic of plurality: Reconciling politics and morality in Hannah Arendt. In History and Judgement. Vienna: IWM Junior Visiting Fellows' Conferences.

Phillip, M. (2014). *Public ethics and political Judgment*. Committee on Standards in Public Life.

Politics. (1996). In *Collins English Dictionary* (7th ed.). Harper Collins Publishers.

Sen, A. (2009). *The idea of justice*. Penguin Books Limited.

Thompson, D. F. (2018). Political Ethics. In H. Lafollette (Ed.), *International Encyclopaedia of Ethics*. Wiley Blackwell.

Walzer, M. (1973). Political Action: The Problem of Dirty Hands. *Philosophy & Public Affairs*, *2*(2), 160-180. http://www.jstor.org/stable/2265139

Wells, C. (2015). *Subjects between ethics and politics* (Unpublished doctoral dissertation). University of Vanderbilt, Nashville, TN.

Why money and muscle still rule in Indian politics: an interview with Milan Vaishnav. (2017, January 16). *Reuters*. Retrieved from https://in.reuters.com/article/milan-vaishnav-book-politics-india-idINKBN1500B

World Inequality Lab. (2018). *World Inequality Report*. Author.

ADDITIONAL READING

Buchanan, A. (2017). Secession. In Edward N. Zalta (ed.), *Stanford Encyclopaedia of Philosophy* (Fall Edition).: https://plato.stanford.edu/archives/fall2017/entries/secession/

Chaturvedi, V., & Sahani, P. (2020). *Understanding Ethics*. MLBD Publishing House.

Galston, W. (2010). Realism in political theory. *European Journal of Political Theory*, *9*(4), 385–411. doi:10.1177/1474885110374001

Jonsen, A. R., & Butler, L. H. (1975). Public ethics and policy making. *The Hastings Center Report*, *5*(4), 19–31. doi:10.2307/3561420 PMID:1158675

Nagel, T. (2005). The problem of global Justice. *Philosophy & Public Affairs*, *33*(2), 113–147. doi:10.1111/j.1088-4963.2005.00027.x

Thompson, D. F. (2017). Designing responsibility: The problem of many hands in complex organizations. In J. van den Hoven, S. Miller, & T. Pogge (Eds.), *Designing in Ethics* (pp. 32–56). Cambridge University Press. doi:10.1017/9780511844317.003

Weber, M. (1946). Politics as a vocation. In *H.H. Gerth and C. Wright Mills (Translated and edited) Essays in Sociology* (pp. 77–128). Oxford University Press.

KEY TERMS AND DEFINITIONS

Balloting: Use of ballot paper to elect a public representative in open democratic fashion.

Bilateral Treaties and Multinational Agreements: International agreements among two nations or a group of nations which they are expected to honour.

Bloody Hands: A phrase indicating use of unethical practices in public life especially by politicians.

Descriptive Approach: Defining morality in terms of religious commands or other prescriptive norms of conduct.

Normative Approach: Defining moral conduct in accordance with values and norms generally accepted by all rational men.

Ombudsman: A person called by this designation and usually appointed by government to consider and resolve grievances of citizens against government or businesses.

Public Policy: Policies made by politicians or the bureaucracy empowered to do so and which effects a group of men or the entire society.

Utilitarian: Assessing moral worth of an action based on whether it has given maximum benefits to the largest number of people in the target group.

Chapter 2
Environmental Ethics:
When Human Beings and Nature Are Not Two

Josep M. Basart
Universitat Autònoma de Barcelona, Spain

ABSTRACT

This chapter introduces some core issues in environmental ethics (EE) from a multidisciplinary perspective. This approach's primary purpose is to make explicit the richness and complexity of the subject, where arguments coming from natural sciences, politics, philosophy, economics, or psychology often intertwine. Moreover, although it is risky, the diversity of views tries to engage interested readers from different fields and non-specialists. After the introduction, the chapter focuses on the characteristics of the different views on non-anthropocentric EE. Next, five of the main ethical frameworks are presented. This is followed by some specific points of view, which are capital in the subject (intergenerational ethics, precautionary principle, deep ecology, environmental justice, indigenous peoples, and feminism). An exploration of consumption and population follows. Finally, some key points on sustainability and human development precede the conclusion of the chapter.

INTRODUCTION

A radical new conception of man's relationship to the rest of nature would not only be a step towards solving the material planetary problems; there are strong reasons for such a changed consciousness from the point of making us far better humans. If we only stop for a moment and look at the underlying human qualities that our present attitudes toward property and nature draw upon and reinforce, we have to be struck by how stultifying of our own personal growth and satisfaction they can become when they take rein of us. (Stone 1972)

Traditionally, ethics has focused on analyzing a specific set of questions about relationships and behaviors affecting rational beings. Under this paradigm, only individual human beings have to be considered

DOI: 10.4018/978-1-7998-4117-3.ch002

by ethics. Nevertheless, since the 20th century other non-human living beings have been recognized by many as deserving moral consideration (e.g. conscious animals or trees). In fact, this widening of the ethical concern has extended itself even to encompass the whole biotic community of our planet. As a result, there is now a far more complex setting for human beings because many new questions have appeared and the answers are still being discussed, with different points of view, inside several subjects (natural sciences, economics, politics, philosophy, psychology and law, among others). In this sense it can be said that the field of environmental ethics is a necessary opportunity to undertake a full revision of the pursuits and responsibilities that human beings assume in relation to Earth, all its inhabitants and to future generations.

This ongoing revision arisen by environmental ethics is, and will be, a huge task. Certainly, ethics is playing a necessary role in it, but ethics alone is not enough. What follows here is an interdisciplinary attempt to introduce some of the most relevant concepts, perspectives, issues and difficulties involved in both the study of its many implications and the proposal of actions and policies that could help leave the dead end we are in.

BACKGROUND

Since the 17th century many philosophers have criticized the ill-treatment of animals or the unnecessary pain that we provoke to them. Based on different reasons, J. Locke (1632-1704), J-J. Rousseau (1712-1778), I. Kant (1724-1804) and J. Bentham (1748-1832) appear among these first voices. Nevertheless, the origin of Environmental Ethics (EE) as a branch of applied ethics, a subfield of philosophy, dates back to the early 1970s and inherits the increasing environmental worries reflected in several works published in the previous decade. Three of them are often highlighted due to their subsequent influence on contemporary debates. Carson (1963) reported how some pesticides used in farming could poison the environment and also, through food, have consequences on public health. White (1967) argued that one of the reasons of our ecological crisis could be found in the justification provided by the Christian theology of human exploitation of nature, as if it had been created to serve the necessities and desires of human beings. Finally, Hardin (1968) exposed how the unrestricted private use of public resources together with the continued increasing of population was going to provoke serious shortages and several environmental damages.

According to B. A. Minteer, EE is "[…] the field devoted to understanding the character and structure of environmental values and clarifying the responsibilities these create for moral agents and for society as whole" (Gardiner 2017, p. 528). Let the concept of anthropocentrism be our starting point for exploring EE. Anthropocentrism defends that only human beings have intrinsic moral value. Therefore, all the other organisms, species and systems have no moral significance to human beings. A distinction can be made between weak and strong anthropocentrism. Thus, the previous definition corresponds to strong anthropocentrism, while weak anthropocentrism accepts that non-humans can possess some inherent moral values, but they always being second-class to human values. In addition, epistemological anthropocentrism recognizes that every perspective on the world is inherently human-centered. Or, in other words, all our concepts and ideas depend on particular characteristics of perception and mental processes in human beings.

In non-anthropocentric ethics the set of moral subjects is more comprehensive, while moral agents remain to be limited to rational beings. Different versions of ethical non-anthropocentrism (sentiocentrism, biocentrism, ecocentrism) argue that there exist some parts or elements in the non-human world that have intrinsic moral values. That is to say, moral significance (standing) is recognized to some non-human entities in such a way that human moral values not always overrule them. These non-anthropocentric visions of nature together with weak anthropocentrism constitute the core of EE.

Sentiocentrism (or sentientism) is the philosophy which defends that all sentient animals, not just human beings, have intrinsic value. Therefore, the capacity of experiencing pleasure and pain is a sufficient condition for an animal to acquire moral significance. This means that their interests must be equally taken into account, without discriminations among the different species. For Singer (1990) sentience is also the necessary condition for moral standing, so implying that stones or trees are not morally significant. Due to the difficulties that this position arises when opposed interests from different species are considered, the notion of gradual sentiocentrism has to be considered. In this case, the unequal complexity of conflicting interests can make a difference. For instance, when an African child is threatened by a hungry leopard all present human beings are expected to take sides with the child by confronting or even killing the leopard. Regan (1983) highlights the presence of consciousness when arguing that sentient beings must be respected. This is so because, like human beings, they are subjects to a life where something else than avoiding suffering is desired. For instance, food and water, health, social interaction, shelter or freedom of movement. At least in mammals and birds all of these conditions are required for living a good life.

Biocentrism claim that moral standing should not be limited to sentient beings. From this point of view moral consideration is deserved by all individual living organisms. Therefore, consciousness is considered to be a faculty that should not be a necessary condition to grant moral relevance. Schweitzer (1923) argued that human beings should not be a hinder to the will to live manifested in all living things. Following this approach, Taylor (1986) defends that all living organisms have inherent value for themselves. For this reason, we have to take into account their moral standing when considering our deeds. When this inherent value is considered to be the same for all living things (egalitarian biocentrism) human action on the world faces serious obstacles. For instance, our dependence on food, clothes and shelter requires the death of animals or plants that otherwise would continue alive. A rigorous respect for their will to live would bring our species close to extinction. It is not surprising that this outcome results inacceptable for almost everyone, so some corrections seems unavoidable. Several options have been proposed to relax somehow this required respect for all living beings. Among them the distinction between basic needs and non-basic ones, lists of principles, complexity of capacities, or hierarchies of interests (Attfield 1983; Taylor 1986; Schmidtz 2011).

Ecocentrism is the philosophy originated in Aldo Leopold works, mainly in *A Sand County Almanac* (Leopold 1949) where the term land ethic was introduced to argue that land is much more than soil. It is a very complex community of life with many circuits and fluxes of energy that have to be preserved: "A thing is right when it tends to preserve the integrity, stability and beauty of the biotic community. It is wrong when it tends otherwise" (Leopold 1949, pp. 224-225). From these words current ecocentrism recognizes intrinsic value in ecosystems in such a way that human moral consideration has to include the biosphere, the whole biotic community in which human beings belong. In doing so, ecocentrism shows a strong holistic aspect which is revolutionary in comparison with anthropocentric ethics.

The emphasis that ecocentrism puts on respecting and protecting the biotic community has arisen the fear that individuals (including human beings) could be sacrificed for the sake of the community. This criticism has been countered by pointing out that human beings interact in different circles of relations (like family, friends, neighbors, work, city or country) with different duties and commitments in every one of them. Ecocentrism is not a replacement to the existing spheres of social ethics, but an addition to them. Another important criticism is how to be aware of all the responsibilities and consequences that our daily decisions produce in spatially separated ecosystems. Due to this complexity it is just not possible to complete a rational analysis that involves an agent and a patient. According to Callicott (1998) the validity of the land ethic does not rely on the agent's rationality and in impartial reasons but mainly on moral sentiments and affections. Thus, with this ethical turn he follows the arguments articulated by the 18th century philosopher David Hume when analyzing the principles of morals.

ETHICAL FRAMEWORKS

There are a number of EE frameworks, some of them enriched by distinct variations. In this section, only an influential example of each one of the main ethical frameworks applied to EE are outlined.

Biocentric Consequentialism

Consequentialism is an ethical theory stating that a conduct or action is right/wrong depending on the good/bad results it produces. For this reason, consequentialism does not dictate beforehand what conducts or actions are right/wrong. Robin Attfield (Attfield 1994, 2003) holds that flourishing of individuals is the main ethical concern. Therefore, from this consequentialist point of view a situation in the world (state of affairs) is right for an individual if it contributes to its flourishing, and it is wrong otherwise. His theory is egalitarian in some aspects and inegalitarian in others. It is egalitarian when it defends equal consideration of equal interests without any advantage for human beings in front of other species. Is inegalitarian when it distinguishes basic needs from non-basic needs and when it establishes that everyone's basic needs should be satisfied before anyone's non-basic needs. Also, because sophisticated capacities developed by individuals make a difference.

Deontological Biocentrism

Deontological ethical theories defend that whether an action is right/wrong does not depend on its good/bad consequences. Instead they hold that it is intrinsically good to respect a certain set of duties or rules and that it is wrong to breach them. Taylor (1981, 1986) argues that in the biosphere each individual organism is a "teleological center of life", seeking its own good in its own way. The well-being of all the individual organisms recognized as having inherent worth (value), conditions our relations with the biotic communities, that is to say we have a prima facie moral obligation to protect or promote their good for *their* sake. This obligation for us as a moral agent derives from taking a previous attitude of "respect for nature" that implies a moral commitment to live by certain principles. In particular, the duties of non-maleficence and non-interference, together with moral qualities as fairness and benevolence.

Virtue-Oriented Approach

Instead of analyzing what is right/wrong or good/bad, virtue ethics put its focus on exploring how to live a fulfilling (flourishing) human life. In consequence, the main issues in virtue ethics are moral character and the quality of human actions. Sandler (2007) holds that there is no general hierarchy of virtues to be used as guidance for action everywhere, because cultural relativism cannot be ignored. For Sandler, virtues come from goals, and action is assisted by these virtues. The flourishing of human beings requires virtues that can satisfy a pluralistic set of goods. These goods include autonomy, meaning in their lives, learning and high regard for goods situated outside the circle of human interests. Even so, his theory is not anthropocentric because some environmental virtues take into account the well-being of nonhuman entities without a particular interest in human fulfilling.

Care

Ethical theories leading western philosophy are mainly based on isolated agents, abstract reasoning, calculations, rule-following processes and impersonal justice. In contrast to them, care ethics highlights practices, empathy, caring, relationships, interdependence and context (Gilligan 1982; Held 2007). The caring approach to social problems often appears in feminist works on ethics. Also, in mainstream EE care ethics has not still received the recognition that it deserves. The ecofeminist Marti Kheel (2008) has studied what she considers to be the handicaps of the theories elaborated to address our actual environmental crisis. She stands for a non-anthropocentric, nonviolent and caring philosophy of environment able to acknowledge the importance of feelings to ethical thinking, clarifying how narratives affect our ethical awareness, respect the worth of both individuals and entities and understand the prevailing logic of domination over women and nature.

Pragmatism

This ethical framework puts its focus on the resultant actions. Needs and practical interests of human beings are the ultimate reference point for evaluations and judgments. Pragmatism is a type of relativism because it refuses any sort of absolutism and promotes the searching of constructive procedures for addressing human problems. Therefore, environmental pragmatists are well-disposed towards the plurality of principles and values when analyzing different contexts in environment problems. They are also interested in shaping effective public policies. Norton (1991) claims that in environmentalist theories there are a growing convergence towards a consensus in policies even though different basic values remain. In front of the two opposed approaches, "aggregators" (those centered on utilitarian resource management) and "moralists" (those rejecting all form of economic analysis), he holds a pluralistic integration where moral principles arise from improved management practices. Moreover, he argues that to take into account the complexity of interrelated, hierarchic natural systems is the key element that will make the integrative approach to be successful.

POINTS OF VIEW

Intergenerational Ethics

Ethical responsibility has its limits. For instance, in general, we are not responsible for the future consequences of our actions when we are incapable of foreseeing them. Nevertheless, the vast knowledge accumulated by sciences together with the astonishing possibilities opened by new technologies has risen the necessity of thinking about the consequences of our actions on future generations. The persisting way of living followed by many millions of people, especially in the economically developed countries, has generated or aggravated some serious problems: pollution (air, water and soil), global warming, natural resources depletion, loss of biodiversity, deforestation, ocean acidification and acid rain, among others. These adversities bring suffering to many people nowadays, but they will probably provoke many damages and difficulties to following generations. It seems that these unwelcome anthropogenic effects on environment generate some kind of non-reciprocal responsibility on us, even when there are different points of view on its extent.

Some crucial questions are still open. For instance: Can we really have duties to unborn people? What rights can we attribute to them? How could we know what future generations will value and prefer? Who is authorized to take action on behalf of future generations? Also, have we any responsibility in front of future non-human beings? However, even taking into account our ignorance about what future generations will prefer, some basic things will be necessary to them: a functioning ecosystem, the biotic conditions required for a healthy life, and the knowledge, skills and means allowing them to overcome their problems and crisis. These minimum requirements could be used as a criterion for deciding what course of actions we may follow or even promote and what activities we should avoid or restrict. To do otherwise would mean the oblivion of the chain of life to which we belong as human beings.

Precautionary Principle

Risks cannot be removed from human existence; they are part of it. At the same time, humanity has enjoyed many benefits from scientific and technological developments, and some of them have facilitated the reduction of risks for humans. Nevertheless, science and technology have also contributed to the generation of new threats to human (and non-human) life. The first type of regulations for dealing with these threats were based on the Polluter Pays Principle, but it alone was not enough to ensure a proper level of environment protection, some additional preventive policy was required. In this context, the Precautionary Principle (PP) arose as the answer to the need of a decision-making procedure oriented to anticipatory risk control.

Different definitions for the PP have been proposed. One of the most influential appearing in an internationally approved declaration is the Principle 15 of the Rio Declaration (1992 United Nations Conference on Environment and Development): "In order to protect the environment, the precautionary approach shall be widely applied by States according to their capabilities. Where there are threats of serious or irreversible damage, lack of full scientific certainty shall not be used as a reason for postponing cost-effective measures to prevent environmental degradation." (UNESCO 1992). Later, the World Commission on the Ethics of Scientific Knowledge and Technology elaborated a document (COMEST 2005) where a clearer definition was presented and discussed. According to it, the main characteristics defining the PP should be six. First, a high degree of uncertainty has to be present; second, a previous scientific analysis is always required; third, plausibility of the risk is enough to consider its application; fourth, it should only be applied to morally unacceptable dangers; fifth, the intervention has to be proportional to the threat, taking into account their good and bad consequences, and sixth, the intervention has to be done before possible harm happens.

Deep Ecology

This movement was first introduced by the influential Norwegian philosopher Arne Næss. In (Næss 1973) a contrast between the shallow ecology movement and the deep ecology movement is presented. In his opinion, the former is oriented towards the environment protection from the perspective of rich, developed countries. With a wider purpose than the protection of the environment, the later denies moral hierarchies in nature and claims the intrinsic value of every living thing. Besides this biospheric egalitarianism, the core of deep ecology contains the view of life as a web of complex interrelations and interdependences, instead of a collection of individual, autonomous objects. In this sense, it denies the rationalistic approach to humanity according to which the existence of a mind in human beings makes them more advanced than a supposedly matter-based nature.

The concept of deep ecology has been revised and modified since its origins (Næss 1989). At the beginning, a certain metaphysical dimension of ecology was developed. According to it, human beings could gradually transform themselves from identifying with a narrow ego until realizing their ecological self. In this last step, the interests of nature coincide with the interests of human beings. Therefore, human love for nature is not different from love for oneself. Through the eighties, deep ecology moved to a second level where the previous, somewhat mystic formulations were avoided and a more pragmatic turn took the lead (Devall / Sessions 1985). A Deep Ecology Platform of eight principles was articulated in 1984 with the purpose of transforming deep ecology into a social movement easier to be understood and followed by the public. These principles were as follows:

1. The welfare of all kinds of life in the planet has intrinsic value, independently of their utility for human beings.
2. All different forms of life are values in themselves.
3. Human beings have to respect all forms of life except for satisfying critical needs.
4. Current interference of human beings in the life of non-human beings is expanding and cannot be accepted.
5. The well-being of human life is possible together with a decrease of the size of the population. Moreover, the well-being of nonhuman life demands this decrease.
6. Present plans and actions have to be revised. These changes will create a new situation much divergent from the present.
7. The goal of these changes has to be that of respecting the diversity of life rather than looking for maximizing economic benefits.
8. The acceptation of the previous principles implies the obligation to contribute to the implementation of the necessary changes.

Environmental Justice

The following results are extracted from a report by the University of Michigan (Center for Sustainable Systems 2019):

- Due to uneven distribution patterns, minority and low income communities have far less access to green spaces than white, affluent communities and have limited resources to maintain the green spaces they do have.

- Food prices are higher and food quality is poorer in areas with higher rates of poverty.[12] The average U.S. household in 2017 spent about 14% of its income on food; low-income families spent more than 30%.
- Low-income households spend twice as much of their gross income on energy costs than the national average, despite consuming less energy.
- Roughly 3% of the country's oil and natural gas reserves, 15% of coal reserves and between 37-55% of uranium reserves are located on Indigenous land. These resources and their associated land are sometimes taken away from Indigenous people once they are discovered.
- Though wealthy, developed nations like the U.S. emit larger concentrations of greenhouse gases per capita, developing nations experience the worst effects of climate change relative to wealthier countries due to their limited resources and ability to adapt.

Due to global businesses or natural phenomena, many environmental effects cross national boundaries. As a result, communities living far away from their origin may suffer their undesirable consequences. Environmental justice is one means to fight against poverty and environmental deterioration, "calling for the legal and social empowerment of the poor, and freedom from the inequities that result from entrenched and unsustainable forms of resource use" (UNDP 2014), bearing in mind that it is both a local and a global matter. Distributive justice and participatory justice are the two main concerns in environmental justice (Schlosberg 2007; Nixon 2011). Distributive justice refers to a fair sharing of society's environmental risks, impacts and benefits. Examples of these burdens are soil, water and air pollution, industrial waste, loss of natural resources, droughts, floods and storms. Participatory justice points to the capacity of free participation, deliberation and decision-making in all the issues particularly affecting groups and communities. This participation requires avoiding all kind of social discrimination by reason of gender, color, culture or socioeconomic status.

The laissez-faire capitalism has provoked the exploitation by northern rich countries of many natural resources situated in southern developing countries (e.g. oil, gas, coal and uranium). These resources are necessary to keep up the unsustainable model of production and consumption in developed countries. Thus, when considering the tons of CO_2 emissions per capita, Australia (16.2T), United States (15.0T) and Canada (14.9T) appear among the four world's most polluting countries, where Saudi Arabia ranks first (16.3T). But, at the same time, countries trying to follow the industrialization model created by western countries are frequently criticized by its rate of CO_2 global emissions, for instance, China (29%) and India (7%). These criticisms seem to ignore that the tons of CO_2 emissions per capita of these countries are smaller than the ones in Australia, United States and Canada: 6.4T in China and 1.6T in India (UCS, 2019).

Public policies focused on the expansion of environmental justice should consider the following issues (UNDP 2014):

- The development of rights-based frameworks to attain goals of environmental sustainability and equity.
- The encouragement of innovations to national legal and justice systems.
- The incorporation of legal and social accountability mechanisms in government systems to empower the poor.
- A genuine attention towards indigenous and tribal communities, and community norms and justice systems.

- The integration of rights-based environmental justice approaches in the private sector.

A significant declaration of 13 principles on environmental rule of law was approved in the 1st International Union for Conservation of Nature World Congress on Environmental Law (WCEL 2016).

Indigenous Peoples

The shared wisdom expressed by indigenous peoples on environment issues appear as a powerful voice that provides guidance for ethical decision-making about policies and actions in relation to environmental concerns. Their understanding of Earth has many points in common with the core principles of care ethics. In particular, how their life expands on a rich web of close, interdependent and reciprocal relationships among all living beings and entities. In addition, they are the owners of a deep ecological knowledge strongly tied to the natural world and consider themselves to be the keepers of the land as well as the protectors of both the youth and the future generations (Krakoff / Rosser 2012).

Indigenous peoples have often shown their consensus on environment matters through published declarations approved in international meetings. The Kari-Oca Declaration presented to the Rio de Janeiro 1992 Earth Summit (Kari-Oca 1 Declaration 1992) demands their fundamental rights and states their proposals on land, biodiversity, development, culture and intellectual property. The Anchorage Declaration (Anchorage Declaration 2009) accorded in the 2009 Indigenous Peoples' Global Summit on Climate Change presents a 14 points list of Calls for Action against climate crisis ("Mother Earth is no longer in a period of climate change, but in climate crisis"), and insists on the "[…] urgent need for collective action".

The Kari-Oca 2 Declaration (Kari-Oca 2 Declaration 2012) presented in Rio de Janeiro twenty years after the first Earth Summit, denounces the worsening of their living conditions; "[…] the violations of the inherent rights of Indigenous Peoples […] have intensified", "[…] our inherent rights to our lands,

territories and resources are increasingly and alarmingly under attack by the collaboration of governments and transnational corporations", "Indigenous activists and leaders defending their territories continue to suffer repression, militarization, including assassination, imprisonment, harassment and vilification as *terrorists*." "Gross violations of Indigenous Peoples' rights to food sovereignty continue unabated thus resulting to food *insecurity*". In their opinion, "Imperialist globalization, […], overproduction, overconsumption and overexploitation, […] capitalist development, colonialism" have a clear responsibility in the actual disaster. Meanwhile, in the 2007 General Assembly of the United Nations a Declaration on The Rights of Indigenous People was approved (UN 2007). Article 43 of this document reads, "The rights recognized herein constitute the minimum standards for the survival, dignity and well-being of the indigenous peoples of the world".

Feminism

Environmental feminism as a social movement dates back to the early 1970s, when feminist philosophers began to explore the links between sexism and exploitation of nature (Diamond / Orenstein 1990). From a North-South perspective Mies and Shiva (1993) developed a deep critique of the effects that industrial development and patriarchal capitalism produces on children, women, and nature. Patriarchy means the excessive power and control held by men. The fight against this sexist oppression belongs to all the diverse feminist approaches to EE, and for this reason ecological feminism (ecofeminism) first introduces a radical question on the nature of humans, namely, what does it mean to be a human being? Many subsequent discussions in EE will be conditioned by what the answer to this question may be. After analyzing patriarchy, feminist philosophers coincide that domination of nature and domination in human societies are closely related, although not all their theories equally explain why it is so. For Plumwood (1993) the origin of these dominations lie in the central role dualisms have in rationalist thought and the separations that they introduce in the world. In particular, some pairs like human/nature, man/woman, mind/body, and reason/emotion. All the first (second) elements in these pairs appear as opposed to the second (first) elements, with the characteristic that each opposition establishes irreconcilable differences into the pair. However, Warren (1990) argues that the differences by themselves should not create any problem. The subordination appears when these dualisms grant superiority (human over nature, man over woman, and so on) and justify control and exploitation. This happens whenever the dualisms are included in an oppressive conceptual framework that imposes domination based on a hierarchy of values. A logic of domination is coupled to this framework with the final purpose of justifying the power of the dominant (men) over the dominated (women or nature).

From the perspective of sustainable development UN WOMEN (2014) proposes three conditions that sustainable development should take into account in order to protect gender equality. First, compliance with human rights standards; second, recognition of the unpaid care work of women; and third, unrestricted and equal participation by women in all issues. Moreover, and with the same purpose, the report offers a set of forty actions to be considered by States, international organizations, trade unions, the private sector, and non-governmental organizations.

RELATED ISSUES

Consumption

Life requires consumption, i.e. the action of using up some resources. For human beings it means the acquisition, use, and eventual discard of material things. Thus, the capacity of consumption depends on the availability of resources, and due to the fact that Earth is finite, some limits on consumption seem to be unavoidable, unless we do not care about its grim consequences for both humans and environment.

The *ecological footprint* measures the ecological assets that a given population requires to produce the natural resources it consumes and to absorb its waste. The *biocapacity* of a region represents the productivity of its ecological assets. When the ecological footprint is higher (smaller) than the biocapacity an ecological deficit (reserve) results. The global hectare (gha) is a unit of measured productivity used to evaluate both the ecological footprint and the biocapacity of Earth. The global hectare is based on the area-weighted average productivity of usable land and water. Recent results show that humanity's total ecological footprint has increased steadily since 1961, and in 2016 it reached 20.5 billion gha whereas biocapacity reached 12.16 billion gha. Thus, the 2016 ecological deficit amounted to 8.34 billion gha, that is to say the 68% of Earth's capacity (Global Footprint Network 2019; Lin 2018). These data show that urgent changes are required. The distinction between unconscious consumers and conscious citizens can offer some indications to choosing more sensible ways of consumption. Nevertheless, besides the importance of thoughtful personal decisions, structural transformations, bold policies and steady commitments to act by governments and institutions, both at local and global level, are also necessary.

A first approach to consumption differentiates between basic (necessary) needs and desires (unnecessary). But a clear separation between them is not always easy because the distinction between necessary and unnecessary is socially conditioned. Moreover, some things were considered a luxury a few years ago whereas today the same things are considered necessities (for instance, running water and indoor toilets in developed countries). At the same time, capitalist economy promotes both types of consumption (necessary and unnecessary) with the underlying supposition that the more you get of what you like the more it will make you happy. Therefore, consumption increases globally, due to different reasons. Among them, because of: people around us consume some products; planned or perceived obsolescence; the identification with a socioeconomic status; dissatisfaction, compulsion, or as a result of successful advertising. The world's largest luxury goods companies generated sales of US$212 billion in the period 30 June 2015-30 June 2016 (Deloitte 2017). Main products sold were clothing, footwear, bags, cosmetics, fragrances, jewelry and watches.

Certainly, life may become very hard when some basic needs are not satisfied. Nevertheless, once these needs are satisfied, an increasing in consumption does not guarantee a rise in life satisfaction. According to the Easterlin Paradox, at a point in time happiness changes directly with income both among and within nations, but over time happiness does not increase as income continues to grow (Easterlin 1974). Recently, Fanning and O'Neill (2019, p. 819) have concluded that "Each wellbeing-consumption pair is strongly correlated at a point in time, but wellbeing does not increase with either GDP per capita or carbon footprint per capita in the long-run".

Population

By the end of 2019, world's population is exceeding 7.7 billion, and projections estimate that by 2050 more than 9.2 billion people will share Earth, whereas thirty years ago (1990) 5.5 billion lived in our planet (UN 2019). This increasing in the number of inhabitants has profound impacts on the environment, with the consumption of resources and the accumulation of waste products being the most significant. However, the challenges presented by these figures do not depend solely on the number of people. In addition to population size, three other factors affecting population should not be ignored: distribution, composition, and consumption (Australian Academy of Science, 2019). These other factors closely related to our ecological footprint are even more critical than the population size. This remark should not be forgotten because there is an easy way to be wrong when addressing the environmental crisis: blaming those southern countries were fertility is higher than in northern countries. In fact, in the early 1970s a simple equation was devised as a means for analyzing the interactions of population, economic growth, and technology in relation to the environment. Expressed as $I = P \times A \times T$ it is known, as the IPAT equation. This formula describes the multiplicative contribution of population size (P), affluence, in terms of level of consumption (A) and technology (T) to the impact (I) on environment. Thus, at first, the IPAT equation made a contribution to understanding the diverse causes of environmental impact. Even so, since its origins many modifications have been proposed to make it more useful in different scenarios or for different purposes (Chertow 2008).

Therefore, caution is necessary when studying the complex relationships between population and impact on the environment. In fact, it is a good starting point to remember that there are no traces of agreement on what Earth's carrying capacity is. The main reason is that this figure depends on many variables over people: diet (vegetarian, omnivorous), residence (city, village), how they move and travel (public transportation, car, walking), consumption patterns, culture, et cetera. Moreover, this capacity is sensitive to the expected (or unexpected) new possibilities introduced by emerging technologies. Distribution of population also has an influence; for instance, the fact that 83,4% of population lives in developing countries (UNDP 2019), and the increasing rural-to-urban migrations. Composition of population takes into account, among other issues, life expectancy and family income. Finally, different patterns of consumption reflect themselves accordingly in the ecological footprint. In fact, Earth's carrying capacity is a dynamic figure, "How many people Earth can support depends in part on how many will wear cotton and how many polyester; on how many will eat meat and how many bean sprouts; [...] These choices will change in time and so will the number of people Earth can support" (Cohen 1995).

SUSTAINABILITY AND HUMAN DEVELOPMENT

In their Introduction to the *Limits to growth* report commissioned by the Club of Rome, the authors stated that "It is possible to alter these growth trends and to establish a condition of ecological and economic stability that is sustainable far into the future" (Meadows 1972, p. 24). This is perhaps the first instance, in a research work, of the term *sustainable* with its present-day implications. Later (UN, 1987) *sustainable development* was a key concept in *Our Common Future* (also known as Brundtland Report) by the World Commission on Environment and Development: "3.30 [...] sustainable development is [...] a process of change in which the exploitation of resources, the direction of investments, the orientation of technological development, and institutional change are made consistent with future as well as present needs". Today, sustainability is one of the more frequent issues in EE, although the meaning of this concept is still controversial and used with different purposes in different branches of knowledge. A first distinction of this concept consists of the contrast between weak sustainability and strong sustainability. The former is common in traditional economics, mostly interested in capital (the capacity to produce goods and services) and in present individual well-being. The capital is human, manufactured, natural or social, and it is assumed that these different types of capital are substitutable within certain limits. The maintenance of the total capital stock is what finally matters. The latter interpretation, adopted by ecological economics, claims that both natural and human-made capital are complementary and, a contribution to human welfare. As such, they should be preserved as they are, without substitutions. In this sense, strong sustainability is concerned with not just present individuals, but also with communities and future generations.

The seriousness of the current threats pending over human and non-human life on Earth asks for the acceptance of the demands assumed by strong sustainability. To begin with, to take into account that sustainability not only refers to economics and natural environment, it also requires the protection of social assets as knowledge, institutions, freedom, values and cultures, to name a few. This characteristic is made explicit in the Preamble of the 2030 Agenda for Sustainable Development approved the 25 September 2015 by all the United Nations members (UN 2015): "The 17 Sustainable Development Goals and 169 targets which we are announcing today [...] are integrated and indivisible and balance the three dimensions of sustainable development: the economic, social and environmental". Heyd (2010) also points to the weight of values present in cultural frameworks when analyzing what factors make it difficult to put into practice effective policies against climate change. Paralleling Kant's well-known first formulation of categorical imperatives for our moral duties ("act only in accordance with the maxim through which you can at the same time will that it becomes a universal law"), Riechmann (2000) states that universalizability is a necessary condition for an action to be sustainable. This condition underlines the importance of recognizing moral limits to what I may do by thinking of what the total effects on the planet would be if everybody did as I do. Moreover, this responsibility includes the expected effects both today and in the future. Also, irreversibility is an alert for unsustainability. The exploitation of finite natural resources should not go on without a previous careful consideration of how related ecosystems are affected and how their exhaustion could limit the options or capacities of future generations.

There is a widespread consensus on the fact that underdevelopment is an undesirable state of affairs for societies, since this word implies incompleteness or insufficiency in some aspects that could be improved. However, its opposite is not always seen as something unwelcome, even when overdevelopment implies a dangerous excess. The pattern of unlimited economic growth followed by rich countries requires a consumerism that is a form of excessive, unhealthy development, which threatens the sheer human well-being it supposedly should provide (Jackson 2017). The GDP indicator appeared first in a US Congress report delivered in 1934 (Kuznets 1934), its purpose was to provide a measure of the nation's productivity. In his document the author explicitly alerted readers (pp. 6-7) to a possible misunderstanding of the indicator: "Economic welfare cannot be adequately measured unless the personal distribution of income is known. And no income measurement undertakes to estimate the reverse side of income, that is, the intensity and unpleasantness of effort going into the earning of income. The welfare of a nation can, therefore, scarcely be inferred from a measurement of national income as defined above". Since this warning, further criticisms have been argued whenever GDP is used to infer human well-being or social progress in a country (Giannetti 2015). They are based, for instance, on the fact that GDP does not take into account some important aspects like distribution of growth, volunteer work, negative externalities or a country's natural capital value. At the same time, crimes, natural disasters and war can increase the GDP. There exist several social indicators that include aspects of progress ignored in GDP and that are used by well-known non-governmental organizations. These alternatives provide an appropriate measure of human well-being in societies. Among them, the Sustainable Development Index, the Human Development Index, the Gender Inequality Index, the OECD Better Life Index and the Happy Planet Index.

Human well-being can be evaluated by considering neither subjective categories nor the material means of enhancing well-being. The capabilities approach sets a quite different framework. It claims that the freedom to achieving well-being has uppermost moral importance, and that this freedom depends on people's real capabilities, that is, their opportunities to be and do what they hold in high esteem. Also, valuable capacities are incommensurable, so that no measure of progress can never be reduced to a single number. In this sense, this approach rejects normative theories that rely exclusively on utility when formulating moral judgments. A. K. Sen and M. Nussbaum are two of the main proponents of this broad perspective (Sen 1984, 2013; Nussbaum 2000, 2011) while Holland (2014) considers it in the context of environmental issues. Nussbaum is close to defining a theory of justice from the definition of 10 central capabilities (needs or values) necessary for all human-beings, while Sen rejects the possibility of a general formulation of social justice because, he argues, the selection of a set of valuable capacities will always depend on a particular context,and should remain open to future changes. The central functional capacities formulated by Nussbaum (2011) are concerned with life; bodily health; bodily integrity; senses, imagination and thought; emotions; practical reason; affiliation; other species; play, and control over one's environment. The capabilities approach has played a part in the definition of some of the main indices used to evaluating different aspects of sustainable development. This has been a significant contribution to a better understanding and a sensitive evaluation of both sustainability and human development.

FUTURE RESEARCH DIRECTIONS

The previous sections of this chapter have shown the special characteristics of EE. Among them there are their tight links to different areas of knowledge. In this sense, research in EE is open to many issues and subjects. The following list includes some topics of special interest in current research:

- Moral status of nonhuman living beings and the intrinsic value of biological diversity.
- Costs and impacts of green technology.
- Renewable sources of energy.
- Ecosystem management.
- Environmental conflicts and environmental justice.
- The role of growth in human well-being.
- Rights of future generations.
- Women's empowerment (both political and economic).
- Overpopulation (concept, moral issues and management).
- Ethical aspects of vegetarianism and veganism.
- Genetically modified organisms in agriculture.
- Sacred dimension of nature.
- Cultural aspects of sustainable development.
- Environmental activism and rebellion.

CONCLUSION

The threats introduced by some environmental challenges that humanity have to deal with today have transformed EE into a singular subject, that is, it has become a clearly purpose-oriented subject. EE cannot simply get involved in an intellectual exercise with concepts and ideas. We need from it some specific guidance that could be useful in dealing with the difficulties and doubts we are facing worldwide.

This chapter has pointed to the complexity of the interrelationships present in EE. It is not just a field of interest for philosophy, rather it is frequently an issue for ecology, politics, economics, psychology and law, among others. In this sense, all of these disciplines have introduced several heterogeneous points of view on this particular branch of applied ethics. And what is more, there are reasons enough to argue that EE has to play a key role in any attempt to define a global ethics for our time. Four of these reasons are: a) the inherent global reach associated to EE; b) the interdependences that it shows between nature and culture (or society); c) the emphasis that it puts on the well-being of future generations, and d) its extension of moral standing to non-human beings. Somehow we need to reconcile ourselves with the environment—people in particular living in richer countries—so that some broken equilibrium could be restored as much as possible. This healing will require deep changes in the way we live and relate to life. Here it seems that EE offers a suitable framework for devising our way back home.

REFERENCES

Anchorage Declaration. (n.d.). Retrieved from https://unfccc.int/resource/docs/2009/smsn/ngo/168.pdf

Attfield, R. (1983). *The Ethics of Environmental Concern*. Basil Blackwell.

Attfield, R. (1994). Reasoning about the Environment. In R. Attfield (Ed.), *Environmental Philosophy: Principles and Prospects* (pp. 135–149). Avebury and Brookfield.

Attfield, R. (2003). *Environmental Ethics: An Overview for the Twenty-First Century*. Polity Press.

Australian Academy of Science. (n.d.). *Population and Environment: A Global Challenge*. Retrieved from https://www.science.org.au/curious/earth-environment/population-environment

Callicott, J. B. (1998). The Conceptual Foundations of the Land Ethic. In M. E. Zimmerman, J. B. Callicott, G. Sessions, K. J. Warren, & J. Clark (Eds.), *Environmental Philosophy: From Animal Rights to Radical Ecology* (pp. 101–123). Prentice Hall.

Carson, R. (1962). *Silent Spring*. Houghton Mifflin.

Center for Sustainable Systems. (n.d.). *Environmental Justice Factsheet*. Retrieved from http://css.umich.edu/factsheets/environmental-justice-factsheet

Chertow, M. R. (2008). The IPAT Equation and Its Variants. *Journal of Industrial Ecology*, *4*(4), 13–29. doi:10.1162/10881980052541927

Coen, J. E. (1995). Population growth and earth's human carrying capacity. *Science*, *269*(5222), 341–346. doi:10.1126cience.7618100 PMID:7618100

COMEST. (n.d.). *The Precautionary Principle*. Retrieved from https://unesdoc.unesco.org/ark:/48223/pf0000139578

Deloitte. (2017). *Global Powers of Luxury Goods 2017*. Retrieved from https://www2.deloitte.com/content/dam/Deloitte/global/Documents/consumer-industrial-products/gx-cip-global-powers-luxury-2017.pdf

Devall, B., & Sessions, G. (1985). *Deep Ecology: Living as if Nature Mattered*. Peregrine Smith Books.

Diamond, I., & Orenstein, G. F. (Eds.). (1990). *Reweaving the World: The Emergence of Ecofeminism*. Sierra Club Books.

Easterlin, R. A. (1974). Does economic growth improve the human lot? Some empirical evidence. In P. A. David & M. W. Reder (Eds.), *Nations and Households in Economic Growth* (pp. 89–125). Academic Press. doi:10.1016/B978-0-12-205050-3.50008-7

Fanning, A. L., & O'Neill, D. W. (2019). The Wellbeing–Consumption paradox: Happiness, health, income, and carbon emissions in growing versus non-growing economies. *Journal of Cleaner Production*, *212*, 810–821. doi:10.1016/j.jclepro.2018.11.223

Gardiner, S. M., & Thompson, A. (Eds.). (2017). *The Oxford Handbook of Environmental Ethics*. Oxford University Press.

Giannetti, B. F., Agostinho, F., Almeida, C. M. V. B., & Huisingh, D. (2015). A review of limitations of GDP and alternative indices to monitor human wellbeing and to manage eco-system functionality. *Journal of Cleaner Production*, *87*, 11–25. doi:10.1016/j.jclepro.2014.10.051

Gilligan, C. (1982). *In a different voice*. Harvard University Press.

Global Footprint Network. (n.d.). Retrieved from https://data.footprintnetwork.org/#/

Hardin, G. (1968). The Tragedy of the Commons. *Science*, *162*(3859), 1243–1248. doi:10.1126cience.162.3859.1243 PMID:5699198

Held, V. (2007). *The Ethics of Care: Personal, Political, and Global*. Oxford University Press. doi:10.1093/oxfordhb/9780195325911.003.0020

Heyd, T. (2010). Climate Change, Individual Responsibilities and Cultural Frameworks. *Human Ecology Review*, *17*(2), 86–95.

Holland, B. (2014). *Allocating the Earth: A Distributional Framework for Protecting Capabilities in Environmental Law and Policy*. Oxford University Press. doi:10.1093/acprof:oso/9780199692071.001.0001

Jackson, T. (2007). *Prosperity without growth: foundations for the economy of tomorrow*. Routledge.

Kari-Oca 1 Declaration. (n.d.). Retrieved from http://nafaforestry.org/ff/download/kari_oca.pdf

Kari-Oca 2 Declaration. (n.d.). Retrieved from https://www.ienearth.org/kari-oca-2-declaration/

Kheel, M. (2008). *Nature Ethics: An Ecofeminist Perspective*. Rowman and Littlefield.

Krakoff, S. A., & Rosser, E. (Eds.). (2012). *Tribes, Land, and the Environment*. Ashgate Publishing Company.

Kuznets, S. (n.d.). *National Income, 1929-1932*. Retrieved from https://fraser.stlouisfed.org/title/971

Leopold, A. (1949). *A Sand County Almanac*. Oxford University Press.

Lin, D., Hanscom, L., Murthy, A., Galli, A., Evans, M., Neill, E., Mancini, M., Martindill, J., Medouar, F.-Z., Huang, S., & Wackernagel, M. (2018). Ecological Footprint Accounting for Countries: Updates and Results of the National Footprint Accounts, 2012–2018. *Resources*, *7*(3), 58–79. doi:10.3390/resources7030058

Meadows, D. H. (1972). *The limits to growth*. Universe Books.

Mies, M., & Shiva, V. (1993). *Ecofeminism*. Zed Books. doi:10.5040/9781350219786

Næss, A. (1973). The Shallow and the Deep Long-Range Ecology Movement. *Inquiry*, *16*(1), 95–100. doi:10.1080/00201747308601682

Næss, A. (1989). *Ecology, Community and Life Style*. Cambridge University Press. doi:10.1017/CBO9780511525599

Nixon, R. (2011). *Slow Violence and the Environmentalism of the Poor*. Harvard University Press. doi:10.4159/harvard.9780674061194

Norton, B. G. (1991). *Toward Unity among Environmentalists*. Oxford University Press.

Nussbaum, M. (2000). *Women and Human Development: A Study in Human Capabilities*. Cambridge University Press. doi:10.1017/CBO9780511841286

Nussbaum, M. (2011). *Creating Capabilities: The Human Development Approach*. Belknap Press. doi:10.4159/harvard.9780674061200

Plumwood, V. (1993). *Feminism and the Mastery of Nature*. Routledge.

Regan, T. (1983). *The Case for Animal Rights*. University of California Press.

Riechmann, J. (2000). *Un mundo vulnerable. Ensayos sobre ecología, ética y tecnociencia*. Los Libros de la Catarata.

Sandler, R. (2007). *Character and Environment: A Virtue-Oriented Approach to Environmental Ethics*. Columbia University Press.

Schlosberg, D. (2007). *Defining Environmental Justice: Theories, Movements, and Nature*. Oxford University Press. doi:10.1093/acprof:oso/9780199286294.001.0001

Schmidtz, D. (2011). Respect for Everything. *Ethics, Policy & Environment*, *14*(2), 127–138. doi:10.1080/21550085.2011.578359

Schweitzer, A. (1923). *The Philosophy of Civilization*. A. & C. Black Ltd.

Sen, A. K. (1984). Well-Being, Agency and Freedom: The Dewey Lectures 1984. *The Journal of Philosophy*, *82*(4), 169–221.

Sen, A. K. (2013). The Ends and Means of Sustainability. *Journal of Human Development and Capabilities*, *14*(1), 6–20. doi:10.1080/19452829.2012.747492

Singer, P. (1990). *Animal Liberation: A New Ethics for Our Treatment of Animals*. Avon Books.

Stone, C. D. (1972). Should Trees Have Standing? Toward Legal Rights for Natural Objects. *Southern California Law Review*, *45*, 450–501.

Taylor, P. W. (1981). The Ethics of Respect for Nature. *Environmental Ethics*, *3*(3), 197–218. doi:10.5840/enviroethics19813321

Taylor, P. W. (1986). *Respect for Nature: A Theory of Environmental Ethics*. Princeton University Press.

UCS. (n.d.). *Union of Concerned Scientists Report*. Retrieved from https://www.ucsusa.org/resources/each-countrys-share-co2-emissions

UN. (1987). *Our Common Future*. Retrieved from https://sustainabledevelopment.un.org/content/documents/5987our-common-future.pdf

UN. (2007). *United Nations Declaration on The Rights of Indigenous Peoples*. Retrieved from https://www.un.org/development/desa/indigenouspeoples/declaration-on-the-rights-of-indigenous-peoples.html

UN. (2015). *Transforming Our World: The 2030 Agenda for Sustainable Development*. Retrieved from https://sustainabledevelopment.un.org/post2015/transformingourworld

UN. (2019). *World Population Prospects 2019*. Retrieved from https://population.un.org/wpp/

UN Women. (2014). *World Survey on The Role of Women in Development 2014: Gender Equality and Sustainable Development.* Retrieved from https://www.unwomen.org/en/digital-library/publications/2014/10/world-survey-2014#view

UNDP. (2014). *Environmental Justice. Comparative Experiences in Legal Empowerment.* Retrieved from https://www.undp.org/content/undp/en/home/librarypage/democratic-governance/access_to_justiceandruleoflaw/environmental-justice---comparative-experiences.html

UNDP. (2019). *Human Development Data (1990-2018).* Retrieved from http://hdr.undp.org/en/data

UNESCO. (n.d.). *The Rio Declaration on Environment and Development.* Retrieved from http://www.unesco.org/education/pdf/RIO_E.PDF

Warren, K. J. (1990). The Power and the Promise of Ecological Feminism. *Environmental Ethics, 12*(2), 125–146. doi:10.5840/enviroethics199012221

WCEL. (n.d.). *IUCN World Declaration on the Environment Rule of Law.* Retrieved from https://www.iucn.org/commissions/world-commission-environmental-law/wcel-resources/environmental-rule-law

White, L. (1967). The Historical Roots of Our Ecological Crisis. *Science, 155*(3767), 1203–1207. doi:10.1126cience.155.3767.1203 PMID:17847526

ADDITIONAL READING

Frasz, G. (1993). Environmental Virtue Ethics: A New Direction for Environmental Ethics. *Environmental Ethics, 15*(3), 259–274. doi:10.5840/enviroethics199315319

Friends of the Earth. (2018). *C40 Cities: Why Women Will Save the Planet.* Zed Books.

Hourdequin, M. (2015). *Environmental Ethics. From Theory to Practice.* Bloomsbury Academic.

Jamieson, D. (Ed.). (2001). *A Companion to Environmental Philosophy.* Blackwell Publishing. doi:10.1002/9780470751664

Pojman, L. P., Pojman, P., & McShane, K. (2015). *Environmental Ethics: Readings in Theory and Application.* Cengage Learning.

KEY TERMS AND DEFINITIONS

Anthropocene: Point of view which defines Earth's most recent geologic time period as being shaped by humans (anthropogenic), based on the overwhelming power that humans have to alter the main process that rule all the ecosystems present in the biosphere.

Anthropocentrism: Philosophical theory asserting that human beings are the most significant entities in the world. It considers human beings as superior to nature and holds that human life has intrinsic value while other beings have instrumental value for the humankind.

Biocentrism: Non-anthropocentric philosophical theory postulating that all living organisms have intrinsic value and equal moral standing. Therefore, human needs and interests are not prior to those of other living entities.

Deep Ecology: A social movement characterized by the recognition of the intrinsic value of all living beings. It also constitutes a criticism of development based on the production and consumption models promoted by industrialization culture.

Ecocentrism: Holistic philosophical theory arguing that natural entities like the ecosystems (with all their elements and processes) have intrinsic value. In doing so it extends ethics to the whole biosphere.

Intergenerational Ethics: It is the branch of ethics that analyzes if current inhabitants of Earth have the moral obligation of conserving life's vital resources for future generations.

Sentiocentrism: Also sentientism. It is the philosophy that all sentient beings, those who can feel and suffer, have intrinsic moral value. Therefore, human beings owe moral consideration to them.

Sustainability: It refers to the quality of living in such a way that it does not endanger current and future potential to satisfy human needs and hopes.

Chapter 3
The Rise of Ethical Concern for Animals as a Major Social Phenomenon

Bernard E. Rollin
Colorado State University, USA

ABSTRACT

The second half of the 20th century represented a major rise in new ethical concerns, including, in the 1970s, the moral status of animals. Until then, analgesia was never used in veterinary medicine, even though many modalities were already known to control pain. This author and others wrote American law that required effective analgesics, which the US Congress mandated in 1985. The research community eventually recognized that failure to control pain in animal research subjects involved not only immoral pain to the animals but also caused invalid research results. Public concern in these areas also spread to farmed animals in intensive agriculture settings, where it was perceived that the profit motive had replaced the husbandry ethic. The increasing number of people owning companion animals also influenced their view of farm animals' moral status and others. The larger number of companion animal owners results from the alienation of human social relations, with animals' replacing the emotional value of human companionship.

INTRODUCTION

The second half of the 20th century represented a major rise in new ethical concerns. From civil rights to attention to gender equality, thinking that was at best nascent in society emerged as full-blown moral concerns. Among these concerns that surfaced in the 1970s, was societal attention to the moral status of animals.

The pioneer in this area was Peter Singer who, drawing on historical utilitarianism, articulated a moral ideal for the treatment of animals, based on their ability to feel pain. Part of this ideal was a rejection of factory farming and animal research – in fact any invasive use of animals not to their benefit (Singer 1975). I had personal concerns about animal treatment since the 1960s,but could not see my way clear

DOI: 10.4018/978-1-7998-4117-3.ch003

to demanding an end to animal use, for I could not see society acquiescing to such a demand. Thus, my own work in animal ethics was directed at making things better for animals, rather than abolition of animal use. The basis of my arguments was respect for what Aristotle called an animal's *telos*, or biological and psychological natures (Rollin, 2016). My practical activities in this area began when I was asked to develop the field of veterinary ethics at the veterinary school where I was teaching. The third week of class, I discovered, much to my horror, that the students were taught surgery by using an unwanted pound dog multiple times. To be specific, students were compelled to do nine surgeries on the same dog. One of my students prevailed upon me to visit the ward where these animals were kept. Upon visiting, I found out that no analgesics, no pain killers, were given to these animals. The ward was horrific with the dogs whining and crying out of uncontrolled pain. As I talked to the nurse responsible for these animals, I found out that analgesia was never used in veterinary medicine, even though a variety of modalities were already known to control pain.

Full of righteous indignation, I immediately approached our surgery department and the Dean, and asked them to consider the image of the school if the reliance on multiple survival surgery became known to the general public, our university being a state-funded Institute. Reluctantly, they agreed and we moved the next academic year to single terminal surgery, a trend that thankfully spread to other veterinary schools. I also launched an attack on laboratory exercises that appeared to have no other purpose than to excise the idealism that drew students to veterinary medicine in the first place.

In the ensuing months, I started to explore my suspicion that the lack of pain control was not only characteristic of veterinary school in their training programs, but also was ubiquitous in the field of animal research. Despite the fact that the Congress had passed the Animal Welfare Act in 1968, that legislation did not mandate the use of analgesia, but rather recorded situations where analgesia might be used at the discretion of institutional research veterinarians. Thanks to fruitful conversations that I had held with our university's veterinarian in charge of research animals, Dr. David Neil, and prominent surgeon Dr. Harry Gorman, I learned that the failure to deploy analgesia was virtually ubiquitous, thanks to the aforementioned loophole in the Animal Welfare Act. Ironically, the first textbook of veterinary analgesia came out of CSU in the early 1960s, yet the authors do not use analgesia! Our group worked hard during the next decade to develop effective analgesic requirements, which were then carried to Congress by Rep. Patricia Schroeder of Colorado.

In 1982, I went before a congressional committee to plead our case for mandatory pain control. The congressman in charge told me that the research community claimed to use copious amounts of analgesia on a regular basis. I knew that was not the case and told him so. He responded that the burden of proof was on me to show the falsity of the research community's claim. Fortunately, I had a friend who was a librarian at the National Agricultural Library and asked him to do a literature search on analgesia for laboratory animals. He called me back and told me he could not find a single paper on the subject! I disseminated this point to Congress, who were sufficiently appalled to vote for mandatory control of pain for animals used in research, a law that went into effect in 1986. To my knowledge, this 1985 Bill was the first congressionally mandated requirement directly supporting pain control in animal welfare. In the ensuing years, with the influence of federal law, the number of papers dealing with animal pain and suffering and its amelioration increased to a level that was unimaginable!

Despite the research community's claim that this legislation would harm animal research, that was patently false. Pain and stress are widely known to deform research results, and failure to control them led to data that distorted the facts and rendered the results of animal research untrustworthy. Neglect of pain in general was also epidemic in medicine and science. One horrendous example not widely known was the failure to use *anesthesia* when performing open-heart surgery on newborn infants. My students and I publicized this until social pressure forced the elimination of this practice. Failure to use anesthesia was defended by unbelievably fatuous arguments. When I attended the pain society meeting on this, I encountered the following remarkable argument: the author of the policy against using pain control in infants asked me if I had been circumcised. When I replied that I had, he said "well you don't remember it, so why does it matter?" I in turn asked him if they had ever examined the subsequent history of children who did not receive anesthesia. It turned out, not surprisingly, that such children tended to develop chronic pain in later life! By the early 1990s, failure to use anesthesia in infants had been abolished.

In addition to pain, there are vast numbers of aspects of animal nature or *telos* that most people do not notice. Yet the violations of these interests have major consequences not only for the animals, but for their functioning as models in science as well. Virtually any aspect of animal nature can, if violated, have myriad consequences for the animals' metabolism, physiology, and normal functioning, as well as for the animal's welfare. There is a vast literature detailing such considerations, as we will shortly indicate. Yet the average researcher knows nothing of this.

In the mid-1980's, I was approached by an editor of an anthology on the ethics of animal research. Knowing that I had a fair amount of knowledge concerning research, he asked me to discuss the most outrageous situation regarding animals used in science. I unhesitatingly explained that one could get an MD/PhD degree in an area using laboratory animals, and never learn anything about the animals one uses except that they model a particular disease or syndrome, and get funded to establish a research program while knowing literally nothing of the biological and psychological needs of the animals in question. This explains the extraordinary anecdote I heard on three different continents from veterinarians whose expertise was laboratory animals. In each case, the veterinarians had been approached by researchers complaining that the scientist had been supplied with "sick dogs" by the veterinarian. The basis for these complaints was the fact that the animals in question all had "fevers" above 98.6 F. What the researchers did not know, of course, is that the *normal body temperature of the dog is 101.5!* In 1989, I and a veterinarian colleague published a two-volume book on proper care of laboratory animals, a book still recommended by NIH for researchers.

Laboratory animals are kept in accordance with human convenience, not out of respect for their biological natures, or *telos,* which in most cases caretakers are not even aware of. Cages are designed for ease of cleaning, not for fitting animal needs. An excellent example is provided by rats and mice, who by nature are nocturnal burrowing creatures. Yet, for human convenience, they were kept in 24 hours of light. So inimical was this to rat nature that they experienced retinal deterioration. Monkeys, which are highly social, were kept in isolation in cages totally devoid of stimulation. Not only were animal natures not respected, they were not even known, and thus no attempts were made to accommodate them. So impoverished were environments for primates, that it took 1986 federal law to demand housing for nonhuman primates that "enhances their psychological well-being."

This sort of neglect was true for virtually every species used in research. Dogs were often caged singly with no attention paid to their socialization with humans. Cats, who are creatures that jump and climb, were not provided with perches. Whether in agriculture or in research, absolutely no attempts were made to respect animal nature, thereby putting their welfare in a highly compromised situation. In animals used in research, absolutely no attention was paid to the stressfulness of environments, despite the patent truth that even tiny variations in environment wrought havoc with physiological and metabolic variables. Increasing the cage size for rodents has been shown to reduce the toxicity of amphetamine by 50 percent. The LD50 test for acute toxicity of amphetamine in rats, that is the amount of amphetamine needed to kill half the rats (provide a Lethal Dose to 50 percent of them), increases by a factor of seven when the rats are caged in groups of 12, compared to when they are caged singly. Virtually monthly, examples are published illustrating the profound effect of housing conditions.

Improper Treatment Distorts Research Results

It is thus evident that ignoring animal *telos* in animal housing has profound effects upon one's research results. Thus, the requirements of good science dictate that attention be paid to animals' natures lest important variables relevant to the object of study may be ignored or distorted. Remarkably, compliance with emerging societal ethics for animal treatment creates a demand for precisely the same attention to animal nature! As we mentioned earlier, it is rare that the requirements of science and societal ethics converge. In order to minimize the stress that results from conflict between the requirements of research and an animal's needs and nature, we should come as close as possible to creating conditions for the animals that meet the dictates of their *telos.* Happily the very same requirement is ever increasingly being dictated by the advancements in moral obligations for animals that are being expressed in societal ethics.

Astonishing as it may be that researchers ignored both ethical and scientific reasons for respecting *telos*, the knowledge that the failure to respect animal nature wreaks havoc with scientifically relevant variables is not new. Despite the historical failure to utilize analgesia in painful experiments, as discussed earlier, simple common sense should have suggested to researchers that an animal suffering pain is not a biologically normal animal. Metabolic variables and physiologic variables are disturbed by the failure to control pain as are immunological parameters and, correlatively, resistance to disease. There is also data showing the converse. Animals who are treated exceptionally well have a significantly higher degree of reproductive success, and rabbits who are treated with exceptional care develop significantly fewer atherosclerotic lesions when fed a high cholesterol diet. The attempt to develop a cage for mice that was considerably more congenial to mouse *telos* beginning in the 1990's was met with extreme criticism on the grounds that it would vitiate, invalidate, and at least call into question all the previous mouse data accumulated from mice housed in cages designed for human convenience, not mouse natures!

Even something as putatively trivial as uncorking a bottle of ether in a laboratory can have profoundly measurable effects on animals. The key point to reiterate is the degree to which the failure to create environments accommodating the animals' *teloi* is as harmful to science as to ethics. Thus, we find in the area of animal research a scientific as well as an ethical reason to learn as much about animal needs and natures as possible, in order to assure the validity of the animal as a model. As is well known, the more we minimize and control extraneous effects on animals, the sounder will be our science. Thus, by accommodating the dictates of an animal's *telos,* we do good while we do well! Thus, albeit unwittingly, we find a sound reason to unify ethical and scientific considerations. And once again, we find a powerful reason for viewing animals in terms of their *telos*, their biological and psychological natures. If we look at animals in a reductionistic way, strictly in terms of physics and chemistry, we not only lose sight of what makes an animal what it is, we guarantee deformation of the variables central and vital to our inquiry!

Attendant on the points we have just made, a startling conclusion emerges for animal research. As indicated earlier, the laws governing animal research are implemented by animal care and use committees consisting overwhelmingly of scientists who are suffused with scientific ideology. In other writings, I have stressed that this system represents an extreme case of the fox watching the henhouse. This is not to suggest that we need to distrust scientists on committees. In my experience, there is a strong tendency on the part of such scientists to abide by federal law. It is rather that scientists tend to see the world through ideological glasses based on the way in which they are both trained and suffused. Thus, they will inevitably miss relevant considerations filtered out by that ideology. In addition, scientific research is largely done with public money in the name of public interest. Yet when ordinary people see the world through common-sense, ethics-based lenses, they see a good deal that scientists do not and about which scientists tend not to care. As we have indicated, ordinary people now care a great deal about the metaphysics of ordinary common sense, and about the ethical issues to which science gives rise. For these reasons, ordinary people should have a significant voice in deciding what research is to be funded and how it is to be conducted.

To the standard complaint that only scientists can understand what the scientists do, there is a simple rejoinder. Science has become so specialized that most scientists are essentially lay people outside their own immediate area. In addition, the purpose of a piece of scientific research can be explained to intelligent lay people by intelligent scientists. If public money is the fuel that drives science, in a democratic society the public should have a significant voice in how that money is expended, both in terms of what areas are researched, and how ethically that research is conducted!

There are some limited but promising indicators that the research community is moving in that direction. In the most recent edition of the NIH *Guide to the Care and Use of Experimental Animals,* there is a directive that social animals should be housed in a manner that respects their sociality. There is also a directive promoting the adoption of adoptable animals at the end of a research project. Furthermore – and this is a highly significant tip of the hat to common sense and social morality – cost to the animals utilized in research project should be weighed against the benefit likely to be produced by the research. As we approach a new era in society wherein the general public concerns itself far more dramatically with both the logic and ethics of research, and no longer fully abides by "trust me, I'm a scientist," the need for incorporating ethics into science becomes far more pressing.

Confinement Agriculture and Animal Ethics

Suspicion of lack of pain control in research and medicine are not the only reasons that concern over animal ethics suddenly blossomed in the second half of the 20th century. Also significant was the end of husbandry as a basis for agriculture. Traditionally, as far back as the beginning of animal domestication 12,000 years ago, animal agriculture was based in exemplary care – putting the animals into environments where they would not only survive, but thrive, and providing them with food during famine, water during drought, protection from predation, whatever medical attention was available, and so forth. This strategy was solidly based in self-interest. The better the animals did, the better the producers fared. Even for those producers who did not particularly value animal welfare, the realization that their own fate and success were solidly linked with the fate and survival of their animals militated in favor of good animal treatment.

The link between agriculture and good husbandry persisted until the early 18th century, when much of agriculture became industrialized and husbandry began to vanish. With the rise of industrial agriculture, husbandry was replaced by emphasis on efficiency and saving money. This grew gradually more prevalent until the 1960s, when such groups as the Brambell commission refocused attention on good care. Concern for farm animal welfare has continued to develop in the ensuing years, until it is of concern to large numbers of consumers in most Western countries. The majority of consumers are in fact horrified by keeping animals in small, impoverished enclosures, unable to perform the most basic activities characteristic of their natures. There are battery chicken operations where chickens are literally living on the backs of other chickens one layer below them.

The beef cattle business has remained to some extent husbandry- based, with the notable exception of feedlots, which crowd many cattle into a relatively small pen, and depend upon antibiotics to prevent disease. This in turn, has helped in a Darwinian way to develop microbial resistance to antimicrobials. The dairy industry is even worse, with dairy cattle being raised in concrete- based confinement. As one of my friends, for 50 years a dairy veterinarian, has affirmed, most cattle never even see a blade of grass! The swine industry is a very similar story, with five large, corporate entities producing 95% of the pork and raising animals in large, cramped industrial facilities also depending heavily upon antibiotics. Similarly, with the broiler industry. Given that children have universally been raised on books and stories depicting animal agriculture as clean and green, it is no wonder that people are horrified by confinement agriculture.

Change in the Social Status of Companion Animals

While all of the reasons we have tendered help explain the rise of animal ethics in society, perhaps the most significant is a change in societal structure during the second half of the 20th century. This has implications for the status of companion animals. Prior to that point in time, companion animals were one step above prized trinkets.

According to repeated surveys, today a minimum of 89% of dog owners and up to a maximum of 99% claim to view their animal companions as "members of the family".

One can conjecture with some plausibility why this societal change in attitude has occurred: We are societally isolated from one another, particularly in urban and suburban locales. *Gesellschaft*, i.e., disconnected aggregates of individuals, has ever-increasingly replaced *gemeinschaft*, i.e., tightly bonded organic social unity. Divorce affects slightly less than half of marriages, leading to a feeling that home is no longer a stable source of comfort and belonging, particularly among children. Furthermore, career opportunities are often nonexistent in small communities, so young people are drawn to urban contexts, thereby eroding the communal sense present in such communities, and also eroding extended families. As a person who spent the first 25 years of his life in New York City, I can attest to how difficult it is to make friends, particularly for young women who inevitably attract "circling predators".

In fact, one of the few legitimate reasons for talking to and interacting with strangers in New York City is if one is pushing a baby carriage or walking a dog. Communities of people taking their dog to parks are able to form groups at the same time, creating significant camaraderie, and I was part of such a community. Even though we didn't know each other's names, and in many cases just referred to each other as "Fido's person" or "Red's owner," we cared for each other. There was one time when the owner of a giant German Shepherd required hospitalization for a week. During that period, many of us took turns taking the dog to the park. New York paranoia was thrown to the winds, and we shared a key to the owner's apartment.

A few years later, I remarked to Phil, the owner of the German Shepherd, that my physician was threatening to put me in hospital because of my recurrent asthma. The next day at the park, Phil handed me an envelope. I looked inside, and found a key. When I inquired him about it, he replied, "It is a key to my cabin in Thunder Bay, Ontario in the country. Go there and breathe". That moment beautifully exemplified the extent to which a dog can serve to break down barriers and create bonds among people.

With human bonding having reached a low point, one can understand why people bond closely with animals. In such a world, the sense of obligation to animals is significantly enhanced, and moral concern for companion animals is also greatly strengthened. Thus, it should be no surprise that the sense of moral obligation to companion animals has been strongly buttressed. And companion animals have become the paradigm example for animals in the social mind.

Service animals have proliferated, ranging from service dogs to peacocks, and universities have been forced to rescind their prohibitions against dogs in dorms and classes, as more and more students bring notes from physicians and psychologists attesting to student emotional dependence on animal companions. Wounded veterans increasingly depend on PTSD (post-traumatic stress disorder) trained animals.

The upshot of all this is clear. Ever- increasingly, animals have assumed the role of friends in our society. I remember a group of my students asking me how I can work on animal issues when there is so much child abuse. I asked them in return how many of them had been betrayed by an animal. The answer was consistently zero. I asked them how many had been betrayed by a friend, or relative, or a lover. The number of such people was consistently high. In this way, I helped them understand the degree to which animals serve as friends who are steadfastly loyal.

In short, a variety of factors have converged to create ethical consciousness of our obligations to animals. Most of these factors are reflective of major societal change. Given that the traditional ethic pertaining to animals is simply devoted to stopping deliberate cruelty, what consistent form will be emerging ethics for animals take? To answer this question, we must recall Plato's dictum about teaching ethics. Plato pointed out, in essence, that in dealing with ethics and adults, one cannot *teach,* i.e. create ethics de novo, one must remind, or, in his phrase cause them to recollect. And, obviously, the ethic we know best is that which has developed to deal with issues of human beings.

Extracting Animal Ethics from Human Ethics

The US societal ethic is embedded in the Constitution. The Bill of Rights, the basis for the American social ethic, is in essence a theory of human nature, of what is essential to functioning as a human being.

Clearly, the essence of the Bill of Rights is a theory of human nature-- we are beings who wish to believe according to our own lights; express ourselves clearly, hold on to our property, be protected from cruel and unusual punishment, and so on. A moment's reflection reveals that one can construct a similar Bill of Rights for different sorts of animals, depending upon their natures.

Ordinary human beings tend to believe that animals have natures. Even as a child, I found that my peers were extremely uncomfortable with zoos as they were constituted-- essentially prisons for animals. This is exactly the same mentality that renders confinement agriculture uncomfortable to many ordinary people. So it is easy to see that the logic behind our societal ethic is not difficult to extrapolate to animals.

Two years ago, I redid my literature search on pain control in animals and now found 12,000 papers. Certainly, things are far better than they were. Physical pain, of the sort subject to control by analgesics, is no small thing. The utilitarian (pain-centered approach to animal ethics) is very laudable, but upon close examination, reveals itself to be difficult if not impossible to implement. For example, it is extremely difficult to see how one weighs emotional pain against physical pain. An animal can suffer by virtue of the threat of a beating, as compared with the actual beating. How does one score the former in comparison to the latter? How does one score *fear*? How does one count the emotional pain of a cow whose baby is removed from her immediately upon being birthed, when we know that under normal extensive conditions she will not wean the calf for 7 to 9 months?

A *metaphysics* is a way of looking at the world. Ordinary common sense finds that objects in the world are qualitatively different-- beautiful and ugly, good and bad, green and yellow, hot and cold, living and nonliving, wet and dry. Modern science considers as real only what is mathematically expressible. There are no words in physics and chemistry to describe the qualitative distinctions taken for granted in ordinary experience. The most articulate spokesperson of commonsense metaphysics is Aristotle, who sees the world of reality as coextensive with the world of our experience, something the scientific revolution explicitly denied.

Unlike the modern scientist, where one set of laws, the laws of physics and chemistry explain all phenomena, in Aristotle's world each kind of thing obeyed its unique set of laws. For Aristotle, all things had a *nature* or *unique function,* which he called its *telos,* what it *did.* Living things were naturally explained by how they performed the *functions* of living things; sensation, locomotion, nutrition, reproduction-- all living things are characterized by these functions but perform them in unique ways. A predator like a tiger finds food and eats in a manner different from an herbivore. A bird moves very differently from a snake. To know an animal is to know its *telos,* which we understand by observing numerous instances of the kind of creature it is. Thus, we understand animals, not by reducing what they do to the laws of physics but by observing how they fulfill the functions of a living thing.

When we begin to think of animals in moral terms, and how we manage them, the most important criterion is to respect their *telos.* If we wish to assure animal well-being, respect for the animals' *telos* is as important, or probably more important, than avoiding physical pain. The pain of a surgical procedure can be rendered tolerable and minimized by the proper use of analgesics. But the suffering engendered by keeping a social animal caged individually is not remedied by drugs. The separation of calf from cow at one day of age is not alleviated by narcotics. A clearer example is provided by coyotes and wolves that are caught in steel jawed traps. These animals have been known to chew off their paws in order to escape. Since the physical pain of being held in a trap is far less horrendous than the pain of chewing off one's limb, we must conclude that being kept immobilized is a far worse experience than very severe physical pain to the animal.

Clearly, if we wish to assure the well-being of research animals, and are doing a good job controlling physical pain, the next step is to make sure the animals' *telos* needs are respected in the way in which they are kept. Primates simply presented with food, surely a very basic drive, will often not eat unless the food is randomly distributed in their cages, requiring foraging. Many animals will withstand pain and injury to escape severe confinement. These examples illustrate that things besides pain matter to animals; other situations may be worse than physical pain; any *telos* violation creates a negative experience.

Further, there is no simple word to express the many ways we can hurt animals besides causing them physical pain; the ways are as countless as possible telos violations. "Positive mattering"—the many states that matter to animals in a positive way-- of course would include all states that are positive for the animal – freedom of movement, pleasure, a sense of security, companionship, play.

CONCLUSION

We are morally obligated to expand the scope of biological science so that it studies all of the ways things can matter in a positive or negative way to animal well-being. We also need to understand which forms of *telos violation* matter most to animals, and how. And we need to determine this without harming the animals.

When we begin to think of animals in moral terms, and how we manage them, the most important criterion is to respect their *telos.* If we wish to assure animal well-being, respect for the animals' *telos* is as important, or probably more important, than avoiding physical pain. The pain of a surgical procedure can be rendered tolerable and minimized by the proper use of analgesics. But the suffering engendered by keeping a social animal caged individually is not remedied by drugs. The separation of calf from cow at one day of age is not alleviated by narcotics.

Further, there is no simple word to express the many ways we can hurt animals besides causing them physical pain; the ways are as countless as possible telos violations. "Positive mattering"—the many states that matter to animals in a positive way-- of course would include all states that are positive for the animal – freedom of movement, pleasure, a sense of security, companionship, play.

To summarize, replacing the notions of pleasure and pain with that of *telos* and the needs and interests flowing therefrom creates major benefits. It helps us to better understand our obligations to animals. It stops us from arraying harms as variegated as creating fear, striking an animal, illness, grief and loneliness on a single axis – pain. It also fits with common sense, which has no difficulty recognizing mental states in animals. Finally, it rationalizes talking of *happiness* in animals, which means satisfying most (or all) of an animal's needs and interests arising from its *telos.*

REFERENCES

Clough, G. (1982). Environmental Effects on Animals Used in Biomedical Research. *Biological Reviews of the Cambridge Philosophical Society, 57*(3), 487–523. doi:10.1111/j.1469-185X.1982.tb00705.x PMID:6982731

Rollin, B. E. (1980). *Animal Rights and Human Morality.* Prometheus Books.

Rollin, B. E. (2016). *A New Basis for Animal Ethics: Telos and Common Sense.* University of Missouri Press.

Rollin, B. E., & Kesel, M. L. (1990). *The Experimental Animal in Bioomedical Research.* Boca Raton: CRC Press.

Singer, P. (1975). *Animal Liberation.* New York: New York Review Press.

Chapter 4

Ethics of Security: From Personal Safety to Cyber Security

Tommi Lehtonen
https://orcid.org/0000-0002-0286-3059
University of Vaasa, Finland

ABSTRACT

This chapter aims to identify and analyse the ethical problems of security, particularly cyber and digital threats. The concepts of security and safety are defined based on existing literature. The chapter addresses the key results and research gaps in the field (i.e., security issues in different areas) and future challenges, both theoretical and empirical. Moreover, the discussion is linked to an analysis of the relationship between utilitarian ethics and deontological ethics, which brings a new perspective to the debate on security ethics in general and cybersecurity. Finally, comprehensive security and absolute safety ideas are discussed, which sheds new light on the complexity of security concerns.

INTRODUCTION

This chapter identifies and analyses the ethical problems of security, with special reference to cyber and digital threats. Amid the growing global threats of climate change and economic inequality, one might find it tempting to define ethics as prescriptions to prevent humans from hurting each other and the rest of the biosphere. From this it can be deduced that ethics is essentially about security, in various forms, and that other important aspects of ethics such as fulfilling one's duties, developing virtues and character strengths, feeling empathy and reinforcing pro-social behaviours ultimately serve the same goal: security, both personal and collective. This view means that the ethics of security is not merely a subfield or special area of ethics, but the core and epicentre of ethics. However, it is possible to remark that emphasising security easily leads to employing ethics as a means to an end, which anchors ethics firmly in one particular paradigm and tradition, namely utilitarianism or consequentialism. To this, the reply can be that nothing automatically forces one to formulate and establish ethics in general, or the ethics of security in particular, merely based on utilitarian premises, but also other theories such as deontological ethics, contractarianism, virtue ethics and care ethics play an important role in the ethics

DOI: 10.4018/978-1-7998-4117-3.ch004

of security (see Table 1). The contribution of different ethical theories to security considerations will be discussed in this chapter.

BACKGROUND

The United Nations Development Programme (UNDP) included the phrase *human security* in its 1994 report, which triggered a debate about the perception of security. That debate has challenged an earlier understanding of security that emphasised the sovereignty of national states, the importance of defence politics and international negotiations and treaties (Martin & Owen, 2015, 1). The present understanding of human security hovers around human development and human rights and even national security, but it is wider than any of them alone (Sen, 2015, 18). In addition to these wide-ranging issues, Amartya Sen points out a much more narrow and personal interest in security: the majority of people are concerned with the security of their own lives and of the lives of their family members (ibid.). Sen emphasises that this general concern has to be addressed and any understanding of security can be integrated with it to the extent that this makes human life more secure. Such an integrative understanding of human security is important precisely because it affects human lives. National security and global security alone, without the integration with an individual's "world of experience", are more abstract entities and, often, more remote from everyday life. These political forms of security are frequently defined in terms of military or ecological preparedness and other modes of national and international problem-solving.

With security being a matter of personal interest, it is worthy to pay attention to the underlying psychological background. Scholars in various fileds have argued that security is a basic human need (Maslow, 1943, p. 376–380; Staub, 2003, p. 2; Doyal & Gough, 1991, p. 214–213). Famously, Maslow ranked security or safety as a psychological need second to the basic physiological needs such as air, water and food. Maslow's hierarchy of needs is often represented as a pyramid in which the bottom represents basic biological drives for survival (Figure 1). Other basic needs, such as the need for safety, love, esteem and self-actualisation, are also universal, but not absolutely necessary for survival. Moreover, a desire to sacrifice oneself for others does not necessarily go against the need for survival or safety, because self-sacrifice can serve the greater good of the survival and safety of a collective, such as a family, clan, race or nation. However, the higher one climbs in the hierarchy of needs, as more specific desires are addressed along the way, the more suspicious one may become of the universality of human needs. Maslow (1943, p. 389) said, "Certainly in any particular culture an individual's conscious motivational content will usually be extremely different from the conscious motivational content of an individual in another society. However, it is the common experience of anthropologists that people, even in different societies, are much more alike than we would think from our first contact with them, and that as we know them better we seem to find more and more of this commonness".

DEFINITIONS

With these background remarks in mind, the concepts of *security* and *safety* need to be defined and analysed. No concept exists in isolation from other concepts and linguistic practices (Gasper, 2015, 33). Therefore, to understand *security* and *safety*, it is necessary to be familiarised with the real-world contexts of users and audiences of those words, purposes within those contexts and accumulated patterns of use

Figure 1. Pyramid showing Maslow's hierarchy of needs
Source: (the author, 2019)

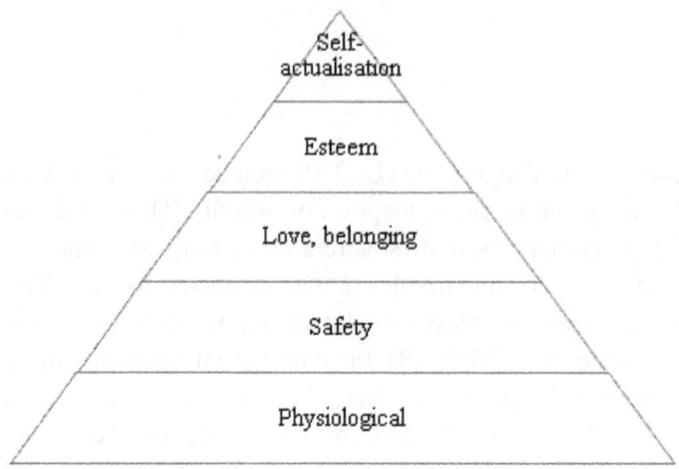

of the words concerned. In what follows, defining *security* and *safety* takes place primarily based on the United Nations characterisations and relevant research literature. Worth special mention is the *Routledge Handbook of Human Security* (2015). The definitional task under discussion is all the more important because the concepts of *security* and *safety* are ambiguous and complex, as will be seen hereafter.

Let us start with some everyday observations about the uses of the terms *safety* and *security*. It is easy to identify examples of expressions where safety means a tool or measure that reduces hazards. Safety can also mean a place or position where there is no danger (Martin & Owen 2015). Thus, in security considerations, the benefit pursued is understood in terms of the pair of safety and danger. For example, family, friends and paid work can be part of one's safety net. For many, a home is a shelter and safe haven.

Security, for its part, is a state where threats and risks can be controlled. Even if such a state can be assessed objectively, security is also a matter of subjective experience. Therefore, one person may feel safe in a situation where another person feels insecure. The objective and subjective aspects of security do not always go hand in hand.

Moreover, security is relative in two ways. First, security is always relative to a real or imagined threat. Viruses, worms and attacks from hackers can potentially threaten any computer connected to the internet. The computer system can be safeguarded from cybersecurity threats with a firewall and virus protection software. Secondly, there are degrees of security. Someone or something can be safe from a risk or danger, i.e., they cannot be affected by it. However, something else can be an even safer option, and the safest option is in many cases the most recommendable to choose, unless one is ready to accept risks such as those related to share trading.

Let us continue our definitional task by leaning on the United Nations 1994 *Human Development Report* and the 2003 report of *Human Security*. In light of them, security can be defined as the condition of being protected from harm or other unwanted outcomes caused by intentional human action. Safety, in turn, is the condition of being protected from harm and injuries, caused by unintentional hazards. Thus, the concepts of security and safety are interrelated but distinguishable based on their being caused by human or natural factors – a topic discussed by philosophers specifically in the context of the problem of evil. Human or moral evil refers to harm and suffering caused by intentional human actions or behaviour, subject to moral evaluation. Natural evil refers to natural disasters and other harm and evil caused by natural phenomena such as disease and dysfunction. However, the distinction between moral and natural evil is not razor-sharp or unambiguous. The necessary preconditions for many accidents include human practices (e.g. transportation, energy production and consumption, residential activities and industrial processes) and technology created by humans even if the accidents themselves are non-intentional, and in that respect, represent natural evil rather than moral evil (Neiman, 2015, 23). In its widest sense, the phrase *human security* refers to the condition of being protected both from intentional and non-intentional harm and from both moral and natural evils.

Des Gasper identifies different definitions of security according to their range. In the widest range, human security includes the promotion and expansion of capabilities that strengthen existing opportunities and create new possibilities for human flourishment (Gasper, 2015, 33). These capabilities include all skills and faculties considered as good and useful and which humans, both as individuals and as members of social and cultural groups, can have innately or can learn to do. Taking this into account, security can be defined as a condition under which human flourishing is maximally possible and maximally sustainable.

SECURITY AS SUSTAINABILITY

The widely accepted definition of sustainable development is that used in the Brundtland Report: "Sustainable development is development that aims to meet the needs of the present without compromising the ability of future generations to meet their own needs" (The World Commission on Environment and Development, 1987, 8). Although this is a vague statement, it has proved durable.

The term *sustainability* is often preferred over *sustainable development*. Development may be perceived as implying growth and, therefore, sustainable development means mitigating the problems caused by, but not necessarily challenging, continued economic growth (Lehtonen, 2016, 191, 193).

The concept of *sustainability* can be used in different ways in relation to security or safety. Environmentalists mean ecological well-being and security when they speak of sustainability. Many businesspeople really mean economic performance and security reflecting the efficiency of businesses when they speak of sustainability. From a more general perspective, sustainability is said to have three major aspects – ecological, social and economic – and it is not possible to achieve a particular level of any of them independently without achieving at least a basic level of all three forms simultaneously (Frigo 2018, 84). This shows that security, as a sum of sustainability, is a complex concept and phenomenon. Thus, it is not possible for subsystems to be secure and sustainable within an unsustainable global system. Security and sustainability are properties of the Earth system as a whole, including the interacting physical, chemical and biological processes and global social, political and economic developments. Therefore, a business or organisation is unlikely to be sustainable if the society at large is not secure.

PEOPLE-CENTRED SECURITY

In the current understanding of security, the 1994 *Human Development Report* and the 2003 UN report of *Human Security* are the major sources. The *Human Development Report* points out that for many people today, a feeling of insecurity arises more from worries about everyday life than from the fear of a global catastrophe, even if many people are concerned about climate change. Job, income, health, one's own children – these are the major human concerns all over the world. Therefore, the report defines security as "safety from the constant threats of hunger, disease, crime and repression". Thus, the report sees security, both individual and social, as the condition of being protected from both moral and natural evils. Moreover, the report characterises security as "protection from sudden and hurtful disruptions in the pattern of our daily lives – whether in our homes, in our jobs, in our communities or in our environment" (UNDP, 1994, 3). Thus, the report takes into account both individual and social dimensions of security.

In the same report, human security is considered a global concern and relevant to people both in rich countries and poor. Unemployment, alcohol and drug abuse, crime, pollution and violations of human rights pose real threats to many people. These threats are persistent and continuously evolving, even though their intensity differs from one place to another. In addition, different security concerns are interdependent. When the security of people is jeopardised anywhere, other nations may get involved through various relationships and dependencies. Famines, pandemics, pollution, drug trafficking, terrorism, ethnic disputes and social disintegration are not confined within national borders. Their consequences can affect all people. Therefore, human security is easier to ensure through early prevention than later intervention (UNDP, 1994, 22-23).

According to the *Human Development Report*, threats can be accumulated and can lead to persisting disparity between ethnic groups or between regions. Violence can follow from social inequalities in power and economic opportunities. When people see their security threatened, they often become less tolerant, as the rise of nationalism and right-wing populism worldwide shows (UNDP, 1994, 23).

The 2003 UN report links human security to the protection of fundamental freedoms. In this vein, the report outlines security as the protection of people from critical and pervasive threats. That protection is achieved by collaboration between social systems representing different interests and values such as environmental, economic and cultural. The social systems concerned include democratic decision-making, the armed forces and environmental protection organisations, among others, and together they establish the requirements for a dignified life and human survival (United Nations Commission on Human Security, 2003, 4).

Both UN documents considered here argue for the concept of security that is people-centred. The reports see that security consists of different interrelated dimensions such as environmental, economic and military. The lack in one dimension does not usually propel a displacement into another. Rather, the lack of security in one dimension is apt to add to insecurity in another dimension. For example, environmental threats can lead to economic threats, which can undermine mood and expectations in a society and can result in depression and poverty. These interlinked factors create a web of risks that threatens human security. Therefore, to analyse the relational ontology and moral psychology of threats is necessary for understanding the requirements for security. With this in mind, a short overview of security concepts is provided from the perspective of different ethical theories.

SECURITY FROM THE PERSPECTIVE OF DIFFERENT ETHICAL THEORIES

According to utilitarianism (or consequentialism), the ethics of security aims at benefits (e.g. peace, well-being and prosperity) and safety necessary for both individuals and societies. In the ethics of duty or deontological framework, the ethics of security focuses on duties, related to the protection of basic and human rights, the following of which is a universal responsibility and which the human reason can recognise as a universal obligation. The Universal Declaration of Human Rights (UDHR) and the United

Nations reports on human security remain in this deontological framework. In contractarianism or a social contract framework, the ethics of security focuses on treaties and conventions made for enhancing security and stability in a society and between countries. Virtue ethics tries to determine and explain the character and communal traits that are necessary for security and that can be created by upbringing and education. Ethics of care, for its part, sees security both as the fruit and foundation of caring relationships and practices.

Table 1. The approach to security in different ethical traditions

	Idea of Security	Why Is Security Important?
Utilitarianism	Security is the goal of ethics and the ultimate benefit of doing good deeds.	Security is required to achieve a variety of other good things (e.g. peace, well-being and prosperity) important for a society.
Ethics of duty	The ultimate ethical duty is to contribute to the security of individuals and communities.	To actively contribute to the security of other persons means treating them as an end in themselves.
Social contract theory	Security is based on legislation, treaties and conventions.	Security is one of the benefits all members of a society can benefit from and one of the hallmarks of a good society.
Virtue ethics	Security requires peaceful and just societies with individuals who have virtues such as practical wisdom, courage, modesty and fairness.	Living in a safe society under good governance enhances the possibility of developing virtues and character strengths.
Ethics of care	Caring brings about security that is the basis for the healthy development of children and adolescents.	Security is both the fruit and foundation of care.

Source: (the author, 2019)

The foremost task of deontological ethics is to demarcate the area of permitted conduct and actions from the area of prohibited conduct and actions. In the field of cyber and digital security, this means building and programming security systems in which the algorithms contain demarcation rules to distinguish between permitted and forbidden "moves". These demarcation rules are comparable to the constitutive rules that define different games (Myerson 1997). In order to realise the demarcation requirement, an algorithm may need to refer to a list of forbidden activities. An advanced deontological system can also learn new rules and situational exceptions based on relevant documents such as legal texts, policy papers and scientific studies (Togelius 2011).

The role of utilitarian ethics in a computerized security system, for its part, is to optimise the decision-making and operations in the area of safe activities and unsafe activities. In this view, utilitarian ethics can be compared to strategic rules that define how the game is played optimally and with a winning strategy. This requires the creation of algorithms that can compute feasible and optimal courses of action in different situations (Boddington 2017; Bonnefon et al. 2015, 2016; Brant 2016). The classic utilitarian maxim paves the way for such an algorithm: "Get the best possible benefit for the largest number of people".

APPROVING RECOGNITION AS A KEY PREMISE OF SECURITY

According to Hegel, human life in society is the struggle for recognition. Although this characterisation has most often been understood to refer to the value and identity of individuals, the concept of recognition is also relevant when considering the safety and security of groups and communities such as ethnic groups and cultural minorities. By recognition, Hegel means identifying a person as having certain rights and responsibilities. By means of the United Nations Universal Declaration of Human Rights (UDHR) (1948), the member countries recognise human individuals and collectives (e.g. people, nations; ethnic, religious and cultural groups; minorities) as having certain rights and responsibilities. Article 3 of the Universal Declaration of Human Rights guarantees security of the person, and the European Convention on Human Rights also mentions and protects that right.

If an individual or a collective has been granted certain rights, such as security and freedom of expression, that is usually taken to imply that the individual or collective in question should assume the related duty to respect the rights of others. Thus, there is a universal duty to respect the security of other persons. However, this is not automatically the case as individuals and groups can easily overlook the fact that everyone else has the same basic rights. This is one reason why the requirements and conditions for recognition are a relevant topic for both individuals and societies.

Axel Honneth points out that the denial of recognition creates social struggles and undermines security. Specifically, the emotional experiences related to attitudes and actions of others towards oneself make one feel that he or she is being recognised or being denied recognition. If recognition is denied, one may feel insecure. Negative emotions can reveal to one that an injustice is taking place or that one is not being adequately recognised. However, negative emotions do not automatically reveal that one is disrespected or unsafe (Honneth, 1995, 138). What they provide is the possibility to identify an injustice or insecurity to be opposed. In this way, the experience of insecurity is a psychological force behind social struggles. Further, experiences of insecurity can provide the basis for political struggles in circumstances conducive for a social movement (Honneth, 2007). Insecurity can also result in retreat, or random resistance. The recognition of universal human rights, security included, attempts to prevent these maladies. This analysis locates the source and justification of social struggles in the lack of recognition (Fraser & Honneth, 2003, 174; Honneth, 1995, 168). Thus, for social and political security, recognition is a necessary condition.

CYBER SECURITY AND NEIGHBOURING CONCEPTS

It has already become clear that security is a complex concept and phenomenon. This complexity also concerns cyber security. In what follows, new light is shed on the major challenges of cyber security by identifying and defining key concepts that hover around the security risks related to digital technology.

The world relies on digital technology and algorithms more than ever before. As a result, digital data creation, use and storing soar. Today, individuals and collectives, including industries and governments, store data on computers and transmit it across networks to other computers. This global data system has vulnerabilities that can undermine the rights and well-being of individuals and damage the objectives of organisations (Eggenschwiler, 2017; Fichtner, 2018).

Moreover, secret and anonymous algorithms threaten privacy and even the rule of law. Imposing fines or taxes, judging somebody guilty or sending people to jail because of the inexplicable, unchallengeable judgments of computer programmes undermines legal systems. Such programmes and algorithms are "black boxes", i.e., systems, whose inputs and outputs are known, but whose internal working is not well, or at all, understood. In view of this, cyber security is not merely a technological challenge but is very much a matter of using information ethically and making morally right decisions based on information.

Cyber security protects computer systems from unauthorised access or being otherwise damaged. Thus, cyber security is, primarily, the practice of defending data systems from malicious attacks (Pernice 2019).

Cyber security can also be characterised as an endeavour that strives for security in an electronic and networked society. Accordingly, cyber security recognises, prevents and prepares for the impact of disruptions in electronic and networked systems. Thus, cyber security aims to ensure the integrity, privacy and availability of information. Such security is comprised of tools, technologies and best practices designed to protect data systems from attacks.

Cyber security is also known as information technology security and electronic information security. Because of these neighbouring terms, cyber security is apt to be confused with information security. Information security is a broader category and aims to protect all information assets, whether in physical or digital form. It focuses on securing the availability, integrity and confidentiality of data. In this way, information security is a key element in the quest for cyber security.

In recent years, cyber security gained wide media attention (for example relating to election campaigns, data privacy compromises and sexual misconduct cases) and the lack of security has created an increasing pressure of litigation. These developments can be attributed to a substantial increase in cyber attacks and their significant impact on individuals and organisations. Cyber attack is a broader concept than data network attack because it can also be done in ways other than through a data network. In view of this, the relationship between cyber attack and data network attack is similar to that between information security and cyber security: the first mentioned is a major concept, and the second one a sub-concept.

As seen, cyber security is a concern of a digital and networked society and organisation. The operational environment of digital systems can be trusted and protected in the state of cyber security. At the same time, the purpose is to secure the functioning of a digital and networked society or organisation. Cyber security includes measures that can be used to manage in advance and, if necessary, tolerate various cyber threats and their effects. Cyber threats are potentially harmful events affecting information systems.

Digital security makes use of the tools to shield one's identity, assets and technology in the online and mobile environments. Thus, digital security refers to various ways of protecting one's internet privacy and email and other digital accounts from intrusion by outside users. The tools one can use to protect his or her identity include antivirus software and web services, biometrics and secure personal devices such as smart cards, smartphones and e-passports. People carry these devices with them daily. The mere fact that these devices are used on a daily basis makes them vulnerable to various kinds of attacks, as people easily forget the risks related to daily activities. In an age of nonstop hacks and data breaches, getting a handle on one's own digital security matters more than ever. Moreover, every organisation has to take care of digital security.

A data breach can have devastating consequences for any business. It can destroy a company's reputation and can cause the loss of consumer and partner trust. The loss of critical data can cost a company its competitive advantage. A data breach can also impact corporate revenues due to violation of data protection regulation. It is estimated that, on average, a data breach costs an affected organisation more than four million dollars. Therefore, it is essential that organisations and societies adopt and implement a strong cyber security policy (Eggenschwiler, 2017; Dutton, 2017).

TYPES OF CYBER THREATS

Common cyber threats include computer viruses, spyware, ransomware, and phishing attacks. Traditional cyber security focuses on the implementation of defensive measures (e.g. strong antivirus software and good security habits such as using strong passwords and not using the same passwords) around a defined enviroment such an organisation's intranet or home. Recent developments like increasing remote work and portable devices have dissolved the digital environment, reduced physical presence and visibility into cyber activity and expanded the attack surface (Eggenschwiler, 2017; Fichtner, 2018; Pernice 2019).

Today, digital breaches are rapidly increasing despite record levels of security spending. Global organisations are moving to a human-centric approach that places focus on changes in user behaviour and the essential skills of a digital citizen instead of digital countermeasures. Based on behaviour analytics, human-centric cyber security focuses on how an end-user interacts with data and applies security controls in various data systems. Ultimately, this approach is designed to identify behavioural threats in order to revel the most serious risks, reducing investigation and threat detection time.

Many critical functions in society, such as money, energy, water and air traffic control, are dependent on the operation of information systems and networks. Faults in systems and networks that support these functions may be caused by security issues such as malware or malfunctioning devices.

THE RELATIONSHIP BETWEEN SECURITY AND SUSTAINABLE DEVELOPMENT

The concept of sustainability was briefly mentioned above. In what follows, a closer look at the relationship between security and sustainable development is taken. The concern over climate change and the depletion of natural resources, among other things, has prompted many citizens to re-evaluate the scope and limits of economic growth and examine more closely the notion of sustainable development. That is largely a security issue in terms of both personal and social concerns. The promotion of social justice and safeguarding of human rights have also been portrayed as demands directed towards societies and businesses (Smith et al. 2010). Such demands are understandable because in the global economy, large companies and corporations have significant political power and resources. Thus, business life plays a large role in both economic security and social equality, both of which are important aspects of sustainable development and of a just society. Therefore, safety at work and secure employment are important aspects of security in people's everyday life. Further, people throughout the world are united in their concern over environmental, social and corporate governance issues, i.e., issues of social responsibility. These issues are closely related to security concerns that arise from climate change and other ecological threats and from increasing social inequeality and the widening of income gaps. Since the concept of sustainable development is often mentioned as a remedy or cure for these calamities, there is every reason to pay attention to sustainable development as a relevant aspect of security.

The concept of sustainable development is vague and often coloured by unexplicated economic, political, and ideological background assumptions. Such background assumptions include the ideology of continuous development, the pursuit of profit maximization, the intrinsic value of life and nature, and, among others, the animal liberation thought. The clarification of background assumptions is important because they function as the presuppositions and motives of sustainable development and determine the objectives of security considerations in a global setting. Then again, the recommendations for action related to sustainable development can be similar irrespective of background assumptions. Thus, the background ideologies of sustainable development are not necessarily mutually exclusive but can be concurrent.

Sustainable development, in the widest sense of the word, is said of any positive change. Therefore, sustainable development looks like a cloud and its edges are hazy. It is not, however, a dry cloud but electrically charged: different aspects of sustainable development resist each other. Consequently, the concept of sustainable development has been criticised for the reason that poorly compatible ingredients such as economic growth and environmental protection are counted to belong to sustainable development. These tensions make sustainable development a complex, if not problematic, premise and approach for security considerations. Generally speaking, security always consists of a variety of aspects and ingredients, which means that achieving good comprehensive security may require compromises in many details and sub-aspects of security.

As mentioned, the most often referred definition of sustainable development is that of the 1987 Brundtland Report. According to it, sustainable development meets the needs of the present without compromising the ability of future generations to meet their own needs. This definition suggests that there should be a joint effort of the present generation to give the next and subsequent generations a healthy and secure society in all respects. Anthropologist and historian Joseph Tainter (2006, 92) criticises this definition as being too wide and general.

According to Tainter (2006, 92-93), sustainability is a matter of solving complex ecological, economic and social problems. Thus, sustainability is based on the ability to solve problems. One might want to add that major problems include global security issues.

Tainter says that sustainability in the long term requires understanding and controlling complex problems. His central, but not completely uncontroversial, claim is related to this: Complexity grows when social problems are attempted to be solved at the same time when costs are going up. Complex social solutions can both support and prevent sustainable development and security. Therefore, Tainter separates (2006, 94) three different results that can arise from attempts to solve big problems: social collapse, simplification, and increasing complexity. Especially the first and the last ones increase insecurity.

The control of growing complexity continuously requires more investment and energy. As the costs of problem-solving grow, the point will finally be reached where no additional investment generates a corresponding gain, but risks and insecurity gain ground (Tainter, 2006, 94). The complexity that is created by solving problems can enhance society's sustainability, but also lead to insecurity and destruction (Tainter, 2006, 100).

Thus, Tainter criticises the definition given by Brundtland and claims that sustainability is an active condition for solving problems – not a passive consequence of reducing consumption, as Brundtland suggests. Therefore, sustainability requires more resources, not the reduction of consumption. Therefore, we must afford sustainability in the same way as we must afford security. On the other hand, the costs of sustainability can endanger and destroy the whole society (Tainter, 2006, 99).

Tainter's view can be illustrated by comparing the whole of ecological, economic and social sustainability to a complex machine. Resources and sophisticated measures are required to haul and renew the machine part by part so that it does not shut down, but keeps moving, safely, all the time.

I have already mentioned climate change as a major threat to sustainable development and global security. The key facts about climate change are well known. Peter Singer (2016, 130-131) summarises them as follows: "Our planet's atmosphere has already absorbed such a large quantity of human-produced greenhouse gases that global warming is underway, with more extreme heat waves, droughts and floods than ever before. Arctic sea ice is melting, and rising sea levels are threatening to inundate low-lying densely populated coastal regions in several countries. If rainfall patterns change, hundreds of millions of people could become climate refugees."

Singer (2016, 131) also points out that an overwhelming majority of scientists in the relevant fields believe the following: the Earth is "on track to exceed the level of global warming at which feedback mechanisms will kick in and climate change will become uncontrollable, with unpredictable and possibly catastrophic consequences."

Moreover, it is often pointed out that the affluent countries have caused the problem, "owing to their higher greenhouse gas emissions over the past two centuries. They continue to have the highest levels of per capita emissions, and they can reduce emissions with the least hardship. There is no doubt that, ethically, the world's developed countries should be taking the lead in reducing emissions" and thereby contributing to global security. (Singer, 2016, 131.)

Singer (2016, 131) also calls attention to the fact that continuing global population growth undermines "the impact of whatever emission reductions affluent countries can be persuaded to make". Thus, besides the use of fossil fuels, global population growth is another major driver of climate change and another major threat to global security.

Singer (2016, 131) refers to four factors that influence the level of emissions: economic output per capita, the units of energy used to generate each unit of economic output, greenhouse gases emitted per unit of energy, and total population. "A reduction in any three of these factors will be offset by an increase in the fourth. [...] globally, economic and population growth continue to be the 'most important drivers' of the increase in CO_2 emissions from fossil-fuel combustion." (Singer, 2016, 130-131.) Thus, also in the context of climate change, security consists of a variety of aspects, which implies that security requires bargaining and compromises.

COMPREHENSIVE SECURITY

I have mentioned the concept of comprehensive security. Comprehensive security can be defined as a state of preparedness for threats and risks to vital functions of society. Thus, the term *comprehensive security* refers to the fact that security consists of a variety of components. It consists of components because there are many threats and dangers. In what follows, some major components are identified and analysed.

An individuals' sense or feeling of security is made up of a variety of matters, some of which are more important than others. For example, getting a job or recovering from a serious illness can significantly increase a person's sense of security. However, the sense of security associated with individual things easily fades over time. One begins to take health and workplace for granted, and they may no longer increase the sense of security, especially if new concerns have appeared on the horizon.

The comprehensive security of a society means an objective state where the whole is more than the sum of its parts. It is about coordinating and developing the various aspects of security, such as cyber, energy and environmental security. It requires decisions and measures that cross the boundaries of administrative sectors and businesses. It also requires organisations and special officials with the task of planning and steering the coordination of various components.

Comprehensive security has its internal and external aspects and has a political dimension and a geographical dimension. Internal security includes social balance, economic and political stability, clean nature, the ability to deal with the refugee crisis, with the threat of terrorism and other security threats, and in principle all that strengthens democracy and the well-being of citizens. Thus, schools and social and health services also represent internal security.

External comprehensive security includes the fight against military threats, peace in the near abroad and the world, the ability to combat climate change and other cross-border environmental threats, the resolution of refugee problems, the prevention of trade wars and international terrorism, and so on. External comprehensive security has traditionally been treated as a security policy issue. The cornerstones of security policy are foreign and defence policy, but comprehensive security also has dimensions of commercial and environmental policy.

Comprehensive security is taken care of so society can maintain its ability to function even in difficult circumstances and to recover quickly from disruptions. Comprehensive security is also provided so individuals and groups can live in peace and feel safe.

One can feel safe if one does not understand what it is to fear. However, one can also feel safe when he or she realises that caution is needed. Such security is related to self-knowledge and confidence.

Security is also a social construct the need for which is created by identifying threats and painting doomsday scenarios. The sale of insurance, alarm systems and survival kits are based on this. It is also the basis of the arms trade.

The authority responsible for comprehensive security is "in danger of shooting himself in the foot" by emphasising threats and security solutions. Knowledge increases pain, and risk awareness increases insecurity. Therefore, the cheapest way to increase overall security is to avoid public discussion of the topic.

This, of course, requires that security is not already shaken and fear has not conquered the mind. Secrecy would then only aggravate the situation. When fear is on, the most affordable way to increase comprehensive security is through calm speech and behaviour. In addition, other security solutions can be available. They come in many qualities and prices. Not all threats have a solution, but it is not always known in advance what threats cannot be solved.

Citizens prefer to hear bad and insecure news rather than "radio silence" from the authorities. Recently, bad news has been heard on climate change, the world economy and international politics. Security is shaken. Still, many people would rather be aware of the situation than to be ignorant. They consider truth and honesty more important than a sense of security.

ABSOLUTE SAFETY

Have you felt safe and secure so that no matter what happens, nothing can hurt you?

According to national and international surveys, security is a major value for both individuals and communities (Pew Research Center, 2019). However, security is relative in two ways. First security is related to a real or perceived threat, and second, there are degrees of security: safe, safer, safest.

Ludwig Wittgenstein talked about the experience of feeling absolutely safe: "I mean the state of mind in which one is inclined to say 'I am safe, nothing can injure me whatever happens.'" (Wittgenstein, 1965, p. 8). Such a feeling is not coming from the home alarm system or the Government Security Committee.

Absolute safety can be felt for a variety of reasons and in different life situations. The feeling may be related, for example, to the experience that one's life is related to a larger whole or to a story larger than oneself. The feeling of absolute security is strengthened if one can think that not all dignity will be lost to death.

People have a lot in common and have more in common than any individual. While each one of us dies, the good in person, who is not ultimately private but common, continues in new individuals. This good, which reincarnates and receives new manifestations, is the cognitive and artistic heritage of humanity. Values and virtues also reincarnate and rise like a phoenix from the ashes. They are worth preserving and essential to the future of humanity. Therefore, it is necessary to direct our energies towards the virtues and to focus on their promotion, in the end for fair humanity.

However, humanity is not eternal. The sun also dies out someday. Therefore, does mortality pull the absolute sense of security under the rug? Wittgenstein drew attention to this and wrote: "To be safe essentially means that it is physically impossible that certain things should happen to me and therefore it's nonsense to say that I'm safe *whatever* happens." (Wittgenstein, 1965, p. 9.) Yet, one of the most prominent philosophers of the last century said he felt absolute security.

So how is it? Does humankind's mortality make the sense of absolute security meaningless? At least it can be said that it requires idealism to think like this. Whatever happens to humanity, it is better that humans have had virtues such as wisdom, self-control, courage and justice. Having these and other virtues is permanently better than never having them. Whatever happens, nothing threatens human totally, for the highest in humans are those enduring values, no matter how imperfectly they are realised in us.

Despite its universe-embracing nature, the experience of knowing absolute security is personal. Only an individual can feel safety or anything in the first place.

SOLUTIONS AND RECOMMENDATIONS

Security is a basic human need and thus an essential part of wellbeing. The majority of people are concerned with the security of their own lives and of the lives of their family members. Any understanding of security must be integrated with this general concern. Moreover, a number of international treaties define security as a basic human right.

Secret and anonymous algorithms threaten privacy and the rule of law. Imposing fines or taxes, judging somebody guilty or sending people to jail because of the inexplicable, unchallengeable judgments of computer programmes undermines legal systems. In view of this, cyber security is not merely a technological challenge but very much a matter of using information ethically and making morally right decisions based on information.

Human-centric cyber security focuses on how an end-user interacts with data and extends security controls into all data systems. This approach identifies behavioural threats in order to reveal the most serious threats, reducing investigation and threat detection times.

FUTURE RESEARCH DIRECTIONS

Big security concerns arise from climate change and other ecological threats and from increasing social inequality and the widening of income gaps. Since the concept of sustainable development is often mentioned as a remedy or cure for these calamities, there is every reason to pay attention to sustainable development as a relevant aspect of security.

CONCLUSION

Security is relative in two ways: security is always relative to a real or imagined threat, and there are degrees of security. In the widest range, human security includes the promotion and expansion of capabilities that strengthen existing opportunies and create new possibilities for human flourishment.

In a utilitarian framework, the ethics of security aims at peace, well-being and prosperity, necessary for both individuals and societies. In a deontological framework, the ethics of security focuses on duties related to the protection of basic and human rights, the following of which is a universal responsibility and which the human reason can recognised as a universal obligation. In a social contract framework, the ethics of security focuses on treaties and conventions made for enhancing security and stability in a society and between countries. Virtue ethics tries to determine and explain the character and communal traits that are necessary for security and can be created by upbringing and education. Ethics of care sees security both as the fruit and foundation of caring relationships and practices. In view of these approaches, security is a complex phenomenon.

Digital breaches are rapidly increasing despite record levels of security spending. Every organisation has to take care of digital security. Global organisations are moving to a human-centric approach that places focus on changes in user behaviour and the essential skills of a digital citizen instead of digital countermeasures.

Security and sustainability are properties of the Earth system as a whole, including the interacting physical, chemical and biological processes and global social, political and economic developments. Therefore, a business or organisation is unlikely to be sustainable if the society at large is not secure.

This research received no specific grant from any funding agency in the public, commercial, or not-for-profit sectors.

REFERENCES

Boddington, P. (2017). *Towards a Code of Ethics for Artificial Intelligence*. Springer. doi:10.1007/978-3-319-60648-4

Bonnefon, J.-F., Shariffs, A., & Rahwan, I. (2015). *Autonomous vehicles need experimental ethics: Are we ready for utilitarian cars?* arXiv:1510.03346.

Bonnefon, J.-F., Shariffs, A., & Rahwan, I. (2016). The social dilemma of autonomous vehicles. *Science*, *352*(6293), 1573–1576. doi:10.1126cience.aaf2654 PMID:27339987

Brant, T. (2016). Who should your self-driving car save in a crash? You or pedestrians? *PC Magazine*. https://uk.pcmag.com/electronics/82490/who-should-your-self-driving-car-save-in-a-crash-you-or-pede

Doyal, L., & Gough, I. (1991). *A Theory of Human Need*. Macmillan. doi:10.1007/978-1-349-21500-3

Dutton, W. H. (2017). Fostering a cyber security mindset. *Internet Policy Review*, *6*(1). https://policyreview.info/node/443/pdf

Eggenschwiler, J. (2017). Accountability challenges confronting cyberspace governance. *Internet Policy Review*, *6*(3). https://policyreview.info/node/712/pdf

Fichtner, L. (2018). What kind of cyber security? Theorising cyber security and mapping approaches. *Internet Policy Review*, *7*(2). https://policyreview.info/node/788/pdf

Fraser, N., & Honneth, A. (2003). *Redistribution or Recognition: A Political-Philosophical Exchange*. Verso. doi:10.1002/9780470756119.ch54

Frigo, G. (2018). *The energy ethic and strong sustainability: Outlining key principles for a moral compass. In Strongly Sustainable Societies: Organising Human Activities on a Hot and Full Earth*. Routledge.

Gasper, D. (2015). *Human security: From definitions to investigating a discourse. In Routledge Handbook of Human Security*. Routledge.

Honneth, A. (1995). *The Struggle for Recognition: The Grammar of Social Conflicts*. Polity.

Honneth, A. (2007). *Disrespect: The Normative Foundations of Critical Theory*. Polity.

Lehtonen, T. (2018). *Sustainable investment and degrowth. In Strongly Sustainable Societies: Organising Human Activities on a Hot and Full Earth*. Routledge.

Martin, M., & Owen, T. (2015). *Introduction. In Routledge Handbook of Human Security*. Routledge.

Maslow, A. H. (1943). A theory of human motivation. *Psychological Review*, *50*(4), 370–396. doi:10.1037/h0054346

Myerson, R. B. (1997). *Game Theory: Analysis of Conflict*. Harvard University Press.

Neiman, S. (2015). *Evil in Modern Thought: An Alternative History of Philosophy*. Princeton University Press.

Nemati, H. (2008). *Information Security and Ethics: Concepts, Methodologies, Tools, and Applications*. Information Science Reference. doi:10.4018/978-1-59904-937-3

Ortmeier, P. J., & Meese, E. III. (2009). *Leadership, Ethics and Policing: Challenges for the 21st Century* (2nd ed.). Pearson.

Pernice, I. (2019). Protecting the global digital information ecosystem: A practical initiative. *Internet Policy Review*, 8. https://policyreview.info/articles/news/protecting-global-digital-information-ecosystem-practical-initiative/1386

Pew Research Center. (2019). *Public's 2019 Priorities: Economy, Health Care, Education and Security All Near Top of List*. https://www.people-press.org/2019/01/24/publics-2019-priorities-economy-health-care-education-and-security-all-near-top-of-list/

Sen, A. (2015). *Birth of discourse. In Routledge Handbook of Human Security*. Routledge.

Singer, P. (2016). *Ethics in the Real World: 82 Brief Essays on Things That Matter*. Princeton University Press.

Smith, A., Voß, J.-P., & Grin, J. (2010). Innovation Studies and Sustainability Transitions: The Allure of the Multi-level Perspective and Its Challenges. *Research Policy*, *39*(4), 435–448. doi:10.1016/j.respol.2010.01.023

Staub, E. (2003). Notes on cultures of violence, cultures of caring and peace, and fulfillment of basic human needs. *Political Psychology*, *24*(1), 1–19. doi:10.1111/0162-895X.00314

Tainter, J. A. (2006). Social complexity and sustainability. *Ecological Complexity*, *3*(2), 91–103. doi:10.1016/j.ecocom.2005.07.004

The World Commission on Environment and Development. (1987). *Our Common Future*. Oxford University Press.

Togelius, J. (2011). A procedural critique of deontological reasoning. *DiGRA '11: Proceedings of the 2011 DiGRA International Conference: Think Design Play*, *6*. http://www.digra.org/digital-library/publications/a-procedural-critique-of-deontological-reasoning/

United Nations Commission on Human Security. (2003). *Human Security Now*. United Nations.

United Nations Development Programme. (1994). *Human Development Report*. Oxford University Press.

Wittgenstein, L. (1965). A lecture on ethics. *The Philosophical Review*, *74*(1), 3–12. doi:10.2307/2183526

KEY TERMS AND DEFINITIONS

Absolute Safety: Wittgenstein's term for the paradoxical state of mind in which one is inclined to say, "nothing can injure me whatever happens."

Comprehensive Security: A state of preparedness for the threats and risks to vital functions of society.

Cyber Security: The practice of ensuring the integrity, confidentiality, and availability of information; a set of tools, approaches and technologies designed to protect networks, devices, programmes, and data from attacks or unauthorised access.

Digital Security: The practice of securing one's identity, assets and technology in the online and mobile environments.

Information Security: The practice of protecting all information assets whether in hard copy or in digital form.

Safety: A tool or measure that reduces hazards; a place or position where there is no danger.

Security: The condition of being protected both from intentional and non-intentional harm and both from moral and natural evils.

Chapter 5
Healthcare Digitalized:
Patient/Counsellor Interaction in the Digitalized Era

Miguel H. Kottow
Universidad de Chile, Chile

ABSTRACT

Ever since medicine became a recognized profession, the relationship between patients and physicians was marked by authoritarian paternalism. With the advent of bioethics in the 1970s, patients' right to participate in decision making led to proclaim autonomy as the primary principle in clinical medicine and biomedical research, practically exercised as informed consent; yet, the issue remains contended and poorly regulated. Healthcare digitalization disassembles persons into clouds of data. Individual decision making is interfered with and replaced by dominant algorithms, supposedly delivering a P4 composite of precision medicine: personalized, preventive, predictive, participatory. Biomedicine develops into medicalization, marketization contractual client/provider relationship, and neglect of personal care for the ill and frail. These trends become dominant in digitalized healthcare as personal healthcare relationships, and ethically unsatisfactory medical services replace the psychosocial, existential elements of health/disease.

INTRODUCTION

From Holocene to Anthropocene

Holocene (*holo* 'whole' + Greek *kainos* 'new') is the name for the era where the earth became warm enough to be habitable, initiating the evolution of living organisms including the human species. While all vegetal and animal species either adapted to changing environmental conditions or disappeared, human beings rapidly expanded their ability to develop tool-making, replacing slow biological adaptation by actively pursuing cultural achievements that mine natural resources to not only satisfy needs but also give vent to ever more sophisticated desires demanding increasingly cost-intensive production. Mankind

DOI: 10.4018/978-1-7998-4117-3.ch005

also was quick to engage in competitive practices and developing warfare as an effective way of securing survival of the fittest, as Hobbes succinctly characterized as *homo lupus homini.* In time, exploitation of nature overshot the mark of sustainability, turning into a destructive exspoliation carried out in plunderous an ravenous ways that could not but trigger socioeconomic inequities. As instrumental development peaks, one third of world population remains undernourished, suffering hunger and multidimensional poverty.

Transforming nature to fulfill human whims exacted the inevitable price of maiming the goose that lays the golden eggs, and of creating shortages that lead to a monumental divide between the have and the have not, the privileged and the dispossessed, disingenuously featured as the difference between rich and poor as if some charitable redistribution might make the gap more palatable.

Uncontrolled expansion and irreversible plundering of nature stress the seams of the Holocene, seriously putting to question the era of wholeness. As wholeness looses meaning, so do its etymological cousins holy and healthy. Nature's integrity suffers increased manipulation by humans, as the era of anthropocene erases the traditional dichotomy between nature and artifice. Many philosophical doubts and questions emerge, but the call for urgency requires facing the question about what is at stake why.

Social rejection of man-made machinery was initiated by the 19th century aggressive Luddite movement objecting against industrialization that would lead to underpaid unqualified labour and unemployment, eventually becoming a systemic feature of profit seeking capitalism and its persistent quest for cheap labour.

Modern humanity lives in and cannot avoid relying on, technological instruments that thrive in times of digitalization introduced in all wakes of life, from day to day chores to highly sophisticated production and use of goods and services. Even those too poor to profit from technical advances, are nevertheless influenced, often negatively, by technology and its consequences. Luddite technophobia is as obsolete as uncritical technophilia bent on "progress" is irresponsibly oblivious to the enormous and to a great extent negative impact it has on world affairs, fuelling growing divides between affluent consumers and majorities that lack access to technical means to relieve their daily toils to secure basic necessities.

In 1978 the Club of Rome warned that the limits to growth are being overstepped, causing additional alarm when a 20 years follow-up report showed that natural resources were being exploited with enormous polluting effects that had gone way beyond the limits of growth. The United Nations issued the Brundtland Report and the idea of sustainable growth (1987); in its traditional meaning sustainability is defined as corporative, national, even global economic growth that can be maintained without running into financial problems, but the ecological perspective sees sustainable global as development that meets the needs of the present without compromising the ability of future generations to meet their own needs. This very bland formulation does not define or quantify present needs, presenting a very vague call for saving resources that future generations will need. As long as a considerable proportion of actually living humans are far from having their basic needs met, it is implausible to save resources for the future, unless massive redistribution is undertaken to cover the present essential needs of the world's population before assessing how much can be saved for the future. Since nothing of this has happened in the last 30 years, s evolving into a mild but persistent minority that criticizes technoscientific expansion, merging with current calls to save the planet from destructive exspoliation, and fuelling crusades to resort to a "small is beautiful" use of technology. Sustainability is reduced to a window dressing effort at proclaiming global justice.

The 20th century has been prodigal in presenting critical views on unfettered positivism and its faith in science as the one and only font of true knowledge. Beginning wth C.P. Snow's "The Two Cultures" pleading for a better understanding between science and "traditional" culture, mainly literature, Günther

Anders “The Obsolescence of Man”, Richard Rorty’s neopragmatism, Martha Nussbaum’s efforts at ·Cultivating Humanity” and, of course, Arne Naess’ Deep Ecology that inspired Van Renselaer Potter to develop his brand of global bioethics which he proposed as a “science of survival”.

Schwartz gained some notoriety with his slogan “small is beautiful”, but had little durable influence. Hans Jonas famously requested technoscientific impetus to slower their pace and adopt strategies of frugal responsibility, well aware that this was a non-starter. Jonas’ pessimism should be a warning to humanitarian arguments against the uncontrolled expansion of technoscience, which intuitively plausible and correct, but have had little or no purchase in slowing the invasiveness of globalization, untamed ‘savage capitalism’ and unfettered consumerism. Isabelle Stenghers writes convincingly about “slow” science to counteract the hysterical proliferation of research activities. Unrest about so-called cuttingedge biotechnology is on the rise, but cannot be efficiently countered by religious beliefs, secular conservationists, environmentalists, conservative bioethics or futurologists announcing dire consequences for the survival of humankind.

The Rise of Medical Scientifism

“The science of medicine” is a well known chapter of Hippocratic writings praising the need of accurate observation of the human body and its diseases, in order to ground well-formed knowledge aimed ad diagnostics and prognosis. Therapy was less of an issue, for the course of disease was determined by the forces of nature –*vis medicatrix naturae*–. The emphasis on a scientific attitude was meant to For the Greeks, medicine was a technical practice – *techne iatrique*– practiced by poorly trained itinerant individuals, who needed to acquire systematic knowledge and adopt ethical guidelines as presented by the Hippocratic Oath. Medicine remained a scarcely regulated practice till the advent of the first codes of medical ethics that preached decency, honesty, appropriate demeanor as well as loyalty allowing medicine to become an established and highly regarded profession in the course of the next decades. Instead of self-complimenting itself as a unique combination of art, technique and science, medicine had comfortably and profitably settled as a social practice with all the advantages of a recognized profession, delivering an important service, receiving handsome retribution in terms of income and prestige, and guaranteed to remain a monopolistic practice that controls and regulates the training and ethics of its members.

The 19th century initiated major interest in scientific activity in the natural sciences, with important breakthroughs in medical knowledge –Pasteur, Semmelweis, Koch–, initiating an era of interdisciplinary research between chemistry, physics biology and clinical medicine. Experimentation with human beings and clinical trials were fuelling research in biology generously financed by gigantic pharmaceutical enterprises, and medical practice embraced what German physician Bernhard Naunyn (1839-1925) had loudly announced “Medicine shall be a natural science or shall not be”, revived by evidence-based medicine that bolstered laboratory-based diagnostics, sophisticated surgical instrumentation, and multidirectional research, fostered by corporative interests neglecting the common good, the marketing of affordable products, and research oriented towards relevant medical solutions rather than profitable developments.

Modern reliance on science as the only reliable way of gaining knowledge, paved the way to a symbiotic alliance between biology and medicine:

Biomedicine is created by the interaction between different professional groups, traditionally distanced from each other by their training and goals. Biomedicine symbolizes the alliance between medicine, biology and also industry (Sebag 2007: 20).

These major disciplines would form an association for mutual benefit without giving up their core purpose and drawing the limits of its competence. Yet by coining the now current term "biomedicine" an apparent interdisciplinarity is suggested that creates a confusion and misrepresentation of purposes. Going from bench to bedside is a complex process, often short-circuited and causing unexpected and unwanted complications, misunderstandings, false expectations

In the year 2000, the Institute of Medicine convened the Clinical Research Roundtable (CRR) to discuss emergent issues in clinical research in response to the Association of American Medical Colleges' publication of "Clinical Research: A National Call to Action." In June of 2003, the CRR, noting that the clinical research enterprise was truly in crisis, published steps to improve the translation and dissemination of clinical research, identifying 2 translational obstacles—from basic discoveries to clinical practice and from the clinical identification of "things that work" to broader application—as primary concerns (p. 254).

Science supplies technology with potent and sophisticated information to be translated into instruments that, in turn, provides scientific research with higher resolution in tracing processes and enormous capacity of storing infinite amount data.

Physicians are expected to "let scientific methods tell us what to believe. We should believe in science and nothing else" (Ghaemi 2017: 789), The practice of medicine is co-opted by scientific medicine, biomedicine, evidence-based medicine, biological reductionism, and the new biomedical fields of genetics and neuroscience. This is the ground were digital medicine flourishes, unleashing hopes, expectation, fears and doomsday premonitions.

Much of medical care and clinical research occurs in hospitals that have undergone huge modifications since their medieval function as asylum for the poor and hopelessly sick, to the tertiary medicine of post-World War II hospitals, now being transformed into huge institutions were clinical and research work are intermingled, giving birth to biomedical platforms (Keating & Cambrosio 2003)

It is generally held, for example, that medical technologies tend to remove the patient from the clinical picture and redirect the clinician's attention from the bedside to the laboratory. In addition, it is maintained that medical technologies have downgraded or eliminated skills embodied in what is termed "clinical judgment", forcing diagnosing physicians to become increasingly subservient to technologies whose development they cannot control (Ibid.: 324).

Beware of Technomedicine

In the wake of these major social changes, that include medical practice and public health policies, philosophers, sociologists and bioethicists have raised the banner of the humanities, too often resorting to hyperbolic talk of "dehumanized" even "inhuman" medicine. Reprehensible actions are performed by human beings, so they cannot be ousted into the realm of inhumanity when, in fact, they fit into Nietzsche's lament that they are "human much too human", meaning that to be human is to be defective, hence the quest for the suprahuman. Only by acknowledging how despicable actions originate in human beings, can accountability be required, moral judgement passed and redress be obligated.

What critics of contemporary medicine are claiming is the loss and distortion of personal interaction between patient and healer. Hence the failure of the humanities to correct the idea that it is necessary to believe in science "and nothing else".

This extreme positivism dismisses the efforts of scholars who resist the demise of everything else but scientific knowledge, beginning with Wilhelm Dilthey (1833-1911) distinction between sciences of nature –*Naturwissenschaften*– and the human sciences –*Geisteswissenschaften*–; Max Weber's (1864-1920) differentiation between formal-procedural rationality –*Zweckrationalität*–, *and* substantive-value rationality –*Wertrationalität*–, a distinction Jürgen Habermas (1920–) renamed 'instrumental or pragmatic rationality', and 'communicative rationality', lamenting that the lifeworld of communicative or value rationality was being colonized by pragmatism.

C.P. Snow (1905-1980) famously pleaded for a unification of science and the humanities, especially literature, and Van Rensalaer Potter (1911-2001) pledged alliance to bioethics as a bridge to the future linking human values to medicine and ecology. Mainstream biomedicine continued to trod the scientific path, taking little notice of holistic perspectives, notably the biopsychosocial view that medical management based on individual bodies was a poor substitute for a an understanding of psychological features and social circumstances that contributed to dysfuncionality and disability.

Holistic approaches understand that excessive demands of adaptation to a stressful lifeworld will cause bodily dysfunction, a collapse of defences against some inborn predisposition, lack of energy and empowerment to fulfil normal tasks, the experience of illness and the setting in of disease. A therapeutic intervention that does not address the psychic and social components of the patient's collapse will at best compensate but not heal unless the existential dimension of disease is acknowledged. Similar efforts were displayed by personalistic approaches stressing that pathography needed to be complemented with biography, that the ill body was, in fact, an ill person requiring care rather than repair. Biomedicine centres on pathography, whereas a personalistic approach focuses on biography and pathography.

None of these approaches has neared its goal of slowing the trend towards a depersonalized scientific form of biomedical practice, and it is more than doubtful that they will gain any ascendancy in current times, where pragmatism dominates over humane communication and value-sharing.

On the verge of fully adopting digitilized practices in biological research and medical practice, the heritage of René Descartes' (1596-1650) view that human nature is a composite of reason –*res cogitans*- and matter –res extensa–. The Cartesian duality of two equally fundamental substances has never been abandoned as it gained philosophical presence as the mind/body interaction in ways that continue to be unfathomable, although its influence on medicine needs to be, and has been, subject to thoughtful analysis by phenomenology and by clinical bioethics concerned with the patient/physician encounter.

Phenomenology of Medicine

The concept of lifeworld was introduced by Husserl and introduced in medical thought by Merleau-Ponty who rejected the Cartesian duality of mind/body to be replaced by the human body as lived –experience, perception– and living –biological organism–. Lived experiences of the world through the body in health and sickness could be exposed by first-person narrative, depicting the person's lifeworld in health and illness. Phenomenologists of medicine believe that biomedicine's scientific third-person approach to disease performs poorly by neglecting the first-person narrative of "dis-*ease*" –the experience of illness– and centering its diagnostic and therapeutic interventions on the nosological dysfunction recognized as a disease (Svenaeus 2019). Reinforcing attention to the patient's lived experience of illness is also strongly advocated by narrative medicine –Rita Charon at Columbia University- but has not grown beyond a scholarly enclave that, being a time-consuming procedure, runs counter to the efficacy demanded by managed medicine relying on technical rather than personal exploration of illness.

Doctor-patient communication follows different patterns of intermingling the voice of medicine and the patient's first-person narrative. In the "Strictly Medicine" form of consultation in acute medical situations, was successfully managed by the physician in complete disregard of the patient's narrative. In the "Lifeworld Blocked" situation, the lifeworld was briefly appeared but was stifled by the doctor, leading to a less satisfactory and successful outcome. Similarly, in "Lifeworld ignored" the patient narrates but the doctor does not listen, the consultation is irritating and less effective, in contrast with the "Mutual Lifeworld" where physicians encourage and interact with the patient's narrative. Patient active participation prompts a more humane and effective medical care, though these advantages peter out in chronic conditions and geriatric care (Barry et al. 2001).

The voice of the lifeworld refers to the patient's contextually-grounded experiences of events and problems in her life. These are reports and descriptions of the world of everyday life expressed from the perspective of a "natural attitude" (Ibid.: 487).

These and other proposals may have academic interest but no influence on actual clinical encounters that are enmeshed in social context, as sociologists keep insisting, showing that "the voice of medicine has developed and retains a tendency to suppress and colonize the *voice of the lifeworld*" (Scrambler & Britten, 2001: 55).

The Doctor-Patient Relationship

Nascent bioethics in Georgetown saw its main task in the dismantlement of medical paternalism, sustaining patients' rights and autonomy by means of an elaborate, and never ending, elaboration of informed consent and its varieties –clinical decision, research protocols, immature, defective or declining mental competence–. The clinical encounter was seen as the centre of medical practice and, therefore, of bioethical concern. Against paternalism, patients must make the autonomous decision be heard, thus explaining why the principle of autonomy always takes pride of place in mainstream bioethics.

First proposals were more of a descriptive nature than propositional models; activity-passivity, guidance-co-operation and mutual participation (Szasz & Hollender): activity/passivity, in which the physician does something to an inert or unresponsive patient; guidance-co-operation, in which the physician tells the patient what to do and the patient obeys; and, finally, mutual participation gave way to a widely regarded four model physician-patient relationship (Emanuel & Emanuel 1992): paternalistic, informative, interpretative, and deliberative, representing progressive patient autonomy and decision competence.

In a review article, Ong et al. (1995) stressed the importance of doctor-patient communication, serving three basic purposes: creating a good interpersonal relationship, exchanging information, and making treatment decisions. To these ends, underlying value like empathy, patient-centred approach, exchange of pertinent information, shared decision-making, outgrowth of paternalistic attitudes were being discussed in the literature at a time where traditional medical practice was moving towards mercantilization and consumerism.

The theoretical concept of patient autonomy became operative in practice as the procedure of informed consent, hailed as the ethical core the clinical encounter, as well as in the recruitment of subjects in clinical research. And yet, after 50 years of academic exposure, informed consent continues to be an unresolved contentious issue, still too often ignored by authoritarian physicians, ad investigators, with a growing tendency to regard biomedical procedures as too complicated to be sufficiently understood by patients, and fraught with uncertainties that, if fully exposed, would undermine trust in biomedical interventions.

The enthusiastic support of patient autonomy, created an atmosphere where the interaction between patients and healers was decontextualized, separating individual actor from their social environment, thus creating a chasm between theory and practice, duly noticed by sociologist.

Recent approaches…tend to present the doctor-patient relationship as an autonomous or self-contained unit of analysis, conceptualizing it moreover in terms of a series of independent and 'de-contextualized' encounters (each one displaying an assembly or mix of predefined positive versus negative interactional or communicative characteristics) (Scrambler & Britten, 2001: 46-47).

The atmosphere of the clinical encounter changes as medicine has grown in scope –medicalization– and turned into merchandise. Consumerism transforms medicine into a commodity to be negotiated by the patient/consumer acting as a reflexive actor "actively calculating, assessing ant, if necessary, countering expert knowledge and autonomy with the objective of maximizing the value of services such as heath care" (Lupton 2004: 199). Opportunity replaces trust as alternative therapies are become increasingly popular, while individuals are expected to place their health and disease prevention in their own hands, exercising autonomous self-care, self-regulation, self-management of risk factors. Criticism of medical institutions and negative personal experiences gnaw at traditional role and status of physicians, who all to often act as dispensers rather than health caretakers. At the price of authority and paternalism, the medical profession was trusted and highly esteemed.

The first part of this essay depicts the continuous and growing repositioning of medicine from the antique *techné iatrique* to a flourishing of the medical profession hailed as a personal interaction between the knowing physician and the suffering patient. Expert clinicians relied on listening to patient's narrative, perform fine-tuned physical examination of the ailing body, and bringing forward informed intuition to reach a diagnostic hypothesis and a proven therapy. the doctor steeped in confidence and know-how, epitomized by such figures as Sir William Osler, Albert Schweitzer, the family physician, the general practitioner offering home visits. The doctor-patient relationship was a personal encounter: It is a safe rule", wrote Sir William Osier, "to have no teaching without a patient for a text, and the best teaching is that taught by the patient himself" (Hunter, 2017: 193). In the computer era's infancy, medicine became an ambitious science-based contemporary biomedicine imbued with a sense of measurable efficiency –managed medicine–reliance on sophisticated technology to reach diagnosis and provide algorithmic therapeutic decision-making, expanding its area of influence –medicalization– as a marketable service. In this process of 'maturation'

The major loss for patients and the core of criticism resulted from the loss of a personal relationship between the ailing and the healers and care providers. Undoubtedly, the demands of efficiency, productivity and speedy resolution of clinical situations has eroded the clinical encounter, whether office appointment or bedside visitation, to a perfunctory minimum. Less interpersonal relationship is richly compensated, so we are told, by technical tests that are more reliable than time-consuming, allowing physicians to listen less, 'run' more tests, and gaze at their computer instead of exercising eye contact with the patient.

At least three factors can be traced to this predominance of functional measurements over embracing the ailing as both organically and existentially damaged. Firstly, medicine has remained faithful to the Cartesian duality of mind and body, deepening knowledge and extending interventions to the organism as a complex network of functions. Mental afflictions are increasingly seen as biochemical or genetic derangements, treated with the growing market of psychopharmaceutical drugs. Secondly, scientific biomedicine dwells on generalization to which the individual case must be adapted, even when its presentation is clinically atypical. Thirdly, socioeconomic contexts are acknowledged, but accepted as determinants that cannot be modified.

Mainstream depersonalized biomedicine strongly resists criticism which is based on concepts and arguments that have lost general acceptance. The unending dispute about what makes a human being a person or the holistic proclamation that humanity and personality are synonymous marks this issue as indeterminate and weak. Catholic bioethics, led by Cardinal Elio Sgreccia (1928-2019) put the person –embodied existence– as a spiritual, relational creation of God fails to convince any but those faithful to the doctrine. Claiming that human nature puts limits to interventions also fails to convince, as these supposedly essential features cannot be described, while the thesis that humans are unexceptional (ontological dualism of material and spiritual modes of being has lost ground, as has the "ontic rupture" that poses an incommensurable difference between man and non-human animals is strongly opposed by animalists (Schaeffer J.-M. (2007). "If God does not exist everything is permitted" (Dostoevsky), can now be amplified to "If humans have no essential nature, no intervention is prohibited".

Finally, the idea that nature will never completely reveal itself, clad in the veil of Isis (Nietzsche, Hadot) is being eroded by the triumphant invasion of science into the most intimate natural processes.

This appear to be the current state of affairs facing the ambitious impetus of digitalization, participating in the demise of personal patient/carer encounters and interactions.

MEDICINE AND (HEALTH) CARE IN A DIGITALIZED WORLD

The Rise of Digitalization

Digitalization has boosted reception, storage, exchange and communication of information fractioned into bits, or other units, that can be subject to complex yet speedy processes of interactions, calculations and application in many areas of human activities, including scientific research. Observers speak of the second industrial revolution, the era of communication, the information age, all agreeing on the major impact these developments have on social values and practices.

Some prefer to speak of the Digital Era in society characterized by an overall increase in information, knowledge and communications, digitalized in the form of bits that can be massively stored, creating huge data banks that allow easy access and flexible use of the numerical accumulation of information.

As modernity eschewed grand narrative, truth became a nonentity, dangerous when wielded by truth-holders prone to fanatically impose their convictions and penalize opponents. No longer aiming at elusive truth, science seeks knowledge by accumulating data, pursuing and creating networks in such enormous dimensions that only digitalization can manage them in a highly efficient manner

"In the world everything is as it is and happens as it does happen. *In it,* there is no value, - and if there were, it would be of no value." Wittgenstein's cryptic assertion may be employed to address science and ask how what "does happen" is faithfully observed, recorded and utilized. Theoretical constructs, fallible inductions, discredited deductions or suspicious hermeneutics (Ricoeur) will not help so, again, efficient data collecting has the floor, though the jury will remain out as to efficacy so long as the world remains unhealthy and in bad shape.

Science's predicament for pursuing knowledge per se, and therefore immune to ethical surveillance, is no longer sustainable, nor does Bacon´s idea remain valid that good science produces knowledge that helps control nature for the good of mankind. Habermas acknowledges that pragmatic rationality is essential to the successful adaptation and survival of mankind, but remains critical that scientific research is fuelled by vested interests, corporative growth and personal advancement. As science penetrates deep into the structure of living beings, it urgently needs to be critically evaluated as to the possible consequences on interfering and transforming human beings as we now know them.

This issue will remain subject to wild guessing as long as applications being offered cannot be evaluated in terms of user experience and response to what is put on the market: will domotics find ample acceptance? Will the public persists in its skepticism and suspicion of genetically modified food? Sophisticated gadgets with guaranteed maintenance and updates become useless if producers go under or are purchased by a giant enterprise with no commitments to honor previous contracts. Buying the newest and finest is costly and risky if expectations are not fulfilled or additional support costs ensue. Second-and equipment may break down before fully paid for, a problem not unheard of in medical environments where equipment rapidly become obsolete as new "cutting-edge" models are put on the market.

The social impacts of DTs are mostly regulated by market strategies, which cautiously secure profit even though safety is endangered. Product recalls are expensive, and big firms have been known to prefer facing litigation charges and defer recalling their harmful cars or medicines, preferring ethical transgression to sales disasters. The ethics of digitalized of a digitalized world are subject to even more speculations and uncertainties than social consequences, but normative precepts cannot wait till harm is entrenched and irreversible. Attentive haste is further complicated by extravagant and unsubstantiated promises and fears: the world will become too toxic for survival, the human species will be replaced by cyborgs, the anthropocenic revolution, like other revolutions before it, will devour its children.

A further consideration blends with social unrest, as it becomes apparent that digital technology will widen the gap between rich and poor already deepened by knowledge disparities, the unequal access to technological appliances, and to many products of recent development: computers, medical devices, These disparities but lacks the power to offer mitigation or solutions, and it remains in the realm of ethics to point out how digitalization entails severe moral issues.

Ethical Issues of Digitalization

Criticism of a digitalized era would remain marginal and a province of the querulous, were it not for its impact on the science of biology. Gone are the days where scientist marvelled at the exquisite morphological and functional design of living beings, led by Aristotelian delight in acquiring knowledge and admiration for those pursuing more knowledge without ulterior motives. The visible harmony of macroscopic features needed to be not only admired but dissected and dismantled in a truly Baconian spirit of knowing nature to better control it for the benefit of humanity. Bacon would have applauded digital technology' increase in speed and breadth of knowledge turnover, though the pragmatic turn towards benefit of searchers and researchers pragmatically changed for to the benefit of searchers and researchers while sidestepping the common weal would disappoint him.

Biomedicine has been most strongly affected as it moved towards the study of basic units –genes, proteins, molecules, enzymes, neural units and networks, nanoparticles–, allowing the accumulation of incommensurable amounts of data.

Enter the era of molecular investigation of chromosomes, genes, mitochondria, proteins, telomeres, and minuscule pathogens –viruses, prions, free radicals–infinitely accelerated by digital scavenging of information, where the volume of data doubles every three years in the wake of the vertiginous expansion of digital technologies (DTs) (Berger, Denner & Rögliner, 2018). The intricacies of biological processes are explored, intervened, and artificially reproduced. Genetic technoscience aims at controlling and selecting hereditary features, neuroscience apply artificial intelligence to map the brains neuronal networks in order to influence behaviour, nanotechnology promises to create organs and make transplant medicine obsolete. Embryos with defective genes will either be eliminated or repaired. In the near future phenotypical features will be enhanced and perfected according to parental wishes.

Social Impact of DTs

The most salient feature of digitalization is its capacity to register, store and make available immense amounts of data provided by sensors that monitor personal preferences, utterances and such trivia as buying practices, watching television programs or language habits. As individuals become transparent, data banking turns opaque and unfathomable. Culling recent publications on these subjects shows concern for the erosion of certain public values: privacy, autonomy, security, human dignity, justice, and balance of power (Royakkers, Timmer, Kool & van Est, 2018). This laundry list points to the underlying concern that quantitative data is piecemeal information of certain aspects of the reality under study: culling data about migration, the economy or biology tends towards an accumulation of molecular knowledge that defies a holistic view of the system being investigated. In the case of living bodies, the Human Genome Project successfully sequencing and mapping the genes of the human genome, only to find that little had been achieved in terms of functionality and interaction of genes, still less about epigenetic factors influencing the phenotypical expression and penetration of specific genes. Whereas genetic medicine is still in its infancy, a multibillion dollars industry is thriving on the commercial development of genomics.

A second major consequence of big data collection is that the positivistic approach favors cause-effect of measurable facts to the detriment of information available from chaotic social and behavioral processes, and tending to dismiss qualitative, ethnographic and phenomenological research of human experience and being-in-the-world. Whereas physical anthropology may be quantitative, cultural anthropology is not. Markham bases her defence of ethnographic research in the digitalized era by Gregory Baton's suggestion for "ecology to be open to dynamics rather than essences of processes of what we end up by labelling 'self', 'other', and 'the social (Markham 2018: 5).

Thirdly, the epistemological value of data is inductive and therefore fallible in terms of predictability. Davie Hume's 'problem of induction' may be overstated, but it certainly is a permanent challenge for clinical medicine, which rarely has firm knowledge about the prognosis of most diseases.

Focusing on the relational aspects of digitalized biomedicine brings up the first suggestions, somewhat magisterially presented as principles:

10 Principles for the Ethics of Digital Health (Stanford Library, 2019)

1. The products of digital health companies should always work in patient's interest.
2. Sharing digital health information should always be to improve a patient's outcomes and those of others.
3. "Do no harm" should apply to the use and sharing of all digital health information.
4. Patients should never be forced to use digital health products against their wishes.
5. Patients should be able to decide whether their information is shared and to know how a digital health company uses information to generate revenues.
6. Digital information should be accurate.
7. Digital information should be protected with strong security tools.
8. Security violations should be reported promptly along with what is being done to fix them.
9. Digital health products should allow patients to be more connected to their caregivers.
10. Patients should be actively engaged in the community that is shaping digital health products.

The triviality of this list is easily recognized when deleting the word "digital" from each one of these recommendations they show their non-specific applicability. Bland common sense is easily countered, explaining how ethical probes into technical achievements has had no persuasive force to avoid some very nasty developments –chemical and biological weapons, nuclear armaments, indiscriminate use of shock therapy, frontal lobotomy– submerged in secrecy or ousted for their maleficence.

A more elaborate presentation on the ethics aspects of digitalization in the patient/caretaker relationship takes issue with three aspects of "patient/counsellor" ethics in the digital era.: clinical medicine, biomedical research, non-therapeutic interventions.

Personalized Genomic Medicine

After completion of the Human Genome Project, a major research program was anounced that would revolutionize medicine by mapping individual genomes in order to anticipate the best medication for each person, and optimizing the dosage, duration and form of administration the patient should receive. Advocates of personalized genomic medicine (PGM) elaborated the twin concepts of systems biology leading to P4 medicine: "systems biology is built around the concept of close integration of computational methods, technology, development and global measurements and analysis of biological systems" (Galas & Hood, 2009: 2). In other words, systems biology results from culling data from diverse disciplines, to be fed into potent digitalized data banks, where interactions, mathematical calculations and algorithms model individual clouds of data calculating the best course of therapeutic decisions for particular medical needs.

System biology is translated into the so-called P4 medical strategy: predictive, personalized, preventive and participatory. Prediction in biomedicine is hazardous, subject to innumerable unknowns –unexpected events, behavioural changes, environmental modifications, intercurrent disorders– that disrupt anticipated causality and introduce an element of unforeseeable chaos. Furthermore, prediction is not necessarily a blessing, as it may cause anguish, interfere with reproductive intentions or otherwise disrupt planning for the future. Personalized medicine, including pharmacogenomics, touted as the main benefit of digitalized data banking, does not refer to the patient, but rather to a stratified group of individuals with common measurable traits –sub-types of cancer, shared genetic traits or response to external agents including medication– yet singular in their lifeworld and values. The notion is growing that experts need to group their patients into precise diagnostic entities that allow a scientifically supported most efficacious treatment, where personal singularities ought not to interfere with the generalities of statistically supported information.

Personalization is piecemeal and must be integrated into a general 'one size fit all' in current medical institutions, practice and medication, that, together with the unavoidable market considerations, have led "to a combination of practices that promoted less rather than more integrated, personalised and seamless care" (Day et al., 2017: 155). P4 claims to detect dormant predispositions that may be effectively prevented from becoming clinical conditions. Preventive medicine and public health measures may be opposed –antivaccine movement– or contrary to local o individual values –rejection of the perils of breast-feeding by HIV(+) mothers–. Finally, participatory biomedical actions are most improbable in digitalized decision processes, since the patient is faced with data rather then personally delivered medical opinions (Lupton, 2014). Patients are expected to not only to access digital information but also to provide input about their experience, filling the role of 'prosumption' –producer and consumer of data–, thus delivering data that will perhaps be employed to improve the delivery of healthcare but also as venues serving the interests of profitable corporative ventures. Data prosumption by the digitally engage patient appears to fundamentally differ from participative patient/caretaker interaction: "In the near future, patients will be segregated into treatment groups by not only determining their genetic mutations, but also by using omic data as an output of disease network activation status." (Duffy 2016: 497). This may be precision medicine, but it certainly is not personalized.

Predictive, preventive, and personalized medicine will transform science, industry, education, and society in ways that we are only beginning to imagine (Hood et al 2004).

To the present day, precision medicine has not delivered beyond identifying a handful of monogenetic diseases. Mainly focused on oncology targeted cancer cases have shown an improved survival rate of merely 2,1 months in a twelve years period (2002-2024) (Maughan 2017). Suspicion grows that precision medicine is more hype than promise, thus raising social and ethical issues regarding high priced diagnostic kits and expensive drugs benefitting investors and affluent consumers to the detriment of deprived populations. Increased research funding is seducing profit-oriented investors, neglecting to increase efforts at solving pressing global healthcare issues –endemic infectious diseases, antibiotics resistance, re-emergence of tuberculosis, dengue, Chagas disease–. Precision medicine focused on genetic research does exactly what it proposes to avoid: focusing on genomics and neglecting to treat the individual as a person whose health and disease depends on epigenetic factors, singular biographical lifeworld events, and socioeconomic circumstances

It is evident that the business plans of every sector of the healthcare industry will need to be entirely transformed over the next 10 years — and the extent to which this will be done by existing companies as opposed to newly created companies is a fascinating question (Galas & Hood, 2009, 1).

At least as important is also the essential, profound change in the way that physicians understand the medical issues (therefore medical school education must be radically changed) and how they relate and interact with their patients. The patient physician relationship will undoubtedly change dramatically (Ibid., 3-4).

Hyperbolic talk in the wake of meagre, and probably intrinsically limited, clinical results, not credible and reckless when suggesting profound changes in the practice of medicine in addition to foreboding dramatic turns in the already fragile patient/caretaker relationship.

The Decline Health Related Care

The age-old distinction between cure by male physicians and care by female nurses and ancillary personnel is becoming untenable as equally qualified share men and women the professional duties of treating and caring for the diseased. And yet, the distinction remains and is deeply embedded in many societies. Since women outlive men they will often be the final carer of the aged and chronically ill in the family. Life expectancy has increased, societies are ageing, and the elderly live longer but with precarious security, dwindling pensions, subject to unhealthy life conditions that contribute to chronic disease with progressive weakness and disability. Social ageing studies sounded the alarm that population in need of protection and care is on the rise.

Hospitals developing into biomedical platforms are showing an alarming tendency to regress to sophisticated treatment schemes that neglect basic nursing care reduced to play a role as comforter rather than active carer (Armstrong, 1983).

Despite the fact that the need for care has grown, the development of an ethics of care notably supported by feminist scholars, the omic approach of digitalized medicine is more concerned with intervention than interpersonal relationality. Dissatisfied with the impersonal way the severely ill were receiving, Cicely Saunders (1918-2005) created the modern hospice centred on humane, personal and symptomatic care of diminished and neglected patients, which developed into what now is known as palliative medicine which for a diversity of reasons falls short from its original care-oriented tasks.

As the pragmatic thrust of biomedicine and digitalization further leads to systems biology and P4 medicine, the provision of relational and personal care become unhinged from productive consumerist societies, where the elderly, the chronically ill and the marginalized are more of a burden than an object of caring concern. Biomedicine's and now digitalized medicine's tendency is to regard "their clients as little more than biological machines" subject to an "authoritarian and oppressive" professional/clients relationship (Porter 1997: 17). Foucault, in turn, was concerned that holistic perspectives increase power and surveillance, an observation that appears to become true as digitalized medicine pushes forward. Reactive criticism advocated a holistic kind of care, based on Habermas' drive to reinstall communicative rationality with its respect for values and personal lifeworld in all interpersonal relationships inspired in recognition and care.

Swelling waves of migrants, demographic increase of the aged, and the impersonality of sophisticated medical services lay bare the need for care, but also the lack of social motivation to deliver it. The WHO reports on ageing (2015) note that a "workforce prepared to provide long-term care" ought to be developed, though acknowledging that governments are financially feeble and that self-care will have provide what security plans are unable to provide. The brunt of care continues to fall on family voluntary cooperation, showing clear signs of fatigue and burnout. Professional caregivers are underpaid and stressed that result in preoccupying lacunae of uncared and neglected people in all wakes of life, and increasingly cause insensitivity and violence against the care dependents. Modern society has been seen as tolerating a culture of neglect.

Robotic Care

Enter robotic caretakers and, with them, hosts of scholars presenting pro/con arguments that cannot agree on what care is. Frugal arguments believe that care is needed for individuals who lack, or have lost, the capabilities of everyday chores including care of their body. As a minimum, robotic caregivers ought to assists and complement the insufficiencies and deficiencies of human care. More ambitious proposals require care to protect human dignity by restoring, perhaps even enhancing, the necessary capabilities required to live a good life, by going beyond limited human efforts, that is, outperforming what caregivers can or will do for standard care requirements (Coeckelbergh, 2010).

Entrepreneurs have sensed a market niche for technical support of care. With varying degree of sophistication, robots are designed to carry out tasks that the disabled have difficulty in tackling, including the project of customized robots that cater to specific needs of particular individuals rich enough to afford them. Technology fans and financial supporters expect robots to eventually cover all the needs of those who will depend on them, whereas critics claim that caregiving robots are too clumsy and limited in their capacities to handle anything but gross household chores (Sparrow & Sparrow, 2006). Technical scepticism is unconvincing, for there is every reason to believe that why machines will continue to refine and amplify their abilities, though it is convincingly argued that personal needs may not be artificially met. Even pet robots cannot replace interpersonal contacts, and the main problem that looms over the needy is isolation and loneliness. Social neglect, increased needs for care, the hype and hubris of technology, and the deceptive idea that dependence can be solved by self-care, are announcing that digita lized health and medical care will fall short of actually developing a caring society. An unexplored psychosocial area is the degree of receptivity that the targeted public might muster to prefer robotic over human care, a question that has not be explored or, at best, by inquiring whether people prefer robotic instead of no ancillary care. As long as human beings depend on relational ethics, they will unwillingly bow to assistance of a robot, but things may change if transhumanism becomes one of a kind with digitalized life.

Non Therapeutic, Enhancing and Perfective Digitilized Biomedical Research

The port of entry to the realm where the utopian idea of a transhuman gained momentum occurred when systems biology entered the fascinating and highly exciting world of neuroscience. Nowhere can the rift between science and humanities be more clearly seen than in cutting-edge genetics and neuroscience, both disciplines marching along oblivious to such revered ideas as human nature, intrinsic value of nature, human exceptionality, rebellion against transcendent order, the despair of theodicy believing in god and asking why he allows so much misery.

Human beings are no longer the fixed product of their genome, for their genetic makeup may be modified, purified, enhanced. According to neuroscientific research showing that brain activity is the product of neural network interactions that are set before consciousness embarks on choice and analyses alternatives. Mental activity is, if you will, a secretion of brain functioning. Science thus has faces modern man with the fourth great loss of centrality: geocentrism was replaced by heliocentrism, Darwin degraded the idea of human as exceptional creation by positioning humanity in the yet unfinished process of the evolution of species, while Freud showed our rationality to be basically dependent on unconscious psychic processes. And now, neuroscience simply degrades the mental into a complex material self- regulating neural dance.

Humanism takes refuge that free will, intentional agency and goal-setting activities are anthropological features that will always distinguish humans from other living beings.

Humanism qualitative arguments and communicational rationality have largely been silenced by the loss of essentialism that served to give sanctity as a defence against intervention. "More broadly, we can use 'ecology' as Gregory Bateson did, to be open to dynamics rather than essences of processes of what we end up labelling 'self,' 'other,' and 'the social' (Markham 2018: 5).

Discussions and debates have been ear deafening and tumultuous as too much emphasis was invested in the anticipation and horrified reaction of artificially creating a transhuman being –cyborg? Posthuman? Overhuman?–

Transhumanism sees the human being as a transient state in evolutionary development. It recognizes the incompleteness and weakness inherent caused by biological evolution to be overcome by aiming at a new existential step that transcends the condition humana…The goal of transhumanism is to reach transcendence within the physical universe assisted by scientific methods. (Nahm, 2013: 17).

Transhumanism pretends to employ biotransformative technologies, in order to "enhance" the human organism, with the ultimate aim being to modify the human organism so radically as to "overcome fundamental human limitations" (Transhumanism FAQ, 2016) and thereby the "human" as such. In other words, to use tranhumanist terminology, their fundamental goal is to become "posthuman." (Porter, 2017: 238).

The transhumanist program is marked by a value problem: which human values should be preserved and enhanced, which excised from the human genome? Who is to decide? Will such currently respected values as "health" or "normal" carry their meaning into a posthuman society? There is an obvious Nitzschean element in transhumanism: "the ideal of the over human is an ideal of self-mastery over the will and over the self as a whole" (Porter 2017: 247). The self-sufficiency of posthuman beings would affect the anthropological trait of relationality changing it from an essential survival feature to a playful and dispensable interaction. Consequently, in a flawless self-master the idea of a patient requiring healthcare would be an image of obsolete times, self-mastery will see no place for patient/counsellor situations. To think about interpersonal relationships in transhuman society us simple a non-starter when posthumanity creates a transpersonal being who rob traditional persons of their autonomy.

Therapy and Enhancement

Moving a ratchet down to a more realistic endeavour of genetics, one encounter a scientific discipline focused on therapeutic interventions in genomes that harbour a severe disease or a predisposition to some serious clinical condition. Surprisingly, or perhaps not so unexpected, the possibility of tampering with non defective genes in order to modify a phenotypical characteristic or enhance functions that are performing within the normal range, opens up a future market of enhancement *a la carte*. Some enhancements are presented as therapeutic for below average growth, learning abilities, or other under par performances. Therapeutic enhancement must remain voluntary and subject to an informed consent that hardly can have the usual patient/counsellor, rather presenting as a client/provider relationship.

Enhancement has further moved away from a patient/counsellor decision making process when suggesting that moral enhancement, for example, ought to be mandatory. Taking a dark view of world affairs, some bioethicists stress the urgent necessity of biomedical technology to install ethical enhancement that by habitual methods is too slow and unpredictable (Persson & Savulescu, 2014). Whichever way instrumental enhancement may go, be it genetic and therefore heritable, or neuroscientific and non transmittable, is doing away with irreversible interventions in the human body that traditionally have been the exclusive privilege of health regaining medical therapies. The patient/counsellor relationship is out of joint.

DISCUSSION

Debates on digitally supported scientific knowledge of, and intervention in, the human body go from very naive progress proposals to apocalyptic declarations that the world will only survive if technology is curtailed, to passionate proclamations that survival will depend on more and increasingly efficient cognition and more precise instrumental intervention in natural processes. The question raised many years ago by philosopher Hubert Dreyfus, whether reflecting on technological limitations might reveal the limitations of technology, will remain unheeded because either side is biased by vested interests to expand technology beyond limits, or limit it in the name of other values too dear to be renounced.

Humanism has lost the support of absolute values and is no match to the self-sufficiency of technoscientific expansion that fuses adaptive advantages with artificial creation of desires that spin the self-impelled roundabout of production and consume, neglecting to address the needs of those unable to mount the glittering market carousel that has sold out on fair distribution of essential services like medical care, education, social security. Reliance on human rights doctrine has been insistently brought forth as an insufficiently heeded argument against inequities produced by globalization. Dampened by reality and official recognition that economic, social and cultural rights may depend on "progressive realisation", UNESCO found it necessary and opportune to issue the Universal Declaration of Bioethics and Human Rights (2005) where autonomy and informed consent appear as a major concern. Bioethics has weakened its critical voice of biomedicine's focus on biology to the detriment of personal life-world, to embrace actual innovations future promises in healthcare.

And yet, force and status of human rights doctrine are too weak to take up the challenges posed by the invasive strides of neuroscience and neurotechnology, as yet unable to separate benefits from risks and complications. New rights have been proposed signalling the risk persons may soon be facing: in "coming decades": the right to cognitive liberty, the right to mental privacy, the right to mental integrity, and the right to psychological continuity (Ienca & Andorno, 2017).

Humanism deprived of essentialism lacks the force to establish strong and unshakeable arguments, unable to establish absolute limits t human intervention in natural processes, weakly and unconvincingly suggesting conceptual armours against extreme and irreversible modifications of anthropological features erstwhile hailed as categorical of human nature.

Digitalized biomedicine is doing away with a personal relationship between patient and health caretaker and, as individuals are expected to resort to data bank information to guide them towards self-care and self-responsibility, assisted by caregiving robots, the role of counsellor dwindles into oblivion. Although tt remains speculative whether such extreme digitalization of healthcare and biomedicine will ensue, actual experience shows marked and sustained tendency to rely on instrumentation to the detriment of a clinical encounter or research protocol with satisfactory interpersonal relationships.

CONCLUSION

Bioethics has been the unparalleled champion of patient autonomy, yet unable to reach beyond fierce theoretical elaboration, that has been questioned by those who rely on deeper and more absolute values especially based on religious beliefs. Autonomy remains the *sine qua non* of ethics and the most precious feature of modern individualism, the last-ditch before genetic and neuroscientific control to machines. It supports such basic notions of social life as responsibility and accountability. Debates on man and machine have already visualized that perfecting algorithms might surpass human autonomy, or that machines will come to be autonomous in terms of choosing their own criteria of what can or should no be done why. Much of this is idle speculation, but it has helped show that human autonomy is in itself a fragile construction.

The WTA/H+ statement embraces the desirability and inevitability of germline and enhancing gene therapies, while also calling for public financing of research and a regulatory process to ensure their safety...Nor should laws specify acceptable and unacceptable gene therapies based on distinctions of "therapy" versus "enhancement," or on the basis of speculative long-term social impacts. The only basis for regulations of gene therapy are traditional clinical risk/benefit ratios for the subjects and their children. (Hughes 2005).

Hughes' piece is a masterly example of slippery slope strategy, as he effortlessly glides from supporting germ line therapy, to enhancement and implied, though not named, transhumanism. The Director of an *Institute for Ethics and Emerging Technologies* justified actual clinical risk/benefit evaluation although neither can even remotely by detailed and quantified "for the subjects and their children".

Fear of this slippery slope towards unwanted artificial transformation of human nature is a serious consideration because humanists have been unable to come up with an unassailable prohibition to tamper with individual genomes or cerebral networks, which are initially presented as cognitive conquests with therapeutic aims, moving on to enhancement, perfection and, finally, transformation. Informed consent, if required, has no empirical basis on which to rely, as "Fools rush in where angels fear to tread" (A. Pope 1711).

REFERENCES

Armstrong, D. (1983, January). The fabrication of nurse-patient relationships. *Social Science & Medicine, 17*(8), 457–460. doi:10.1016/0277-9536(83)90051-5 PMID:6879243

Barry, C. A., Stevenson, F. A., Britten, N., Barber, N., & Bradley, C. P. (2001). Giving voice to the life-world. More humane, more effective medical care? A qualitative study of doctor-patient communication in general practice. *Social Science & Medicine, 53*(4), 487–505. doi:10.1016/S0277-9536(00)00351-8 PMID:11459399

Berger, S., Denner, M.-S., & Röglinger, M. (2008). *The Nature of Digital Technologies – Development of a Multi-layer Taxonomy*. www.researchgate.net/publication/330881198

Day, S., Coomes, R. C., McGrath-Lone, L., Schoenborn, C., & Ward, H. (2017). Stratfied, precision or personalized medicine? Cancer services in the 'rea world' of a London hospital. *Sociology of Health & Illness*, *39*(1), 143–158. doi:10.1111/1467-9566.12457 PMID:27460935

Duffy, D. J. (2016). Problems, challenges and promises: Perspectives on precision medicine. *Briefings in Bioinformatics*, *17*(3), 494–504. doi:10.1093/bib/bbv060 PMID:26249224

Emanuel, E. J., & Emanuel, L. L. (1992). Four Models of the Physician-Patient Relationship. *Journal of the American Medical Association*, *267*(16), 2221–2226. doi:10.1001/jama.1992.03480160079038 PMID:1556799

Galas, D. J., & Hood, L. (2009). Systems Biology and Emerging Technologies Will Catalyze the Transition from Reactive Medicine to Predictive, Personalized, Preventive and Participatory (P4) Medicine. *Interdisciplinary Bo Central*, *1*(2), 1–4. doi:10.4051/ibc.2009.2.0006

Ghaemi, S. N. (2017). Biomedical Reductionist, Humanist, and Biopsychosocial Models in Medicine. In T. Schramme & S. Edwards (Eds.), *Handbook of the Philosophy of Medicine* (Vol. 2, pp. 772–791). Springer. doi:10.1007/978-94-017-8688-1_38

Hughes, J. (2015). Transhumanist Position on Human Germline Genetic Modification. *Institute for Ethics and Emerging Technologies*. https://ieet.org/index.php/IEET/more/hughes20150320

Ienca, M., & Andorno, R. (2017). Towards new human rights in the age of neuroscience and neurotechnology. *Life Sciences, Society and Policy*, *13*(1), 5–22. doi:10.118640504-017-0050-1 PMID:28444626

Lupton, D. (2014). The commodification of patient opinion: The digital patient experience economy in the age of big data. *Sociology of Health & Disease.*, *36*(6), 856–569. doi:10.1111/1467-9566.12109 PMID:24443847

Markham, A. N. (2016). *Ethnography in the Digital Internet Era From fields to flows, descriptions to interventions*. http://jeremydean.org/blog/wp-content/uploads/2016/02/ethnographyinthedigitalera-2016finaldraft.pdf

Maughan, T. (2017). The Promise ind the Hype of 'personalized Medicine'. *The New Bioethics: a Multidisciplinary Journal of Biotechnology and the Body*, *23*(1), 13–20. doi:10.1080/20502877.2017.1314886 PMID:28517988

Nahm, T. (2013). 1 Transhumanismus: Die Letzte Grosse Erzählung. In Reader zum Transhumanismus. Berlin: BoD.

Persson, I., & Savulescu, J. (2014). *Unfit for the future*. Oxford University Press.

Porter, A. (2017). Bioethics and Transhumanism. *The Journal of Medicine and Philosophy*, *42*(3), 237–260. doi:10.1093/jmp/jhx001 PMID:28499043

Porter, S. (1997). The patient and power: Sociological perspectives on the consequences of holistic care. *Health & Social Care in the Community*, *51*(1), 17–20. doi:10.1111/j.1365-2524.1997.tb00091.x

Royakkers, L., Timmer, J., Kool, L., & van Est, R. (2018). Societal and ethical issues of digitalization. *Ethics and Information Technology*, *20*(2), 127–142. doi:10.100710676-018-9452-x

Schaeffer, J-M. (2007). *La fin de l'exception humaine.* Paris: Editions Gallimard.

Scrambler, G., & Britten, N. (2001). System, lifeworld and doctor-patient interaction. In G. Scrambler (Ed.), *Habermas, critical theory and health* (pp. 45–67). Routledge.

Sebag, V. (2007). *Droit et bioéthique*. De Boeck & Lancier.

Smith, C., Baveja, R., Grieb, T., & Mashouri, G. A. (2017). Toward a science of translational science. *Journal of Clinical and Translational Science, 1*(4), 253–255. doi:10.1017/cts.2017.14 PMID:29657860

Sparrow, R., & Sparrow, L. (2006). In the hands of machines? The future of aged care. *Minds and Machines, 16*(2), 141–161. doi:10.100711023-006-9030-6

Stanford Libraries. (2019). *Guiding Principles on Ethics in Digital Health Produced During a Seminar at Stanford Libraries*. https://library.stanford.edu/node/156021

Svenaeus, F. (2019). A Defense of Phenomenological Account of Health and Disease. *The Journal of Medicine and Philosophy, 44*(4), 459–478. doi:10.1093/jmp/jhz013 PMID:31356662

Chapter 6
Health Professionals' Attitudes to Obese Patients and Ethical Problems Experienced

Meryem Türkan Işık
Faculty of Nursing, Mersin University, Turkey

Gülay Yıldırım
Department of History and Ethics, Sivas Cumhuriyet Üniversty, Turkey

ABSTRACT

Obesity is one of the health problems that threaten humanity considerably. In our country, considering the right to healthcare of each individual, earned by birth, they have a right to receive an equal and just healthcare. Obese individuals may suffer from negative attitudes of health professionals in providing protective health services and inpatient treatments. In this chapter, the aim was to draw attention to ethical conflicts between obese individuals and health professionals in the process of their healthcare and raise awareness of these problems.

INTRODUCTION

World Health Organization (WHO) reports that obesity in society has nearly increased three fold ever since 1975 and in 2016 more than 1.9 billion adults (39%) were overweight and 600 million of (13%) them were obese (WHO 2018). Obesity, developing in individuals by excessive fat mass in proportion to body size, has become one of the biggest problems of the world (WHO 2018). It was estimated that obese population in Turkiye is by 19.6% of the general population (TSI 2018). All these findings indicate that obesity-prevalence has been increasing over Turkiye as well as the world (Turkish Statistics Institution TSI, News. Number: 31/2017, 7 June 2017.accessed: https://www.tuik.gov.tr/basinOdasi/haberler/2017_31_20170607.pdf; Satman et al. 2010)

Obesity has created significant concern in health services and been emphasized in the provision of health services more and more. Obesity is described as an illness in health policy literature. According to

DOI: 10.4018/978-1-7998-4117-3.ch006

the standard disease term in medicine philosophy literature, obesity is not a disease but is a crucial risk factor for diseases. According to social disease understanding, obesity is considered not only as a public health problem but also as an illness (Hofmann, 2016). Although obesity is a disease, it is accompanied by many physical diseases such as hypertension, Type 2 diabetes mellitus, cerebrovascular accident, muscle-skeleton pain, sleep apnea, breast-ovary-endometrial-prostate-and-colon cancers (Kalan, I.,& Yeşil, Y. 2010,Hofmann, 2016, WHO 2018).

In social terms; obesity is thought as a lifestyle, eating addiction (Berdah, 2010), lack of character, weak will power and inability to take personal responsibility (Greener, Douglas, & van Teijlingen, 2010; Puhl, & Heuer, 2009). These properties of obesity are the reasons for health professionals to develop negative emotions against obese patients such as prejudices, stigma, discrimination and social exclusion and to spare less time for obese individuals. These emotions lead to insufficient medical approaches, avoidance from patient-centered practices, problems in allocating limited resources, a decrease in quality of life. Besides; these negative emotions also cause individuals to be unwilling to benefit from health services or to discontinue treatment; as a result they are hindered to utilize legal right to health services and patient safety is risked. To create a safe zone for hospitalized obese patients to move is one of the problems experienced. To create an area where obese patients can safely move is seen as a concern at health institutions. Health professionals should identify physical inactivity for obese individuals, regard it as one of their health risks and be aware of treatment approaches of this population (Barr, Cunneen, 2001). Similarly; it should be kept in mind that in clinical settings, health professionals should be patient advocates that do not harm patients, provide benefits, protect patient's autonomy, justice principle and put non- judgmental attitudes into practice.

Besides, it is seen that obese individuals suffer more from conditions that threaten their psychological health as compared to normal weight individuals, are more dissatisfied with their bodies and abstain from social relations more. By threatening life, these psychological problems reduce obese individuals' quality of life, impair their body images, result in social isolation, depression and anxiety and decrease their self-esteem (Besler et al. 2010, WHO 2018). As a result of increasing chronic diseases among obese individuals, it is known that they attend health facilities more frequently.

In this respect; obesity is both a risk factor for chronic illnesses, a crucial public health problem that aggravates the process and an economic burden (Obesity and the Economics of Prevention Fit not Fat ed. Sassi F. 2010 OECD Publishing.). If obesity is handled -particularly- in food supply and marketing system as a social determinant, a wise and rehabilitative health policy that will secure health of obese population can be designed. In obesity treatment, biomedical solutions are used, too and it is known that these solutions may pose as serious side effects. Instead of dealing with social problems with biomedical methods; measures taken in line with health policies can contribute more to obese individuals' health status and welfare (Hofmann, 2016).

In conclusion, obese individuals may undergo negative attitudes of health professionals in the provision of protective health services and inpatient treatments; which may –in turn- produce ethical problems. In this chapter, the aim was to draw attention to ethical conflicts occurring between obese individuals and health professionals in the process of their health care and to raise awareness of these problems.

BACKGROUND AND CONTEXT OF STUDY

Obesity Bias and Discrimination Concepts in Health Practices

Despite high prevalence of obesity, obese individuals are often subjected to obesity-bias, negative attitudes and stigma at settings such as education, work, family and health services (Tanneberger et al. 2018; Diedrichs and Barlow, 2011; Puhl and King 2013). Turkish Language Association (TLA) defines bias as "preformed positive or negative judgments, prejudice, preconceived opinion, preconception against someone or something by abusing a certain situation, event and figure" (TLA, 2019).

Theoretical approach of stigma originates from the assumption that overweight and obese individuals are responsible for their physical look caused by personal weakness, laziness, indiscipline and their life conditions. Another theory suggests that excessive emotional bonds with obese patients may produce more burdens on personal resources and thus increase or decrease stigma (Tanneberger et al. 2018). Bias, stigma, discrimination and social exclusion -demonstrated against obese patients by society- aggravate severity of these individuals' possible physical and psychological problems (Diedrichs and Barlow, 2011; Puhl and King 2013). One of the venues where obesity bias occurs often is where health services are provided.

Obesity is a medical condition that may be associated with prejudices in health professionals. This prejudice may result in disrespectful or poor care to be given to obesity patients (Obstetrics & Gynecology 2019). Obesity is accompanied by numerous chronic diseases; as a result of which obese individuals utilize health services more and meet health personnel at different health settings more. Therefore; these individuals are more likely to undergo stigma, discrimination, poor social support (Tanneberger et al. 2018; Wott, & Carels, 2010). Since they are subjected to these negative attitudes; their quality of life is negatively affected. Obese individuals are more dissatisfied with their body as compared to normal-weight people and tend to be abstaining from social relations more. In addition; it is known that obese individuals suffer from social isolation and lack of confidence in interpersonal relations more as well as numerous physical and mental problems such as anxiety, depression, reduced self-esteem, muscle-skeleton pain, sleep apnea (Yen, Liu, Ko and Wu 2014). In a study done, it is reported that as perceived social support increases among overweight and obese individuals so do their self-esteem, body perception and satisfaction with life (Akman, 2019).

In the study carried out with 145 overweight and obese individuals who applied to two special dietary counseling centers in Ankara, the mean scores of obese individuals were stigmatized several times in their life, sometimes in the Dutch Eating Behavior Questionnaire subscales, sometimes for restrictive eating, sometimes for emotional eating, and sometimes for external eating. The participants were found to have mild depression for the CES-Depression Scale and sufficient self-esteem for the Rosenberg Self-Esteem Inventory. As a result of the study, the frequency of the participants with high weight was found to be more common with stigmatizing situations. The frequency of encountering stigmatizing situations and internalization levels of weight bias are high due to the participants' excess weight. Who frequently encounter with stigmatizing condition is felt higher levels of depression. With the frequency of encountering stigmatizing situations, self-esteem perception is lower. The emotional eating and external eating levels of the participants with higher obesity bias internalization score and internalization levels are also high (Cihan, 2019).

Obese individuals state that they feel subjected to prejudices and biases when they receive health services and do not want to continue the treatment and become unwilling to benefit from health services due to negative attitudes against them (Diedrichs and Barlow, 2011; Puhl et al.,2009; Altun and Ercan, 2016; Tylka, et al. 2014; Hand et al. 2013; (Tanneberger et al. 2018; Alkan et al. 2016).

Sometimes, obese patients tend not to report all the symptoms or to hide the disease itself and this worsens their conditions further and prevents their right to health. Those, not experiencing a healthy process in the treatment of the disease, affect the rest of the society negatively due to negative externality (Alkan et al. 2016; Wott, & Carels, 2010). Patients, being subjected to obesity prejudice and unpleasant attitudes by health professionals, stated that they may withdraw from preventive health services, health screenings and cancel medical appointments or delay medical treatments (Aldrich, Hackley, 2010). Obese individuals underline that they feel prejudiced while receiving health services and are generally reluctant to continue treatments due to these negative attitudes (Puhl & Heuer, 2009).

In a review study; it is reported that many studies on the relation between obesity prejudice and stigma and health service quality argue that health professionals bear strong negative attitudes and prejudices against obese individuals. There are strong evidences that emotions caused by prejudices and negative attitudes felt by obese individuals impair their interpersonal relations with health professionals and decision-making process. In case of these attitudes, obese individuals' treatment process and quality of health care services they demand may deteriorate (Phelan and et al. 2015).

In studies reported by health professionals, some widely seen biases reported by health professionals are in the form of such negative attitudes that obese patients are lazy, weak minded, incompatible with the treatment and unsuccessful (Diedrichs and Barlow, 2011; Puhl et al.,2009; Altun and Ercan, 2016; Tylka, et al. 2014; Hand et al. 2013). Generally; health professionals state that they are not ready enough to treat obese patients. At the heart of these views are rooted such assumptions that obese patients do not make efforts to lose weight, are discordant and do not demonstrate enough motivation to change life styles (Tanneberger et al. 2018). Therefore; these factors may cause health professionals to be less patient-centered, not to build positive communication and to spend less time to give health information to obese individuals (Tanneberger et al. 2018). Instead of accepting limited efficacy of the existing weight management methods, health professionals are inclined to accuse patients of maladjustment and lack of self-discipline. On the other hand, disappointment that clinicians experience while trying to help patients lose weight may too intensify this prejudice (Altun & Ercan, 2016). In a study; it was found that obese young women are exposed to stigma more as compared to old women. 53% of obese women told to have received unpleasant remarks from doctors about their weight and 75% of them refused diet as a coping method due to obesity stigma (Puhl, Brownell, 2006).

In the study carried out at Johns Hopkins University Faculty of Medicine, it aimed to develop students' attitudes towards obesity through an innovative ethical session in the "Obesity, Nutrition and Behavior Change" section. The data were applied to 59 students in the medical school (2012-2017) before the ethical session, and the Implicit Relationship Test was applied to measure the bias of the students about the implicit weight, the causes of obesity, as well as the video clips showing negative deviation of weight. As a result of the Implicit Relationship Test to measure the implicit weight bias of all students, it was revealed that 70% were prone to bias, 18% were neutral and 12% were obesity-biased. While most students think that obesity is a disease or behavioral disorder, the vast majority still think it is caused by ignorance, and a small number of obese people are lazy. In the follow-up questionnaire, 30% of the students improved their obesity bias attitudes after the ethical session, 40% of the students struggled to discuss their personal struggles with weight, implicit weight bias, and the Implicit Relationship Test and the results of the questionnaire were useful, more than 70% of video clips were useful and obesity It was determined that discussing personal experiences and beliefs in an ethical framework and discussing popular media developed positive attitudes towards obesity (Geller, Watkins, 2018).

In a study where dietician students' attitudes towards obese individuals were investigated, it was identified that the students viewed obese patients' diet quality and health status worse as compared to those non obese individuals (Puhle, 2009). In another study on nurses who worked at surgical clinics, majority of the nurses reported that they would not give care at obesity surgery. Of nurses' unwillingness to provide care; backache, fear that patients may fall off and inability to meet patients' care needs can be listed. It was also explored that most of the equipments used for obesity care by intensive care nurses who cared 1-3 patients in a month, did not have any training on caring obese patients, worked with unqualified personnel in nursing care procedure and underwent emotional problems were not sufficient for care (Altun et al. 2017). In another study, "your weight may harm your health" sentence was used instead of "obese" term and relative effect of beliefs on patients was investigated and as a conclusion, the term 'obese' worsens the problem further while it was less disturbing for obese patients (Talor 2009).

Stigma related to obesity at health facilities may be occurring due to physical setting and medical tools. On the one hand; small or uncomfortable or unreachable clinical equipments and tools for medical specialists (chairs, weighing machines, blood pressure cuffs) that are not fit for obese individuals with bigger body sizes may create an unpleasant setting for obese patients. Since equipments that are not suitable for obese individuals at health facilities complicate health professionals' efforts and works, their prejudices and negative attitudes as well as feeling of unrest experienced by obese individuals will increase; as a result they will stay away from receiving health services. As unsolved health problems of obese individuals, undergoing these situations, worsen in time, so does health-cost to care obese patients. On the other hand; health professionals' bias reduces quality of communication with obese individuals, outcomes of health services and patient satisfaction; too (Tanneberger et al. 2018).

Additionally; according to the results of a study it was argued that nurses were aware of weight stigma, made efforts to give an emphatic and respectful care to obese patients as compared to other health professionals and demonstrated less prejudice and stigma than other professional groups (Işık et al. 2019; Puhle, 2009; Jay et al., 2009). To the contrary; in another study it was reported that nurses who believed that weight could be kept under control with individual efforts discriminated against obese patients at clinical practices. In a study that was done with overweight or obese women (n=498) in relation to barriers to routine gynecological cancer screenings; it was reported that 52% of the women faced at least one barrier to receiving appropriate services for weight health and most of them complained that they became late to get health services as their weight increased. Reasons for the delay are negative attitudes of health professionals, maltreatment, disgraceful recommendations for losing weight, reluctant weight loss advices and dysfunctional tables, aprons, equipments (Amy, Aalborg, Lyons, Keranen, 2006). In a study done with intensive care nurses (n= 159) in relation to implicit and explicit attitudes towards obesity intensive care patients, obese individuals were found to be weak willed, a bit ugly and "lazy" (Robstad, and et al. 2019). According to results of a study conducted with obese women with a BMI ≥ 30 (n=150), low self-esteem in obese women was found to be associated with body dissatisfaction. Besides, it was also emphasized that the self-esteem in obese women and body image was found significantly to be correlated with socio-cultural attitudes, body mass index and education (Armağan, 2013). In a study conducted in a university nursing faculty (n = 278), it was determined that lifestyle behaviors of students did not affect obesity bias level (Kasar, Akyol, 2019).

Another negative attitude towards obese individuals is discrimination (Durso and Latner, 2008). There are numerous different definitions as to discrimination. In the definition of Turkish Language Association (TLA), discrimination is defined as the adoption of unjust or prejudicial treatment against certain group of people in order to separate them from others in a deliberate and purposive manner (TLA, 2019).

Another definition puts discrimination as direct or indirect obstruction or restriction of minority individuals or groups by the majority of the society without legal grounds from using rights and freedoms benefited by those accepted as the majority of the society. In literature, there are many theories that clarify causes of discrimination. According to Social Dominance Theory, holding some parts of these theories; one or several groups in each society are more dominant than other groups. Dominant group always considers itself superior, others inferior and isolates other groups and shows an inclination to seize their rights. Other theory that retraces roots of discrimination is Social Identity Theory. According to Social Identity Theory, people tend to classify and categorize objects, things, events and other people since birth. This process starting with stigma is nourished with negative prejudices and ends in isolation and discrimination. Stigma -used to rule, to marginalize, to disgrace or to praise subgroups- puts stigmatized people in a certain stereotype and causes them to be perceived as "the other" by everybody. Prejudice is one's thoughts formed in mind against a person or a group and accepted without inspecting whether they are true or not. A prejudiced individual shows discriminatory behavior while thinking about the other (Alkan et al. 2016).

In some studies that investigate health workers' attitudes against obese patients, it was reported that 40% of the physicians demonstrate negative attitudes, 45% of the nurses would not provide care services to obese patients if they were granted the option to give or not to give care to them (Jay et al., 2009) and 52% of these nurses were unwilling to provide care to obese patients (Poon and Tarrant, 2009; Usta and Akyolcu 2014).

In a descriptive study undertaken to determine level of nurses' prejudice against obese patients; it was also emphasized that most of the nurses tend to be prejudiced against obese patients (Işık et al. 2019). Prejudice tendency makes health professionals abstain from serving obese patients and disturbs services and results in situations that increase health cost. When implicit and explicit attitudes of doctors about weight were examined to explore their negative attitudes, it is seen that most of them are prejudiced against obesity (Sabin, Marini, & Nosek, 2012). In a study that investigated how nurses perceived obese patients at children hospital, it was identified that nurses demonstrated prejudice against pediatric obese patients and had negative views on their supportive roles in caring obese patients (Garcia, Amankwah, & Hernandez, 2016).

Ethical Problems Experienced by Health Professionals in the Care/Treatment Process of Obese Patients

Health professionals should provide an effective and holistic health care to obese patients. Awareness of prejudice and discrimination by health professionals against obese individuals should be raised, obese individuals should be given health services as good as possible and ethical sensitivity of health professionals should be improved.

We can examine the problems experienced by health professionals while caring obese patients under two titles: First, basic nonethical problems experienced by health professionals who work at health facilities are lack of sources, materials, insufficient number of health staff and economical problems caused by high cost of care (Forhan and Salas 2013, Altun and Ercan 2016; Altun Ugras G et al. 2017). Second, basic ethical problems experienced by health professionals who work at health facilities are related to justice and equality principle, problems of distributing limited sources equally and ethical problems related to the principles of autonomy beneficence, non-maleficence (Hand, et al. 2013; Royal College of Nursing.2003). We expect that health professionals be patient advocates who demonstrate attitudes that do not harm but provide benefits to patients in terms of medical ethics, protect patient privacy/ autonomy and justice principle and are non-prejudiced in clinical settings (Royal College of Nursing, 2003; Cobanoglu 2007; Ecevit Alpar, et al.2013; Ryan, 2019).

In Turkiye, patient rights of those who are in need of benefiting health services are put under protection according to Turkish Constitution, international agreements, laws and other regulations and Patient Rights Regulation (1998) legalized to protect patients in every aspect just because they are human. By these laws, health professionals are banned from discrimination of any kind among patients due to any reason. According to these legal regulations; health professionals should accept that the condition of obese patients is a disease, should respect for their patient-rights and therefore should approach them kindly (Royal College of Nursing, 2003; Ryan, 2019).

All the patients as well as obese patients should access to care services offered by health professionals in a fair manner, should not be abstained from their rights and their access to rights should not be restricted.

Health professionals are responsible to restore patients' health and to accept patients as they are. To fulfill this responsibility requires that patients be dignified, understood and respected. Professionals should make sure that minimum standards are met to give patients service. Otherwise, beneficence principle of ethics is violated and professionalism values are neglected. Considering the above mentioned study-results of stigma; it may be argued that beneficence principle of ethics is broken because sensitive and quality health services to everyone are not provided.

Yet, studies state that obese individuals are generally harmed due to collateral damages in the course of receiving health services because they abstain from getting health services by feeling embarrassed due to health professionals' negative attitudes. In addition, such simple procedures as taking vital signs or wearing apron pose a problem for them, too.

Stigma or discrimination may be rooted in conscious or unconscious personal beliefs or attitudes of health professionals about patients' personal behaviors and health effects. These motives may affect allocation of health resources.

At emergency services, the situation is the same. Emergency services are clinical settings where difficulties caused by care of obese patients intensify. Growing burden of obese patients, diagnostic assessments and difficulties occurring in treatment process are barriers to population of these patients while they are accessing to care. Due to restrictions caused by medical treatments and care given by emergency physicians to obese patients and limitations at hospital setting; some ethical problems occur. In addition to good qualities, emergency doctors should have the same knowledge and skills to stabilize obese patients as in acute patients and clinical settings at hospitals should be supported. For obesity management, necessary equipments and tools should be used and sufficient number of emergency personnel should be employed, in case of any drawbacks in this sense obese patients should be transferred to more equipped hospitals (Venkat, & Larkin, 2014).

Medical Ethical Principles and Obesity

There is a need for ethical sensitivity and ethical concern so that obesity related problems can be noticed. These issues highlight health professionals' purposive interest in human relations. Ethical sensitivity underlines being aware of and putting an emphasis on ethical dimensions while providing health services whereas ethical concern underlines the importance of detection of ethical problems by health professionals and thus existence of an ethical concern among these professionals. When an awareness of these two dimensions is strengthened, ethical dimension of the difficulties experienced by obese individuals in the delivery of health services can be uncovered; which may be a driving force for action plans of solutions to be developed in case of any concerns that health professionals have.

In the meanwhile, ethical decision-making process comes into play as action plans to be developed for the delivery of health services of obese individuals. During this ethical process, ethical principles that guide health professionals are often used. Ethical values provide a source for the rules that determine behavior on social level as well as a criterion for thinking-processes that determine behaviors in mind. The term "principle" -which directs and sets criteria for basic approaches and behaviors specific to different frameworks and accounts for the most general, basic and abstract rules- can often be used as ethical principle (Yıldırım and Kadıoğlu, 2007).

In providing health services, health professionals and academicians adopt the following principles: justice, beneficence, non-maleficence, accountability, fidelity, autonomy, and veracity. Basic medical ethics principles not only establish a general framework for the rules but also illuminate decision-making process not set by rules. There is no a permanent hierarchy among these principles and to determine the principle that will be adopted as a guide in order to solve the conflicts, it is necessary to make evaluations specific to the problem because the person that experiences the problem and the values are different and principle of uniqueness of each individual affects it. In the context of this medical ethical principles, an approach to a healthy / obese patient should be demonstrated.

Autonomy Principle

Autonomy is defined as one's competence and ability to think freely on his own, to make decisions about himself and to act accordingly. According to this definition of autonomy; individual chooses rational and free decisions and actions to his own values and without any pressure. Thus, as an autonomous individual he realizes his autonomous choice. The action should be purposive, informed and voluntary for an autonomous action to be realized. Therefore, individuals should be informed of the relevant issues and make the due decisions. After individuals have made their decisions, their autonomy should be respected while they are putting their decisions into actions. An approach that respects for autonym requires a high level of communication and effort between health professionals and their patients. From patients' point of view, it requires sharing responsibility of medical actions with health professionals. However, even if paternalism principle decreases its validity, this principle has not been entirely quitted yet (Ersoy N., Aydın E. 1994). Health professionals are all healthy / patient should care about the individual's treatment and care process in the principle of autonomy. Since the obese individual is unique in the process of being in the health institution, information about his condition should be mentioned in detail and in a way that he can understand the possible risks. Health professionals should include the obese individual in the entire treatment and care process, protect their privacy, and keep their secret.

Justice Principle; This principle guides medical practices. It is generally suggested that health is a basic human right and everybody should benefit from health services equally and limited resources should correctly be utilized. In this sense; justice principle argues that -just like basic rights such as freedom to choose and liberty to think freely- it is necessary to create equality of income, equality of opportunities and equality of power and responsibility. In particular; disputes on whether justice principle or other principles of fair division should be used in allocating limited and scarce medical resources make up ethical clashes (Aydın E., Ersoy N. 1994). Healthcare professionals should understand the importance of the principle of justice in obese patient care. It should provide care services in the context of the principle of justice against this vulnerable group by internalizing the process of providing health services in institutions and hospitals where health services are provided.

Non-Maleficence Principle

This principle has been one of the oldest basic principles of medicine. Hippocrates emphasizes this principle by saying "First do no harm". According to this principle; health professionals are banned to do harm not only to patients but also to third parties and to society. The necessity to identify what the harm is and according to whom the harm should be identified is important points in this principle. Sometimes, it is required to do a smaller harm in order to prevent doing another life-threatening harm. A risk-benefit assessment should be done in non-maleficence principle and decisions should be made according to the choices that are the most valuable (Oğuz Y. 1997; Aydın E. 2006). Health professionals are obese patient care process in the patient's treatment and care not to harm in making ethical principles should be kept in mind. In line with this principle, obese patients are provided to receive health care services without disturbing their trust in healthcare professionals.

Beneficence Principle

Philosophically, it may be argued that this principle requires beneficence as high as possible for individuals as many as possible (Akarsu, 1982). Medical practices with a high potential of beneficence may at the same time have a high risk to harm. The significant two principles of modern medicine education –principles of autonomy and beneficence- sometimes contradict each other as the most difficult ethical dilemmas in cases where either is reasoned as a separate action choice. Putting patient beneficence above all else in health-practice becomes another ethical problem that emerges as a result of paternalism (Özcömert G.H., Özçelikay G., Şar S., Asil E. 2000). Healthcare professionals should offer treatment and care to the obese individual / patient in line with the principle of benefit, which is one of the most important medical ethics principle. Health care professionals are found in health care policy action in the direction of benefit, it may violate the autonomy of obese individuals. Obese individuals / patients may face a paternalist attitude in the process of receiving a different expression treatment and health care.

SOLUTIONS AND RECOMMENDATIONS

Health professionals' demonstrating bias and discriminative attitudes prevent obese patients from receiving a care of high quality; which creates risks for patient safety. In order for this group of patients to receive a care of high quality that they need; it is important to determine health professionals' biases and discriminative attitudes so that the right to get a care of high quality can be realized by obese patients. However, it is expected especially from health professionals that they should show affectionate, honest and non- discriminative attitudes. All the health professionals need to be educated on weight bias and stigma in health so that they can approach patients in a non-prejudiced way (Royal College of Nursing, 2003; Hand, et al. 2013).

When a group or an individual are stigmatized and suffer from discriminative acts, they are hurt at social level. In addition to improving public health outcomes of obese individuals, health professionals should stop discrimination based on weight and play a role in building a culture change towards tolerance. That health professionals blame obese individuals for their excessive weight may result in a higher chance for these individuals to show a tendency towards obesity. Similarly; unjust treatment to obese patients should not be rationalized. All the interdisciplinary health education programs should be encouraged to include weight based stigma in their curriculum in order to increase awareness of weight based discrimination and to promote equitable health care for obese and overweight patients. In general, such structural discriminative components as standard-sized patient aprons, equipments or weighing machines too small to be functional for obese patients should be minimized. Finally; interventional studies that aim at reducing weight stigma and discrimination should be disseminated (Hand et al. 2012). In a study done in Australia with health students, a short training program held in order to minimize weight discrimination among health students yielded evidences that these training programs were successful in reducing weight bias (Phillippa 2011;Fruh, et al.2016).

In order to improve clinicians' discriminative attitudes against obese people; useful strategies are listed below:

1. Empathy for obese patients should be enhanced in order to develop a perspective.
2. Perceived norms about negative attitudes and stereotypes against obese patients should be changed.
3. Use of emotion regulation techniques that will create positive effects should be promoted.
4. Such emotion regulation tools as meditation or deep breathing can be used to reduce prejudice.
5. Briefings stating that genetic, environmental, biological and psychological and social factors play a role in weight gain and loss should be organized.
6. It should be remembered that such positive, simplified messages for obesity patients as "Eat less", "Move more" are less threatening. Instead, make concrete recommendations (for example; instead of only saying "You should lose weight." say "Start an exercise program" "Eat at home" etc.). A study done with women demonstrated that individuals who internalized negative weight stigma were inclined to eat more extremely and were less likely to go on diet. Health service providers should know that most of those fighting against overweight and obesity try many measures to lose weight.
7. By reducing focus on body weight and instead of focus on body weight, it should be stated that obesity is a risk factor. By focusing on the diseases and screening the conditions; behaviors and acts that will enhance health and wellbeing should be taught or instead of losing weight, other benefits of physical activity and healthy eating should be underlined; which may pave the way for a behavior change by reducing threat.
8. Patient-centered communication strategies should be implemented.
9. By giving evidences as to the fact that diversity of body sizes is valued, create a feeling of an identity security. Positive and non-stereotypic images of overweight and obese people should be inserted in clinical advertisements and brochures (Phalan, et al.2015; Fruh, et al.2016).
10. Provide a health service by paying attention to the fact that patients have underwent negative experiences with other health professionals about their weight and by dealing them sensitively.
11. Discover not only weight but also all the other reasons of the problem.
12. Instead of emphasizing only weight, put an emphasis on behavioral change, too.
13. Accept difficulties of life style changes.

14. Do not forget that small amounts of weight losses may result in significant health achievements.
15. Create a supportive health setting with waiting rooms where there are big chairs without chair-arms, medical equipments in appropriate sizes, comfortable patient aprons and pleasant reading-materials for patients.
16. As health professionals, be aware of your own biasness (Sharon 2016; Fruh, et al.2016).

Health professionals can be recommended to ask themselves some questions that will help them discover the existing/non existing obesity prejudices. These questions may be as follows:

By judging only from weight basis, do I make assumptions as to a person's character, intelligence, professional success, health status or life style?

Can I work comfortably with people of any shape and weight?

Can I give suitable feedbacks in order to encourage healthy behavior change?

Am I sensitive to needs and worries of obese individuals?

Do I treat individuals or only the condition? (Sharon 2016). In line with the answers given to these questions, health professionals should make self-criticism, discover their attitudes in this respect and take necessary measures.

FUTURE RESEARCH DIRECTIONS

Qualitative studies may be recommended to help healthcare professionals understand existing / non-existing obesity bias and raise awareness of obesity bias.

It may be recommended to carry out quantitative studies with wider participation to help healthcare professionals understand existing / non-existing obesity bias and raise awareness of obesity bias.

Studies should be carried out to evaluate the quality of health care provided to obese patients in all health services.

CONCLUSION

Obesity endemic has been increasing over the world and our country day by day. Overweight and obesity individuals become unwilling to seek for medical assistance for their any kind of health problem due to perceived discrimination of health professionals. Health professionals prejudiced against obese people may hinder our country's effective battle against obesity epidemic. Weight bias should be discussed at health facilities and it should be focused on how obesity-affected patients can lose weight significantly and efficiently. Health service providers should have knowledge on how medical approaches that do not include obesity prejudices and stigma can be given. Besides, it should be kept in mind that weight bias and lack of communication between obesity patients and health professionals result in unfruitful health outcomes, vicious circle of prejudice, impairment of health and rising medical visits. This vicious circle should especially be discussed since it brings about violations of ethical problems.

REFERENCES

ACOG Committee Opinion Summary. (2019). Ethical Considerations for the Care of Patients With Obesity. *American College of Obstetricians and Gynecologists, 133*(1), 231-232.

Akarsu, B. (1982). *Ahlak Öğretileri*. Remzi Kitapevi.

Akman, G. (2019). *Fazla kilolu ve obez bireylerde beden algısı, benlik saygısı ve algılanan sosyal desteğin yaşam doyumuyla ilişkisinin incelenmesi* (Unpublished Master's thesis). University of Maltepe, Istanbul, Turkey.

Aldrich, T., & Hackley, B. (2010). The impact of obesity on gynecologic cancer screening: An integrative literature review. *Journal of Midwifery & Women's Health, 55*(4), 344–356.

Alkan, A., Erdem, R., & Çelik, R. (2016). Discrimative attitudes and behaviour in field of healthcare: A conceptual review. *Hacettepe Sağlık İdaresi Dergisi, 19*(3), 365–390.

Altun, S., & Ercan, A. (2015). *Üniversite öğrencilerinin obeziteye ilişkin önyargılarının belirlenmesi (Yayınlanmamış Yüksek Lisans Tezi)*. Başkent Üniversitesi Sağlık Bilimleri Fakültesi Beslenme ve Diyetetik Anabilim Dalı.

Altun, S., & Ercan, A. (2016). Obezite Önyargısı. *Turkiye Klinikleri J NutrDiet-Special Topics, 2*(1), 17–22.

Altun Uğraş, G., Yüksel, S., Erer, M. T. I., Kettaş, E., & Randa, S. (2017). Are Nurses Willing to Provide Care to Obese Surgical Patients? *Bariatric Surgical Practice and Patient Care, 12*(3), 116–122.

Amy, N. K., Aalborg, A., Lyons, P., & Keranen, L. (2006). Barriers to routine gynecological cancer screening for White and African-American obese women. *International Journal of Obesity, 30*(1), 147.

Armağan, A. (2013). *Obez kadınlarda benlik saygısının beden imgesi üzerindeki etkisinde biçimleyici değişkenlerin incelenmesi* (Unpublished doctoral dissertation). University of Okan, İstanbul, Turkey.

Aydın, E. (2006). Tıp Etiği. Güneş kitapevi, 10-11.

Aydın, E., & Ersoy, N. (1994). The Principle of Justice In Medical Ethics. *Turkiye Klinikleri J Med Ethics, 2*(2), 61–63.

Barr, J., & Cunneen, J. (2001). Understanding the bariatric client and providing a safe hospital environment. *Clinical Nurse Specialist CNS, 15*(5), 219–223.

Berdah, C. (2010). Obesity and psychiatric disorders. *Annales Médico-Psychologiques, 168*(3), 184–190.

Besler, T., Bilici, S., Buzgan, T., Çakır, B., Çakır, B., Çom, S., Güler, S., Hasbay, A., & Irmak, H. (2010). Türkiye obezite (şişmanlık) ile mücadele ve kontrol programı (2010-2014). In TC Sağlık Bakanlığı Temel Sağlık Hizmetleri Genel Müdürlüğü. Ankara: Kuban Matbaacılık. Sağlık Bakanlığı Yayın.

Cihan, G. (2019). Obez bireylerde damgalanma hissi ile yeme davranışları ve depresyon arasındaki ilişkinin incelenmesi (Yayınlanmamış Yüksek Lisans Tezi). Başkent Üniversitesi Sağlık Bilimleri Enstitüsü, Fakültesi Beslenme ve Diyetetik Anabilim Dalı, Ankara, Turkey.

Çobanoğlu, N. (2007). *Tıp Etiği. In Özerklik*. İlke Yayınevi.

Diedrichs, P. C., & Barlow, F. K. (2011). How to lose weight bias fast! Evaluating a brief anti-weight bias intervention. *British Journal of Health Psychology*, *16*, 846–861.

Durso, L. E., & Latner, J. D. (2008). Understanding self-directedstigma: Development of theweightbi-asinternalizationscale. *Obesity (Silver Spring, Md.)*, *16*(Suppl. 2), 580–586.

Ecevit Alpar, Ş., Bahçecik, N., & Karabacak, Ü. (2013). *Çağdaş Hemşirelikte Etik*. İstanbul Tıp Kitapevi.

Ersoy, N., & Aydın, E. (1994). Autonomy And The Prıncıple of Respect For Autonomy In Medıcal Ethıcs. *Turkiye Klinikleri J Med Ethics*, *2*(2), 71–74.

Forhan, M., & Salas, X. R. (2013). Inequities in healthcare: A review of bias and discrimination in obesity treatment. *Canadian Journal of Diabetes*, *37*(3), 205–209.

Fruh, S. M., Nadglowski, J., Hall, H. R., Davis, S. L., Crook, E. D., & Zlomke, K. (2016). Obesity stigma and bias. *The Journal for Nurse Practitioners*, *12*(7), 425–432.

Garcia, J. T., Amankwah, E. K., & Hernandez, R. G. (2016). Assessment of weight bias among pediatric nurses and clinical support staff toward obese patients and their caregivers. *Journal of Pediatric Nursing*, *31*(4), e244–e251.

Geller, G., & Watkins, P. A. (2018). Addressing medical students' negative bias toward patients with obesity through ethics education. *AMA Journal of Ethics*, *20*(10), 948–959.

Greener, J., Douglas, F., & van Teijlingen, E. (2010). More of the same? Conflicting perspectives of obesity causation and intervention amongst overweight people, health professionals and policy makers. *Social Science & Medicine*, *70*(7), 1042–1049.

Hand, W. B., Robinson, J. C., & Creel, E. (2013). Ethical Issues Surrounding Weight Bias and Stigma in Healthcare. *Online Journal of Health Ethics / the University of Mississippi Medical Center*, *8*(2).

Hasta Hakları Yönetmeliği. (1998). *Resmi Gazete*. Retrieved from https://www.mevzuat.gov.tr/Metin.Aspx?MevzuatKod=7.5.4847&sourceXmlSearch=&MevzuatIliski=0

Hofmann, B. (2011). *Ethical aspects of bariatric treatment of adult obesity*. Institute of Health Economics.

Jay, M., Kalet, A., Ark, T., Mc Macken, M., Messito, M. J., Richter, R., Schlair, S., Sherman, S., Zabar, S., & Gillespie, C. (2009). Physicians' Attitudes About Obesity And Their Associations With Competency And Specialty: A Cross-Sectional Study. *BMC Health Serv Res.*, *9*(106), 1-11.

Kalan, I., & Yeşil, Y. (2010). Obezite ile ilişkili kronik hastalıklar. *Diyabet ve obezite, 78*.

Kasar, K. S., & Akyol, A. (2019). Hemşirelik Öğrencilerinin Sağlıklı Yaşam Biçimi Davranışlarının Obezite Önyargı Düzeyine Etkisi. *Hemşirelikte Eğitim Ve Araştırma Dergisi*, *16*(2), 79–86.

Obesity and the Economics of Prevention Fit not Fat. (2010). OECD Publishing.

Oğuz, N. Y. (1997). *Tıp Etiğinde Temel İlkeler*. Etik Bunun Neresinde, Funda Matbaacılık.

Ozcomert, G. H., Ozcelikay, G., Sar, S., & Asil, E. (2000). The Assessment of Benefıcence And Non-Malefıcıence prıncıples From The Perspectıve of Pharmaceutıcal Ethıcs. *Turkiye Klinikleri J Med Ethics*, *8*(2), 101–104.

Phelan, S. M., Burgess, D. J., Yeazel, M. W., Hellerstedt, W. L., Griffin, J. M., & van Ryn, M. (2015). Impact of weight bias and stigma on quality of care and outcomes for patients with obesity. *Obesity Reviews*, *16*(4), 319–326.

Poon, M., & Tarrant, M. (2009). Obesity: attitudes of undergraduate student nurse and registerednurses. *Journal of Clinical Nursing*, *18*, 2355-2365.

Puhl, R., Wharton, C., & Heuer, C. (2009). Weight bias among dietetics students: Implications for treatment practices. *Journal of the American Dietetic Association*, *109*, 438–444.

Puhl, R. M., & Brownell, K. D. (2006). Confronting and coping with weight stigma: An investigation of overweight and obese adults. *Obesity (Silver Spring, Md.)*, *14*(10), 1802–1815.

Puhl, R. M., & Heuer, C. A. (2009). The stigma of obesity: A review and update. *Obesity (Silver Spring, Md.)*, *17*(5), 941–964.

Puhl, R. M., & King, K. M. (2013). Weight discrimination and bullying. *Best Practice & Research. Clinical Endocrinology & Metabolism*, *27*, 117–127.

Royal College of Nursing. (2003, Apr. 22). *Definingnursing*. Retrieved Jan. 23, 2012, Retrieved from https://www.rcn.org.uk/__data/assets/pdf_file/0008/78569/001998.pdf

Sabin, J. A., Marini, M., & Nosek, B. A. (2012). Implicit and explicit anti-fat bias among a large sample of medical doctors by BMI, race/ethnicity and gender. *PLoS One*, *7*(11).

Sikorski, C., Luppa, M., Glaesmer, H., Brähler, E., König, H. H., & Riedel-Heller, S. G. (2013). Attitudes of health care professionals towards female obese patients. *Obesity Facts*, *6*(6), 512–522.

Tailor, A., & Ogden, J. (2009). Avoiding the term 'obesity': An experimental study of the impact of doctors' language on patients' beliefs. *Patient Education and Counseling*, *76*, 260–264.

Tanneberger, A., & Ciupitu-Plath, C. (2018). Nurses' weight bias in caring for obese patients: Do weight controllability beliefs ınfluence the provision of care to obese patients? *Clinical Nursing Research*, *27*(4), 414–432.

Türk Dil Kurumu (TDK). (n.d.). Retrieved from http://tdk.gov.tr/?option=com_gts

Türkiye İstatistik Kurumu. (2017). Retrieved from https://www.tuik.gov.tr/basinOdasi/haberler/2017_31_20170607.pdf

Tylka, T. L., Annunziato, R. A., Burgard, D., Daníelsdóttir, S., Shuman, E., Davis, C., & Calogero, R. M. (2014). The weight-inclusive versus weight-normative approach to health: Evaluating the evidence for prioritizing well-being over weight loss. *Journal of Obesity*, 18.

Usta, E., & Akyolcu, N. (2014). Cerrahi hemşirelerinin fazla kilolu/obez hasta bakımına ilişkin bilgi ve uygulamalarının değerlendirilmesi. *Florence Nightingale Hemşirelik Dergisi*, *22*, 1–7.

Venkat, A., & Larkin, G. L. (2014). Ethical Dilemmas Posed in the Care of Obese Patients in the Emergency Department. *Online Journal of Health Ethics / the University of Mississippi Medical Center*, *10*(1), 5.

Wott, C. B., & Carels, R. A. (2010). Overt weight stigma, psychological distress and weight loss treatment outcomes. *Journal of Health Psychology*, *15*(4), 608–614.

Yıldırım, G., & Kadıoğlu, S. (2007). Etik ve tıp etiği temel kavramları. *CÜ Tıp Fakültesi Dergisi*, *29*(2), 7–12.

ADDITIONAL READING

Fruh, S. M., Nadglowski, J., Hall, H. R., Davis, S. L., Crook, E. D., & Zlomke, K. (2016). Obesity stigma and bias. *The Journal for Nurse Practitioners*, *12*(7), 425–432. doi:10.1016/j.nurpra.2016.05.013 PMID:28408862

Garcia, J. T., Amankwah, E. K., & Hernandez, R. G. (2016). Assessment of weight bias among pediatric nurses and clinical support staff toward obese patients and their caregivers. *Journal of Pediatric Nursing*, *31*(4), e244–e251. doi:10.1016/j.pedn.2016.02.004 PMID:26948091

Hofmann, B. (2016). Obesity as a socially defined disease: Philosophical considerations and implications for policy and care. *Health Care Analysis*, *24*(1), 86–100. doi:10.100710728-015-0291-1 PMID:25822670

Phelan, S. M., Burgess, D. J., Yeazel, M. W., Hellerstedt, W. L., Griffin, J. M., & van Ryn, M. (2015). Impact of weight bias and stigma on quality of care and outcomes for patients with obesity. *Obesity Reviews*, *16*(4), 319–326. doi:10.1111/obr.12266 PMID:25752756

Robstad, N., Westergren, T., Siebler, F., Söderhamn, U., & Fegran, L. (2019). Intensive care nurses' implicit and explicit attitudes and their behavioural intentions towards obese intensive care patients. *Journal of Advanced Nursing*, *75*(12), 3631–3642. doi:10.1111/jan.14205 PMID:31566770

Tanneberger, A., & Ciupitu-Plath, C. (2018). Nurses' weight bias in caring for obese patients: Do weight controllability beliefs ınfluence the provision of care to obese patients? *Clinical Nursing Research*, *27*(4), 414–432. doi:10.1177/1054773816687443 PMID:28052689

KEY TERMS AND DEFINITIONS

Bioethics: Scientific field where ethics and biology science are combined. Bioethics determines how ethical decisions should be made.

Discrimination: Having positive/negative actions or attitudes against a group or an individual.

Empathy: Ability to understand feelings and thoughts of another by identifying yourself with them.

Health Professionals: Occupational groups consisting of doctors, nurses, midwives, pharmacists, dieticians, and others who strive to improve health, protect, prevent disease, and heal in the current disease situation.

Medical Ethics: A science philosophy that determines health professionals' actions as good or bad or right or wrong.

Obesity: One's having too much weight as compared to current height (30≤).

Prejudice: Marginalize/otherize a group or an individual.

Stigma: Action to humiliate an individual or a group due to their characteristics or differences.

Section 2

Ethics in the Media and Marketing World: Post-Truth Era

Chapter 7
Media Ethics:
Evaluation of Television News in the Context of the Media and Ethics Relationship

Özlem Arda
https://orcid.org/0000-0003-2565-1830
Istanbul University, Turkey

Zuhal Akmeşe
Dicle University, Turkey

ABSTRACT

This chapter provides an overview about media ethics that is very important for the news. Today, the rapid development and diversification of mass media tools have also accelerated the works in the field of communication ethics. Media responsibility, issues occurring in media, and public utility issues have come to the central position of communication ethics. Looking at the ethical codes in the media, it is seen that a large part of them are created for printed media, and the information about television is limited. The purpose of this study is to focus on the ethical issues that arise starting from the production stage of the television news with a holistic perspective in the context of the relationship between media and ethics by considering the ethical codes in the media and to offer solutions devoted to an ethical understanding of journalism. Within the scope of this study, the qualitative research method included the content analysis for the news about Princes Diana and Prince Harry as samples.

INTRODUCTION

Ethics is a system that reminds the important rules of moral system and should be valued every area. When looked at the dictionary meaning of the word "ethics" it is seen the corresponding expression "*Science of value judgments with regard to morality, virtue and making distinction between good and evil. All the morality principles which form the basis of one's behaviour. Moral, moral-related, morality,*

DOI: 10.4018/978-1-7998-4117-3.ch007

customs science, moral science" (turkcesozlukler.com). The concept of ethics is defined as the branch of philosophy that searches for the quality and origin of the concepts categorized as good-bad or right-wrong, which are among the moral values existing among individuals. Looking at the origin of the word ethic, it is assumed that it derives from the word *ethos* in Greek and also comes from the word "ethicus", which is a Latin word (Tutar & Yılmaz 2003, p. 53).

It is seen that this word is used in different ways. Its first use in Greek means habit, custom or moral law. Ethical values do not emerge by an authority mechanism, these values emerge as a result of experiences created by the human relationships and human mind during the course of time. In this respect, when we look at the communication ethics due to the relationship between the media and ethics, it is possible to express that the professional ethics, i.e. a whole set of rules in a certain profession area, can only be realized when the people performing this profession adopt these values both mentally and behaviourally and it cannot be valid as a result of external pressure and imposition (Girgin 2000, p. 144). The focus of our study is the relationship between media and ethics as well as the communication ethics in the context of professional ethics and evaluation of television news in ethical context. The fact that information flow has become very important nowadays and the information flow is transferred very fast regardless of the time and space barriers, the mass media tools are diversified in parallel with the developing technology and the information is spread instantly on digital platforms, and the potential to reach millions of people at the same time has increased the importance of communication ethics and made it one of the issues that should be seriously discussed.

The media, which has such an effective and decisive potential in social life shall take into account certain ethical rules and inform the masses in an accurate manner while producing a content that is considered as a source of information, especially news, and does not leave its potential under the auspices of certain interest groups only when the media professionals performing actively in this area adopt and comply with the certain ethical rules (Eroğlu & Atabek 2006, p. 1). The evaluation of television news in the context of ethics, the relationship between media and ethics and awareness of the production processes of television and television news, which is one of the most widely used mass media tools, are important for understanding the study.

The main focus of this study is to draw a frame of the ethical issues in media as television news. It is the target of this chapter to underline the relationship between media and ethics by considering the ethical codes in the media and besides to offer solutions devoted to an ethical understanding of journalism. There are various subheadings as "The Concept of Ethics and Using Ethics – Morality Concepts", "The Relationship between Media and Ethics and Ethical Violations in Media", "Television and Its Functions", "Media and News Production", "Television News and News Production Process", "Ethical Violation During New Production Process", "News Examples Involving Ethical Violation", "News about Private Life of Princess Diana" and "News about Private Life of Prince Harry" for a holistic view.

The Concept of Ethics and Using Ethics – Morality Concepts

The concept of "ethics" is considered to be the guiding and commanding rules of human life in all communities, while the specific ethics rules depend on the religious beliefs, social norms, historical conditions, and economic philosophies of the culture it is in all times. For this reason, it is important to consider the system in which the standards are applied when referring to ethical principles (Matelski 1995, p. 11). Ethics is a discipline that deals with the rules and values that form the basis of the individual and social relations of a human being, which is essentially a social entity, within the discipline of philosophy and examines these rules and values on the basis of good-evil, truthfulness and falsehood (Kolçak 2012, p. 3). Values can be defined as standards that state "in what situation" and "what kind of behaviour is appropriate" in the community. Since beliefs remain abstract in establishing a social order alone, human behaviour needs some criteria that determine the right-wrong, good-bad, just-unfair, normal-pervert according to the belief system. Values provide us with these criteria and standards. In this sense, the values can be defined as the standards that we refer to when we judge our behaviours and choose our purpose in life and also that show the right and wrong while determining socially shared goals and behaviours (Bozkurt 2012, p. 93).

It is observed that the word "morality", which is thought to bear the same meaning is used instead of the word "ethics" extensively in everyday life. Although they are used interchangeably and thought to have the same meaning, there are differences between these two concepts. While morality covers the rules of behaviour that people must follow in society, ethics should be considered as the science of morality that investigates human behaviour within the frame of morality. In the Turkish dictionary, ethics finds its equivalent meaning as "*custom science*", "*the behaviours that a professional group must follow*", "*ethics science*", "*moral, moral-related*" (Uzun 2007, p. 11). In this sense, it is possible to state that the concept of ethics is concerned with the underlying causes of behaviour rather than the behaviour itself. Uzun states that ethics has two methods: normative and descriptive:

Ethics has two methods: normative and descriptive. In the descriptive method, the forms of action and behaviour in a community are investigated in terms of effective values and validity demands that apply to the community, and this method includes judgments for the whole moral law that the majority considers as binding. The normative method, on the other hand, is a definitive method rather than describing the existing one and tries to develop criteria that allow moral evaluation of actions before going for a determination. The most appropriate method in terms of communication problems is the normative method. Two of the most common normative ethical models are generally classified as teleological and deontological. Understanding the theory behind these approaches helps both revive the moral imagination and develop the analytical skills needed for ethical decision making" (Uzun 2007, p. 14)

When the media is taken into consideration in general, media ethics is considered as a sub-category of professional ethics and in this respect the scope of the concept of ethics is important in terms of determining the scope of media ethics and establishing a specific framework. Ethics briefly finds its counterpart with "moral philosophy" or "defining morally acceptable individual, institutional and social values and using these values as the basic criteria for evaluating human behaviour". Although the issue of ethics in the media is among the constant issues on the agenda, whether or not the ethical rules are ignored in all media productions and to what extent the media complies with these principles have always been a matter of discussion.

Communication ethics, on the other hand, describes the professional ethics of those working in communication-related professions, namely in the media sector. In this context it is described as a field defining the professional ethics of people and institutions who produce and engage in news in news agencies, newspapers, radio and television organizations, or on internet platforms, which are expressed as new media, as well as those who create content in these media other than the news and those whose productions take part in the media even if they are not members of the media (Uzun 2009, p. 17). After examining the meanings and scopes of moral and ethical concepts in general, it would be more appropriate to make a general situation assessment by evaluating the concept of ethics on television and communication ethics in this respect, professional ethics and the criteria determined by the national and international organizations with regard to the mentioned profession.

The Relationship Between Media and Ethics and Ethical Violations in Media

Mass communication is expressed *as the transmission and reception of information and entertainment through mass media tools and media technology. The communication process is shaped by very few resources, many buyers and a limited amount of feedback* (Lull 2001, p. 241). Today, the concept of media is used to meet all the mass media tools. The word 'media' which is used for defining the mass media tools in the field of communication is the plural of the Latin word 'medius' and the English word 'medium' which means 'tool, environment, context' in English (Nalçaoğlu 2005, p. 51). The word finds its equivalent in the Turkish Language Association Dictionary as "communication medium, communication tools" (Media, www.tdk.gov.tr 2020). Today, the mass media tools which have become a force that seriously affect, change and transform the social life and the content offered to the masses through these tools intentionally reveal the ethical violations for a number of reasons.

The media is *in a direct and indirect relationship with the power and capital groups due to having a wide coverage. These relationships often play an effective role in shaping content productions.* The network between media organizations, shareholders and the states bring together the use of mass media as ideological devices of the state, and therefore the transfer of the ideology and perspective of the state, i.e. the ruling power, to the masses with these media contents produced. This situation becomes decisive in the process of creating all media content. As media institutions generally have a government-dependent structure in terms of political support in general; *the adoption of taxes, interest rates, labour policies and anti-trust laws also concern the media organizations as well as all the other commercial organizations* (Chomsky & Herman & Peterson 2004, pp. 62-63). Since mass media tools are an important part of property relations in this sense, their use is subject to certain legal regulations, and the property relations of these organizations shape the nature of the product, including the way the media contents are produced (Yaylagül 2013, p. 17). Due to the reasons arising from all these networks, there is a relationship based on interdependence between the media organizations and the governments. Considering the parameters such as media-capital-state-ideology, the reproduction of state ideology through mass media and the functioning of mass media as ideological devices of the state, it is possible to state that mass media tools are moving away from their main purposes, they are shaped in line with these mentioned parameters and especially the produced content reveals ethical violations in this sense.

When making an evaluation on the media productions in the context of media ethics relationship, it is possible to list the ethical violations that come to the fore in the media as the manipulation of content, circulation of unrealistic news, presence of disinformation elements, structuring of contents in an unrealistic, biased and exaggerated manner, marginalizing certain groups, labelling, showing contents containing discrimination elements, and showing contents promoting and normalizing violence, the use of profession for the purpose of interest, biased attitude and violation of the principle of impartiality, tabloidization, exceeding the limits of criticism, insults, etc. in the contents devoted to certain communities, violation of privacy and confidentiality (Işık 2018, p. 158).

When the reasons for the occurrence of these violations are examined, it is seen that media organizations act with some concerns (economic, political, ideological) and capital and property relations come to the fore. In order to eliminate and minimize these violations, platforms are created at local, national and international levels and universal professional ethical principles are created by professional associations performing in the field of media on a global basis and non-governmental organizations and similar institutions that pay regard to public benefit. However, despite all the reactions, the media organizations do not necessarily pay attention to this ethical violation but instead they ignore it for the sake of their interests.

Television and Its Functions

With the developments in communication technologies and with the introduction of television into our lives, we now have the opportunity to watch the world from our homes. Television is a Latin origin word meaning "seeing away", which consists of the combination of the words "tele" and "vision" in the simplest sense (Mutlu 1991, p. 15). The word meaning of television actually refers to both the function and effects of this tool. With the inclusion of this tool into our lives, we watch all the events of the world we live in from the place where we sit. In a sense, all the information is coming to our feet without having to go anywhere. It is enough to sit at home and watch television to learn about our environment and the world in general (Aydoğan 2004, p. 41). In this sense, it is possible to tell that television has made the "vision" of the world possible with these features (Lazar 2001, p. 87). This is important for understanding the limits of the power and impact of the mass media tools. The passive masses watch the world through these tools, are informed about the events happening in the world and their environment through this tool, meet their information needs through this tool, are becoming aware and use the information contained in the messages conveyed by these tools as a resource.

Considering the functions of mass media tools, it is accepted that one of the most important functions of these tools are their functions of "informing and making aware". Considering that *the main mission and aim of mass media tools are to reach the widest possible audience, to ensure the highest consumption of broadcasts and programs that reflect the products and ideology that it markets with its messages* (Rigel & Batuş et al., 2005, p. 262), it is possible to express that television is the mass media tool that has the greatest opportunity to realize this goal. Because television both constitutes the most important news source of the public and the role of television channels as an independent news source can easily be exaggerated (Chomsky & Herman et al. 2004, p. 46) and these advantages increase the credibility of the television. The feature of television that makes the viewer addicted to it and transmits the message through the audio-visual channels to appeal to the audience's multiple senses brings it to a superior and privileged position than all other mass media tools.

Media and News Production

News constitutes one of the most important productions of mass media tools. News are the structures in which the media reconstructs the reality. Bernard Roscho mentions that there are two main factors in the production process of news and the formation of news discourse. These are the institutions studies and the preferences of the editors (Kars 2010, p. 131). The process of preparing the content presented in the news through mass media tools basically depends on a large capital investment and division of labour. As a result of this structure, certain content and images are transferred to the society / audience through technological tools such as television, radio, printed media and cinema. Since mass media institutions are organized as companies or public institutions, they undertake the function of transmitting certain messages to the public in television programs or in different forms in the manner that they are organized and as guided by the forces that control these structures. The purpose of these institutions may be to generate income according to their organizational styles, direct the masses to certain patterns of thought and behaviour, or cover many of these goals. When we consider television and its functions specifically with regard to Mass Media tools, it is possible to state that television effectively exists in all areas of life (culture, art, daily life practices, lifestyles, etc.) and the functions of television are shaped according to the changing needs and conditions of the societies.

In general, the functions of mass media tools coincide with those of television also as a mass media tool. However, the fact that television technology and narration features are different highlights some of these functions. It is possible to express these functions as informing, contributing to education, entertaining, informing and socializing which all have aspects with regard to community. The ability to instantly convey an event occurring anywhere in the world to the audience in an audio-visual way highlights the importance and instrumental superiority of television. Viewers have the opportunity to watch the news about a place on the other side of the world they can never go by only sitting in the hall of their houses. They reach the information about the world in social, cultural, economic, political, artistic and other fields through the television indeed. There is a great awareness about all areas (Akmeşe 2017, p. 67). The television performs its reporting function with the news bulletins it offers to the audience. However, the news goes *through many processes starting from the production process until the moment they are conveyed to the audience.* These ethical violations mainly occur during this process. Selection of an event as news, editing the news and including it in the news bulletin is completely shaped within the scope of the policies of these media organizations, the government-capital relations as well as state policies. This situation raises the problem of turning the news into fiction and leads to ethical violations.

Television News and News Production Process

Image and sound which are two basic elements of television also form the most essential features that distinguish television from other mass media tools. When the studies and regulations *in the field of communication ethics are examined, it is seen that the ethical problem and solution suggestions related to these problems generally focus on the printed media.* However, in reality, the issue of ethics in television journalism cannot be evaluated separately from ethical issues in printed media. As a matter of fact, the technical developments experienced in the course of time and the ability to use the image and sound elements of the television together have made it more privileged than other mass media tools, and even after a certain period, the reporting function, which is one of the main functions of the mass media tools started to be realized through television and the rules of journalism started to be shaped by the television itself. In this respect, Ignacio Ramonet states that a new turnover was entered in the history of journalism and that television has taken power in the field of news, among the other mass media tools especially after the 1991 Gulf War. After this process, he emphasizes that newspapers, in other words, the printed media in general, are now pursuing the rules set by the television, and television, which has become the leader in news media, has replaced the newspaper (Ramonet 2000, p. 29). When we return to the subject of ethical violations, the utilization of moving images and audio-visual elements together by television which has acquired such a decisive role in the field of news, unlike any other mass media tool (radio, newspaper, magazine, etc.), brings up the ethical problems arising from the images in the field of television journalism. In this context, our focusing on the production processes and features of television news will be useful for detecting the ethical violations that emerge.

The selection of the events that will be the subject of television news, the writing of the news belonging to the selected event, the language used in the news and the presentation of the news differ from the news presented in other media. Since the television broadcast stream is limited to 24 hours, there is a certain time period for all the programs broadcast on television. For this reason, the fact that there is a certain time limit for the published news and limited time periods are allocated for the news make it necessary to make a choice among the news. While organizing a news bulletin, the most important event of the day or the news of the event desired to be brought to the fore is given first place in the bulletin and all the news in the bulletin must be listed in a certain logic series as a whole. To summarize, television news; uses spoken language, there is no need to be literate to receive the news, the news text is conveyed to the audience with the narrator using the spoken language as if looking at the viewer through the lens and accompanied by the images. Unlike the newspaper or printed media, there is no chance to read again or look back on the news given through television. It is given instantly. The duration of television news is at most 90 seconds due to the specified limitations unless there is a very extraordinary situation. This situation brings the use of images that are constantly flowing. There is a transition between different images. This fragmented information moves people away from information. Instant reception prevails narrative dynamics (Kars 2010, p. 167). This situation somehow necessitates that news are presented in an entertaining framework by disregarding the education, thinking and informing functions. Television "displays the news to be viewed." In addition, the fact that all television programs are presented in an entertaining manner and that it even appears in the news that seems to be the most real is an indication that the top ideology of any discourse on television is entertainment (Postman 2010, pp. 101-102). Television news goes through a number of filters during the production process. First of all, selecting the event that is considered worthy among thousands of topics and including it in the news bulletin constitute the first stage of production. The selected event goes through many stages until it meets the audience and undergoes some interventions at each stage. These interventions include processes such as manipulating the news, giving incomplete information on the subject and reconstructing the news. The basis of these interventions is that broadcasters consider many parameters, especially government policies.

We witness that television, which especially uses motion picture and sound together in the news production process, concentrates on some unethical images while producing news. Sometimes the detailed images of accidents, tragic events, images containing violence and pornography and images that breach a person's private life can be used in order to highlight the news. Making such news with the concern of rating causes serious ethical violations. Also, in terms of image use; especially the use of images that do not belong to the news and uncontrolled use of the images such as animation archive images, hidden camera records, etc. reveal another dimension of this problem.

Briefly, any misinformation about the event, person or situation on the news as well as any discrimination, marginalizing, presenting the content in a way that harms the dignity of the person, presenting a content that violates the limits of privacy and private life, and using a content that alienates the individuals due to their languages, religions, ethnicities and the groups they belong to are all considered as ethical violation.

Ethical Violation During News Production Process

The apparent purpose of the news circulating in all media environments is to inform the masses and make them aware of the subject. The function of informing the public, which is accepted as one of the main objectives of mass media tools, is moved away from this goal as a result of some interventions during the news production process, and the masses are actually tried to be directed by the news given under the name of informing. This brings both the fictionalizing of the news and ethical violations. Similar filters and network of relations stand out in the core of these interventions in the news production process. Primarily the government policies as well as the media-state, media-shareholder relations, economic, political status etc. causes the restructuring of the news and being formatted in line with these parameters. In the media content, manipulation is put in place in order to direct the masses through news that are believed to be "the most real, most neutral and correct", to present the events with the dimensions desired to be put forward and to make the audience being exposed to these messages in order to make them think in the same way.

Manipulation is a French origin concept. In the Turkish dictionary; it finds its equivalent as orientation, influencing, choosing content, changing information through removing or including, directing people's thoughts (https://sozluk.gov.tr/). When the relevant definitions are evaluated, it is seen that manipulation involves the functions of directing and intervening in certain mental processes and consciousness.

Directing the masses lays under the presentation of media contents through manipulating. This method, which is used to direct the masses to certain attitudes and behaviours, may emerge in different ways in many areas. The manipulation which is carried out by interfering with the nature of the reality finds equivalent with the fictionalizing the discourse in written and visual media content, highlighting some points of the news while ignoring others intentionally, and including the dimension desired to be seen and shown within the frame both fictionally and visually. Today's technological developments have made it much easier to perform manipulative operations on both content, images and documents. The sources, evidences and images used in the news, briefly all the data that constitute the main material can be reflected in many different ways by digital interventions. This situation, in fact, makes it necessary to approach critically and suspiciously even to the concepts of "real", "right", "neutral" and "objective".

Indeed, Baudrillard's (2016 p. 15) statement about manipulation becomes supportive at this point; "The hyper-real era, where we stepped in from the threshold today, shows how such a favourable environment is provided for manipulation-based perception management: All kinds of imaginary and real distinctions are now over, and the truth will never come back. It is now a hyper-real era that repeats itself through models rotating around the same orbit and does nothing but produce a simulation of difference" (cited by Çakmak 2019, p. 1135). These statements are important in explaining the present day where digital technologies are widely used. As a matter of fact, Baudrillard, who emphasizes that many events and facts that we assume as "real" consist of an illusion, expresses this dilemma in which modern man falls in an impressive way.

The concepts of **Disinformation** and **Misinformation** are other interventions that lead to the occurrence of ethical violations. These concepts include the meanings of distortion of source information, a negative situation belonging to the information, or briefly, intervention in the information that constitutes the source of the phenomenon or event. Disinformation is used as a term of intentionally interfering with the content and misinforming the masses. There is a conscious intervention in disinformation. Unreal information is included and used for manipulation purposes (Çakmak 2019, p. 1129). Although disinformation and misinformation are different concepts, these concepts are often confused and even if they are used interchangeably, they are actually two different concepts. While misinformation is the incomplete, false and partial transmission of actual information, disinformation on the other hand means that the incomplete, false or incorrect information is spread to manipulate the masses (Yegen 2018, p. 104). In this context, interfering with the content in order to serve the interests of certain power groups and sovereignties by disconnecting from the context of the events and facts that are the subject of the news reveals ethical violations. It is a known fact that the content is subjected to manipulative processes with these methods in all production of the media.

When making an assessment on the television news production process, we see that there is a situation of influencing, directing the masses and creating a perception about a subject with the interventions that started even at the very first stage of the event that is the subject of the news. These interventions take place in all dimensions in the process of restructuring the news.

Aim and Methodology

The purpose of this chapter is to focus on the ethical issues that arise starting from the production stage of the television news. In the context of the relationship between media and ethics by considering the ethical codes is very important issue for journalism. Within the scope of this study, the qualitative research method included the content analysis was applied through two news. The news was selected according to their concept as called 'private life' of the people from the United Kingdom Palace. The news about Princes Diana and Prince Harry were selected as samples. This news has very strong ability to represent ethical violation of media.

News Examples Involving Ethical Violation

News About Private Life of Princess Diana

When it is considered the parameters that play a role in the background of media-ethics relationship and ethical violations in media, it is possible to state that all media productions are shaped by manipulative interventions and that all the content and formats produced serve a specific purpose. When it is evaluated this situation in terms of television news, it is possible to reach similar results. The news bulletins, which are shown as the equivalent of the informing function of the media, also come to the fore as the media productions that undertake the mission of informing the masses. In contrary to the general opinion, the news programs, which are identified with the concepts of objectivity, impartiality and accuracy that the masses adopt as "real", and which are also considered among the most important elements of mass media, are the programs in which fiction and manipulation are applied most intensely. The restructured news bulletins are one of the most effective weapons used by the media to direct the masses, change their attitudes and behaviours at the desired level and impose the ideologies they emit to the masses.

When the television news is evaluated within the context of the concept of ethics, undoubtedly the common ethical codes of all mass media tools are also necessary and valid in television. However, the feature of television to use audio and image together makes it more privileged and more dangerous than the printed media in terms of ethical violations.

Its widespread use and potential to reach the masses brings with it the danger that a false news can lead to very traumatic consequences. While an ethical violation made in a newspaper or magazine involves only the people who have reached that issue and read the relevant news, the fact that television reaches millions of people immediately causes that the news in circulation may have a much more serious impacts and results on the society.

When looking at the news that have left a mark on the world agenda and include ethical violations, the example of "Princess Diana" comes to the fore in terms of creating an agenda all over the world. It is one of the most important ethical violations as the information about the private life of the princess is carelessly leaked to the press and the press organizations brought the news to the forefront by ignoring the ethical rules and the principle of public interest due to certain economic and political benefits such as rating. Diana's private life was brought to the forefront with different dimensions and different related scenarios were produced which is a very good example of the violation of ethical rules in terms of privacy, intruding into a person's private space and negatively affecting the life of the person concerned.

There are many examples in the world where the private lives of politicians, artists and celebrities are exposed. Bringing forward the issues in the private lives of these people rather than their professional positions and public responsibilities and therefore trying to blockade their professional lives intentionally is not a correct approach in terms of a normal moral understanding and conveying this information to the masses in the quality of news is not appropriate in terms of media ethics.

News About Private Life of Prince Harry

The example of 'Princess Diana' has resonated around the world and has been questioned about media ethics. Privacy was one of the most notable ethical violations here. After years, a member of the Royal Palace of England was again seen making a media coverage as part of the media ethics problem.

According to the *The Telegraph*'s news it can be seen ethical violation clearly: *The publication of naked pictures of Prince Harry and the Duchess of Cambridge shows that the internet is an "ethical vacuum" which the press must rise above, Lord Justice Leveson has warned* (Naked prince news ethics 2012a).

Stephen Mangan's news report from London, Reuters, on August 23, 2012 to understand how private life has been serviced in media: *"Naked Prince Harry photos published by UK's Sun newspaper", "The Sun tabloid on Friday published photographs of Prince Harry naked in Las Vegas, becoming the first British publication to defy a request from the royal family's lawyers. On Thursday, newspapers in Britain did not publish the images of Queen Elizabeth's grandson naked with an unnamed woman while on holiday in Las Vegas, following a request from St James's Palace, the official residence of the prince, through the Press Complaints Commission to respect his privacy. But, almost half of the front page of Friday's Sun newspaper shows a photograph of the naked prince covering up his genitals with his hands while an apparently naked woman hides behind his back in a Las Vegas hotel room. The Sun, part of the British arm of Rupert Murdoch's News Corp, said millions of people around the world had already seen the pictures and its readers had a right to see them. David Dinsmore, managing editor of the Sun, said in a video on the paper's website that the decision to publish was not taken lightly, but the issue had become one of 'the freedom of the press' "* (Naked prince news ethics 2012b).

SOLUTIONS AND RECOMMENDATIONS

The subject of ethics is one of the issues that should be emphasized with utmost attention in all areas. For this reason, both the state and the institutions and organizations operating in that field carry out studies on professional ethics and set some national and international rules and principles. These principles can be formed by both public institutions and non-governmental organizations which are composed of the people who are active in this field. It is a well-known fact that television uses image and sound together which differentiates it from other mass media tools. In this sense, it is necessary to pay attention to both the common principles determined for the press institutions and some sensitive points in terms of television in order not to cause ethical violations. It is important to consider ethical codes at every stage from the selection of the relevant event as news until it is delivered to the audience during the news production process. It is important to comply with the ethical rules of press and it should be taken into consideration whether the news will be beneficial in terms of public interest, whether the institutions, organizations, individuals and groups in the related news will be damaged due to this content and whether the issue is worthy of becoming the news while evaluating the subject in the quality of news. Especially in terms of image use, the false, fraud, irrelevant images, hidden camera recording etc. should not be preferred unless it is compulsory. If it contains public interest and if it is obligatory to use it, the question of whether this material will harm other people should be considered. It is important that the news is presented in a realistic way without manipulating people's feelings and thoughts.

FUTURE RESEARCH DIRECTIONS

Ethics is an important concept in all areas where mass media production is. Content production should be carried out according to ethical principles. Today, there are ethical violations in television journalism, which indicates the problem in the field of journalism. Since there is no concrete design of images such as ethics or physical being, ethical principles should be shaped as a set of concrete rules.

As technology progressed, new communication environments had their own production processes and trends. In this digital age, news and ethics are more troubling. In digital news production, new code of ethics is needed. It is essential to create digital ethics codes for digital journalism, both nationally and internationally.

Big data, as well as other concepts, has changed the framework of the concept of ethics. Television journalism should not be considered separate from big data in the age of life; but at this point, another problem arises. Media ethics, yet, do not have an inclusive structure in all the manufacturing processes of the digital world.

According to Lewis and Westland (2014); *big data is a social, cultural, and technological phenomenon -a complex amalgamation of digital data abundance, emerging analytic techniques, mythology about data-driven insights, and growing critique about the overall consequences of big-data practices for democracy and society. The developments of big data potentially have great meaning for journalism's ways of knowing (epistemology) and doing (expertise), as well as its negotiation of value (economics) and values (ethics).*

On the other hand, there are different issues as "Fake News" and "Post-Truth" notions. Fake news is servicing by the reporters all around the world and it is seen that 'fake' is taking place of reality. In this context journalism ethics is more essential element for the news. Also, this the age of post-truth and the people need truth more than ever. Further research can focus on these problematic fields.

CONCLUSION

Today, the concept of ethics forms one of the study subjects of many scientific fields. Morality refers to value judgments that we call "good" or "right" or "bad" or "wrong" in human relations. Moral philosophy is a discipline of philosophy that deals with the moral dimension of human life. In other words, moral philosophy is the branch of philosophy that examines values, principles and judgments in human life.

Within the framework of globalization policies, many different strategies aimed at increasing consumption are used today with the implementation of ideological, cultural and economic practices. One of these strategies is the creation of popular icons and the conversion of them into advertising vehicles

(Akmeşe 2017, p.134). So it can be thought that positively, international ethics rules can be applied according to the media concept. Also, today, social networks also serve as an important tool for the reach of messages to relevant interlocutors in the field of international politics and diplomacy (Yaşar 2020, p. 13) and social media ethics will be another discussion for future.

Every professional group in the community has a function in the name of social benefit. Each professional group emerges as a result of certain accumulations in order to meet a need in society and to provide a certain benefit to that society. It is a well-known fact that the media / press is accepted *as the fourth power* in the society after the legislative, *executive and judiciary powers* based on its functions of informing, notifying and educating the society. It is even possible to state that the media has shaped and transformed the society with all its dimensions due to its impact and orientation potential. In this sense, the media-ethics relationship is important. The fact that a profession group with such a strong influencing potential can move away from ethical principles and serve the interests of a class, community or group may lead to chaotic situations in the society. In this sense, media ethics comes to the fore. Media ethics principles come to the forefront as one of the most important issues in the world. Media ethics rules are generally established by organizations that are active in this profession on a global basis. These principles are also supported legally. The media and its members, which undertake the function of informing and notifying are responsible to the society in this sense. As mentioned before, it is important to support these ethical rules with an internal basis. Media members' having this awareness and adopting these rules rather than limiting the moral approach with pressure and rules as well as legal obligations and punishment is more important for preventing ethical violations. Otherwise, punishments and sanctions cannot go beyond being restrictive.

Although there are different forms around the world, ethical rules can generally be summarized as conveying the source objectively and impartially without deforming the information, not humiliating and alienating the people due to their language, belief, ethnicity, race and nationality, not violating the principle of privacy, avoiding use of news for profit that will damage and pose danger for the society, not occupying the public agenda with unreal news and not manipulating the news. Today, it is seen that ethical violations occur in many areas as the media remains under the auspices of certain capital groups and governments and has turned into an ideological device that serves the interests of these groups. As a matter of fact; the statement of Ramonet (2000): "Scepticism. Distrust. Difficulty in Believing. These are the dominant feelings that citizens have for the media" actually summarizes the modern man's view of the media very well.

However, getting news is one of society's most fundamental rights. Mass media meets the need for human knowledge. It is a fundamental requirement to think and design and serve news in all media on the basis of public benefit and public service.

There are various media ethics formations and rules available nationally and internationally. Sanctions are not required for those who perform their profession in ideal conditions and level, and it personally heralds the ideal profession. However, in the circumstances that are beyond the ideal, serious sanctions of media ethics rules are within the framework of ethical principles if a deterrent is found by the professionals.

As in the case of Princess Diana in the past, Prince Harry in the near future and also now there are many ethical violations in journalism every day. Privacy of private life, refugee images, reporting as hate speech, violence, ideology of power, personal interest relationship are the most encountered problematics in media.

REFERENCES

Akmeşe, E. (2017). Evrensel Bir Simgenin Sıradışı Öyküsü: Chevolution. In H. Kuruoğlu & A. F. Parsa (Eds.), *Belgesel Filmde Zamanın Ruhu: Belgesel Filmde Değişen Anlam ve Anlatım* (pp. 117–137). Detay Publications.

Akmeşe, Z. (2017). *Televizyonda Çerçeveleme ve Toplumsal Bakış Açısının Oluşturulması*. Yayınlanmamış Doktora Tezi, İstanbul Üniversitesi Sosyal Bilimler Enstitüsü.

Aydoğan, F. (2004). *Medya ve Popüler Kültür*. Mediacat.

Bozkurt, V. (2012). *Değişen Dünyada Sosyoloji Temel Kavramlar Kuramlar*. Ekin Publications.

Çakmak, F. (2019). Medyada Dezenformasyon Sorunsalı ve Manipülasyonun Gücü: İdeoloji ve Söylem Açısından Seçim Haberlerinin Siyasal İletişimdeki Rolü. *Selçuk İletişim Dergisi.*, *12*(2), 1127–1154.

Chomsky, N., Herman, E. S., & Peterson, D. (2004). *Medyanın Kamuoyu İmalatı, Medyanın Tekelleşmesi Kitlelerin Yönlendirilişi ve Zorunlu İtaat* (A. Köymen, Trans.). Chiviyazıları Publications.

Eroğlu, E., & Atabek, N. (2006). İletişim Fakültesi Öğrencilerinin "Etik Eğitimi"ne ve "İletişim Etiği"ne İlişkin Görüşlerinin Değerlendirilmesi. *Küresel İletişim Dergisi, 2*. https://dergipark.org.tr/tr/download/article-file/177843

Girgin, A. (2000). *Yazılı Basında Haber ve Habercilik Etik'i*. İnkılap Kitapevi.

Işık, M. (2018). *İletişim ve Etik*. Eğitim Publications.

Kars, N. (2010). *Haberin Tarihi, Kuramları, Söylemi ve Haberin Tarihi, Kuramları, Söylemi ve Radyo-Televizyon Haberciliği*. Derin Publications.

Kolçak, M. (2012). *Meslek etiği*. Turkey Murathan Publications.

Lazar, J. (2001). *İletişim Bilimi* (C. Anık, Trans.). Vadi Publications.

Lewis, S. C., & Westlund, O. (2014). Big Data and Journalism: Epistemology, expertise, economics, and ethics. *Digital Journalism*, *3*(3), 447–466. doi:10.1080/21670811.2014.976418

Lull, J. (2001). Medya İletişim Kültür. Ankara, Turkey: Vadi Publication.

Matelski, M. J. (1995). *TV Haberciliğinde Etik* (Ö. Düzgören, Trans.). Yapı Kredi Publications.

Mutlu, E. (1991). *Televizyonu Anlamak*. Gündoğan Publications.

Naked prince news ethics. (2012a). Retrieved from https://www.telegraph.co.uk/news/politics/9712859/Leveson-Naked-Harry-and-Kate-pictures-show-internet-is-an-ethical-vacuum.html

Naked prince news ethics. (2012b). Retrieved from https://www.chicagotribune.com/lifestyles/ct-xpm-2012-08-23-sns-rt-us-britain-prince-nakedbre87l08c-20120822-story.html

Nalçaoğlu, H. (2005). *Medya ve Toplum İlişkisini Anlamak Üzere Bir Çerçeve. In Habercinin El Kitabı Dizisi: Medya ve Toplum*. Turkey IPS İletişim Vakfı Yayınları.

Postman, N. (2010). *Televizyon Öldüren Eğlence* (O. Akınhay, Trans.). Ayrıntı Publications.

Ramonet, I. (2000). *Medyanın Zorbalığı* (A. Derman, Trans.). Om Yayınevi.

Rigel, N., & Batuş, G. (2005). *21. Yüzyıl İletişim Çağını Aydınlatan Kuramcılar.* Su Publications.

Tutar, H., & Yılmaz, M. K. (2003). *Genel İletişim Kavramlar ve Modeller* (5th ed.). Nobel Publications.

Uzun, R. (2007). *İletişim Etiği: Sorunlar ve Sorumluluklar.* Gazi Üniversitesi İletişim Fakültesi Basımevi.

Uzun, R. (2009). *İletişim Etiği: Sorunlar ve Sorumluluklar.* Dipnot Publications.

Yaşar, İ. H. (2020). *Sosyal Medya ve Siyaset.* Orion Academy.

Yaylagül, L. (2013). *Kitle İletişim Kuramları.* Dipnot Publications.

Yegen, C. (2018). Doğru Haber Alma Hakkı ve Sosyal Medya Dezenformasyonunu Doğruluk Payı ve Yalansavar ile Tartışmak. *Erciyes İletişim Dergisi, 5*(4), 101-121.

ADDITIONAL READING

Akdal, T. (2017). Haberciliğin Etik Kodları Bağlamında Gazetecilerin Polis Adliye Haberlerine ve Haber Öznelerine Bakışıyla İlgili Bir Derinlemesine Mülakat Çalışması. *Ulakbilge*, Cilt 5, Sayı 18, Volume 5, Issue 18. DOI: . doi:10.7816/ulakbilge-05-18-08

Çaplı, B., & Tuncel, H. (2010). *Televizyon Haberciliğinde Etik.* Fersa Matbaacılık.

D'Ancona, M. (2018). *Post-Truth: The New War on Truth and How to Fight Back.* Ebury Publishing.

García-Avilés, J. A. (2014). Online Newsrooms as Communities of Practice: Exploring Digital Journalists' Applied Ethics. *Journal of Mass Media Ethics: Exploring Questions of Media Morality, 29*(4), 258–272. doi:10.1080/08900523.2014.946600

Işık, M. (Ed.). (2019). *İletişim ve Etik* (5th ed.). Eğitim Yayınevi.

Kenneth, S. (2001). What's Right/Wrong with Journalism Ethics Research? *Journalism Studies, 2*(1), 133–150. https://www.tandfonline.com/doi/abs/10.1080/14616700119246?journalCode=rjos20. doi:10.1080/14616700119246

Mhiripiri, N. A., & Chari, T. (Eds.). (2017). *Media Law, Ethics, and Policy in the Digital Age.* IGI Global. doi:10.4018/978-1-5225-2095-5

Özdemir, Z., & Çetinkaya, A. (Eds.). (2019). *Dijital Çağda Habercilik.* Der Yayınları.

Patching, R., & Hirst, M. (2014). *Journalism Ethics: Arguments and Cases for the Twenty-First Century.* Taylor & Francis Ltd, Routledge.

Sánchez Laws, A. L., & Utne, T. (2019). Ethics Guidelines for Immersive Journalism. *Frontiers in Robotics and AI, 6*, 28. doi:10.3389/frobt.2019.00028 PMID:33501044

Ward, S. J. a. (2010). *Global Journalism Ethics.* McGill-Queen's University Press.

White, A. (Ed.). (2017). Ethics in the News. EJN Report on Challenges for Journalism in the Post-truth Era. Ethical Journalism Network: United Kingdom.

KEY TERMS AND DEFINITIONS

Disinformation: It refers that false information spread in order to deceive public, people. False information which is misdirected purposely by an organization to an opponent power or the media.

Ethical Codes: It refers moral principles or the branch of knowledge dealing with these. Ethical codes are connected with beliefs and principles about what is right and wrong, morally good or correct. The codes of conduct which are expected to adopt by the workers as reporters, journalists. Every sector has its own ethical codes as in media.

Fake News: It refers false reports of events, written and read on websites. False or misleading stories pervading fast via media as internet or other tools for damaging reputation of someone or something else making money.

Manipulation: It can be defined as 'influence'. It refers that controlling someone or something to your own advantage, often unfairly or dishonestly.

Media: "Media" refers the different types of mass communication and some of them are television, radio, the newspapers as a collective noun for the press or news reporting agencies.

Misinformation: It refers that wrong, false, or inaccurate information, or the fact that people are misinformed especially that which is deliberately intended to deceive.

Morality: It refers principles relating to right and wrong or good and bad behaviour. It is a frame that refers goodness or rightness for personal or public life.

News: It is information about the events for the time being that can be audio, visual or audio-visual formats broadcasted on the radios, televisions, newspapers or internet.

Post-Truth: It refers unreal things that contains irrelevance of facts. It is related to an erosion of the common.

Violation: It refers the action of violating someone or something that is depriving of a law, agreement, principle, or something.

Chapter 8
Advertising Ethics in the Context of the Print Media Industry

Tina Tomažič
Faculty of Electrical Engineering and Computer Science, University of Maribor, Slovenia

Katja Udir Mišič
University of Maribor, Slovenia

ABSTRACT

In this research, the authors discuss the mass media from the point of view of economic interest versus corporate social responsibility. The authors prepare a high-quality sociological comparison of the three most important Slovenian print journals from the point of view of articles that contain covert advertisements. The chapter indicates that it is more desirable for the media industries to make a profit than to be a socially responsible company. The results of this research provide insight into covert advertising in Slovenian daily newspapers and show several unique features that characterize an objective picture of a daily newspaper.

INTRODUCTION

Media are, in most countries, the primary source of information, and one of the more important creators of public opinion because of their power and influence. Meanwhile, media formed the environment that directs our everyday activities. There is no doubt that value, quality, and excellence are very important media concepts, and Corporate Social Responsibility (CSR) is a constituent part of all these dimensions.

It is very important to comprehend that the strategic mission of a company and its basic goals must not be viewed just from an economic point of view, but should also contain some quality humanistic values and recognizable ethical aspirations. Loyalty to the sole idea of profit maximizing can be a big problem, unless we succeed in defining it as only a strategic motive, and not the total goal of our existence. In order to attain this we should transform the internal organizational cultures of our companies.

DOI: 10.4018/978-1-7998-4117-3.ch008

According to Bivins (2004), profit is certainly a reasonable enough goal, but it should not be the only goal, especially given the opportunities we place on our media. Our opportunities, to a large degree, also shape the goals of the media. However, each communication has a crucial set of goals in common. Which of the sets is used at any given time depends on the medium and the principle to which the communication is being put. The more common of these goals are information dissemination, persuasion and entertainment. Each of these approaches, in the context of communication, can be used to support others, or can overlap with others.We can set an example by an advertisement being entirely informative, even though its fundamental goal may be to persuade. Entertainment can be used to present information or for a more convincing process. For example, in most public relations campaigns, communication with information is usually preceded by communication aimed at changing behaviors or attitudes. Furthermore, also, Tomkiewicz (2017), considers that, where business is concerned, the basic goal is to increase profits, but it should not be the only goal. Csigene Nagypäl (2014) declared that ethical issues are usually more important than expected corporate benefits. According to Rojek-Adamek (2018), the broad opinion applies ethic as relatively close to social responsibility. Furthermore, Porter & Kramer (2006), argued that CSR stems from ethical incentives and business needs in a competitive business environment. On the other hand, Remisová et al. (2013) stated that Corporate Social Responsibility can be seen as a source of competitive advantage and opportunity, so they should not just be focused exclusively on financial results.

Given the rapid technological development, today, the mass media are an influential force in our daily lives. New Information Technologies are having a considerable impact on countries' economies. Compared to the pre-internet era, markets now operate in unique ways, and both companies and consumers behave in different ways. According to Howard (2011), the economy, especially in the developed world, has become both information-rich and networked. Porter & Kramer (2006) claim that the predominant approaches to CSR are extremely fragmented, and also separated from strategy and business, thus providing much greater occasions for companies to benefit society.

According to Zentes et al. (2017), contemporary society expects companies, inter alia media companies, to behave sustainably and responsibly. Such an approach has also influenced the progress of business ethics perceptions, such as CSR, as presented below, in the context of public administration, which is often characterized by rigidity and a conservative approach. But, according to Kovac and Jukic (2017), public administration, especially in Eastern Europe, has been reformed declaratively to comply with European Standards and trends.

THEORETICAL BACKGROUNDS

Corporate Social Responsibility (CSR)

Corporate Social Responsibility reports have been the subject of empirical research for many years. Namely, social responsibility is a business method that deals with how companies` actions and decisions influence society, and also considers the fact that companies have many other responsibilities in addition to economic and financial ones, namely ethical, philanthropic and legal responsibilities. Social responsibility is evaluated as a behavioral model related to a particular social situation or socio-economic status. The size, level of resources obtained (human resources in particular are important), the results obtained and their impact on market conditions play an important role in the socio-economic situation.

Socio-economic status also depends on the degree of globalization and on the desire for globalization, on the impact of intervention policies and regulations within the scope. This is true at the corporate level. In addition to these, there are also other factors, such as at the local, national and global levels. It is also about the social prestige it enjoys and, last but not least, for the power of the corporation to gain the desired status. Socio-economic position is characterized by regulating instructions, and by efficient behavior which, however, may be more or less consistent with others. The greater the corporation's size, market position, ability to persuade, and power to exert pressure, the larger and more diverse the target group will be, the larger and more different its beliefs, although also more regular and destructive will be the consequences of any violations. In summary, Corporate Social Responsibility means a smart investment that benefits the company and the community as a whole. It is the direct or indirect extent of empathy shown by an organization in regard to stakeholders during the organization's activity within financial or non-financial frameworks. It also contributes towards: Improving and evolving a positive relationship with the community (consumers, suppliers, experts and other stakeholders); influencing targeted audiences, important leaders, media, etc.; positioning the company at higher levels within society; creating a favorable corporate image and, therefore, better exposure for the company; greater customer loyalty to the company and its products; motivating employees within the company, as also in terms of achieving business excellence, and providing real competitive advantages. It seems that companies with a bad reputation want to change their negative image through Corporate Social Responsibility activities (Costa&Menichini, 2013; Dinu, 2011; Fifka, 2013; Gangone & Gănescu, 2014; Gulyás, 2009; Miron, Petcu & Sobolevschi, 2011; Yoon, Giirhan-Canli & Schwarz, 2006).

Supporters of CSR within companies advocate four arguments to make their case: Moral obligation, operating permit, sustainability, and reputation (Bivins, 2004). The moral attitude requires that companies have a responsibility to be good citizens and to "do the right thing". It calls on all those involved to "achieve commercial success in ways that honor ethical values and respect people, communities, and the natural environment". Sustainability emphasizes caring for the environment and the community. In fact, every business needs the explicit or tacit permission of governments, communities, and many other stakeholders, to make a profit. Many companies use their reputation to initiate Corporate Social Responsibility. In this way, they want to enhance the company's image, strengthen the company's brand, raise morale and raise the value of its shares. According to Lee & Nguyen (2013) we should also understand social responsibility via different items about the ad's societal impacts.

Following a stakeholder approach, companies are not only responsible for making a profit, but also for their social impact on society. Businesses need to operate sustainably. Therefore, it is crucial that companies measure their success, both in terms of financial performance (eg profits or return on investment), as well as their impact on the wider economy and society as a whole. From this perspective, the success of a company must be measured on the basis of several criteria. There are four significant expectations that society has of organizations (Zentes et al., 2017):

- The economic sphere ("be profitable") in terms of the necessary financial basis;
- The legal sphere ("obey the law");
- The ethical sphere ("be ethical"), marking a "higher" position of responsibility within the pyramid; and
- The philanthropic sphere ("be a good corporate citizen").

Sustainable or socially responsible management of these companies expects many benefits. Financial measures are deficient, therefore new measures are needed to cover the impact of current strategies and measures, also taking into account sustainability of business growth. Everything we can measure can also be managed. Management's attention, therefore, should not be concentrated merely on financial results (Leo´n-Soriano, et al. 2010). According to Csigéné Nagypál (2014), Corporate Social Responsibility is defined as a self-defining concept, while decision-makers can simply misinterpret sustainable development. Corporate Social Responsibility should be an assemblage of conceivable strategies, policies and activities that can all be considered to be, by some means, socially responsible, and that can be motivated for all purposes and for a variety of purposes, from pure humanitarian action to acting strategically to improve companies` reputations (Buckler, 2017). The social reputations of companies are closely associated with company implementation, because stakeholders have focused increasingly on CSR initiatives in recent years (Lee&Jung, 2016).

Covert Advertising – An Example of Corporate Social Irresponsibility

Covert advertising has been around for a long time within the mass media. For instance, in 1883, Benjamin Day lowered the price of his newspaper the New York Sun to 1 penny. This decision transformed the relationships between publishers, journalists, their readers, and also advertisers (Patterson&Wilkins, 1994). Market experts, within a limited advertising budget, are always looking for new persuasive and efficient advertising channels and strategies (Peng et al. 2014), such as covert advertising. The authors of this paper define covert advertising in the context of ordered and paid content, which is created as an advertisement, displayed as editorial content within a particular media. This type of contributions, because they are labeled as advertising messages, are, however, presented as journalistic work, and therefore they should be considered as completely biased and unethical. Given that the public trust such contributions, they can be considered as common examples of misleading the public. According to Tomažič (2016), the most persuasive and efficient communication is one were the consumer does not recognize the content in a real commercial sense. According to Van Reijmersdal, Tutaj&Boerman (2013), the source and intent of the message are hidden by covert advertising. The credibility of the media is undermined greatly by the publication of hybrid messages. Stankevičienė & Čepulytė (2014) stated that the development of new, ethical or socially responsible practices is conditioned by the inclusion of social responsibility in the daily practice of companies. According to Eteokleous, Leonidou& Katsikeas (2015), CSR has acquired enormous attention in the decades during the 2000s, which is evident in many papers and researches under the said topic., marked by many at first as social, which, however, eventually led to developmentof the term "Corporate Social Irresponsibility". As Gulyás (2009) added, one of the media-specific CSR activities is also to ensure responsible advertising, but covert advertising this certainly isn't. Media companies are, thus, usually seen as lagging behind other industrial sectors in terms of the involvement and performance of CSR, which is also reflected in the low score of the CSR index.

One of the main keys to success lies in effective advertisement (Cymbala&Owczarczuk, 2011). New media and new marketing strategies are fundamentally distinct from traditional advertising, especially when speaking about brand incentives, use of brand ambassadors, product placement, content marketing, text-messaging, and also false websites to motivate the audience to interact with brands under misleading and fake appearances. In so far as the consumer is misled deliberately by this strategy into mistaking advertising promotion for program content, such a strategy is a form of deception, and, thus, unethical (Sandberg 2014; Spence&Van Heekeren 2004). The media's complicity regarding this strategy is equally as ethically faulty as that of the advertiser. Editorial content is remembered better, and receives more attention than commercial messages. Advertisers therefore play an important role in shaping and constructing the politics of editing practices within significant sections of the media. According to Sprott (2008), there are ethical dilemmas and possible public policy concerns regarding those companies that fail to disclose their company as being responsible for the marketing of their goods and services.

A period of information abundance was formed with the advent of cable television and then the Internet. However, the mass audience (focused specifically on the print media) has shrunk, making the strategy of growing attractiveness a more attractive specialized audience (Singer, 2013). Newspapers were also faced with a crisis, but in a different form. In the USA, for example, circulation began to decrease in the late 1980s. This happened before the expansion of the Internet, when graphic browsers were not yet available and when there was no consumer-oriented online news. Although the arrival of the Internet and online news may have contributed futher to the reduced use of print media, there is a fact that young people in particular have became more reluctant to buy and read newspapers, even before they had access to Google News. The share of subscription revenue has been very low for quite some time - for example, about 90 percent in the 1990s, so this has not yet been the worst for the newspaper industry. In fact, more than a decade after the start of circulation, total subscription revenues have increased, and began to decline slightly in mid-2000 Although the circulation decrease showed a developing crisis of news` consumption, it did not in itself create a threat to the future of the newspaper industry (Sparks, 2013). However, newspapers could no longer exist from the proceeds of advances, therefore, their existence was based only on the advertising space they offered to advertisers, as only that brought them profit.. Thus, their own economic survival was ensured. Having newspapers was attractive to advertisers and the public became a way to bring revenue to advertisers. According to Manduchi & Picard (2009), advertising also represents the largest source of revenue for many newspapers. As Ivancin (2012) argues, advertising media planners are purchasing an audience. They want access to people they believe can influence the purchase of their products, or the use of their services, and, of course, shopping in their stores. This target group is, in most cases, decisive, and not the content of the program. Advertising therefore constitutes a contractual transaction and not any statement of values or opinions.

The amount of information obtained should vary, depending on the journalist (Leszczynska, 2010). A top management team should consistently encourage them, earning trust by supporting and defending editorial independence (Ratković-Njegovan&Šiđanin, 2014). New information technologies meet not only classic advertising strategies, but also newer, more favored and more afficient strategies, such as covert advertising. Given the flood of advertising, traditional advertising is becoming less effective and, moreover, it is doing so due to excessive advertising. Advertisers are, therefore, looking for all possible means to achieve pre-set goals in order to achieve maximum impact. Accordingly, as a consequence, this has led to the growth of covert advertising. Reading through the literature, we found the following causes of covert advertising: (1) Profitable media; prior to the transition of the media to industry, their tasks were to inform and entertain the public, thus helping the public to form opinions. (2) Ineffective regulation by governmental structures. In Slovenia, there are three laws (Media Law, Law on Protection of Competition, and Consumer Protection Law) and also Regulation Acts against covert advertising. Almost no one deals with these laws, not even the courts or govermnet agencies – no one is in sharge of reviewing the media. (3) Moreover, in contrast to informing the public, the modern media have been gradually been desensitized to human feeling and problems, and, finally, (4) The supply and demand of covert advertising within the market.

METHODOLOGICAL CONCEPTS

Objectives, Goals and Research Question of the Research

The media operate within the economic system. They do not derive most of their profits directly from the consumer, but indirectly through advertising. In this they differ from most other companies. In addition to public and social safety, the media contribute to economic safety. All media feel the importance of economic development, as economic values follow the criteria of the free market, but they should also respect the social responsibility. Economic development also has a strong impact on the media. In commercial media, economic incentives are the most important, but even strict public media should appreciate the regulations of the free market. Responsibility is also very important. According to Bivins (2004), the idea of social responsibility arose as a means of highlighting companies. Their sense of commitment to the public was decidedly lacking. The realization that everything a company does affects the vast majority of people has contributed to greater accountability. In the social responsibility model, companies are seen as public operations; so their rights are, in fact, privileges - and privileges come only at the expense of restitution in the form of agreed responsibilities.

As well as several dynamic changes within the media arena, in this respect the authors propose to:

- Explain the covert features of advertising in the print media as typical examples of Corporate Social Irresponsibility by the media. Advertising is an essential part of the modern information society . Advertising enables companies to sell products and choose media sources and consumers. They also provide a quality option between different products and services. The advertisers have a lot of opportunities, tools, methods, and approaches to hand, and, with them, they try to attract the attention of potential consumers. They often advertise using new, legally and ethically non-permissible methods of advertising because of more competition on the market. The purposes of these methods are to attract awareness and improve sales; one of them is covert advertising.
- Develop and define the categories of covert advertising, so that they can be measured.
- Prepare high-quality comparisons concerning the three most important Slovenian newspapers (Večer, Dnevnik, and Delo) within the years from 2008 to 2012 inclusive, from the perspective of articles that contained covert advertisements. We investigated whether the covert advertising appeared in the more important Slovenian print journals, and how often.

Based on this, we developed the following research question:

If the number of texts with covert advertisements within the Slovenian press is concealed from year to year, does that mean the media industry is more interested in making a profit than being corporate socially responsible companies with excellent or even just good reputations?

Methodological Concept

In this research, we analyzed, and compared covert advertising from the years 2008 and 2012 as typical examples of Corporate Social Irresponsibility regarding the media. In the first part, we used the descriptive approach: The compilation method, which is a compression process regarding the observations, knowledge, attitudes, conclusions, and results of other authors. Help compilation can be based on a number of concise insights, opinions, and conclusions for reaching new, independent, and generalized conclusions, and the comparative method, i.e., comparing the same or similar facts to determine their similarities and differences. This method allows for the reaching of new and different generalized conclusions that enrich existing knowledge. In the empirical part, we used content analysis. Content analysis was performed in qualitative and quantitative terms as a scientific research method (see also Brantner&Herczeg, 2013; Conley&Tosti-Kharas, 2014; Hartz&Steger, 2010; Jakubowska, 2017). As argued by Bonini & Morello (2014), content analysis is one of the most widely used research methods in the field of Media Reporting in various communication and media studies. Our research involved three Slovenian daily newspapers (with supplementary issues): Delo, Dnevnik, and Večer. These newspapers were selected based on their readability, quality, understanding of public issues, and impacts. All three newspapers were analyzed from 3rd January to 9th January, 2008, and from 26th November to 1st December, 2012, which meant six days per year. We must mention that newspapers in Slovenia are not published on Sundays. The only exception is the newspaper Nedelo. However, we did not analyze it as there is no comparable newspaper with which to compare it.

The newspapers we chose were read in their entireties, and we looked for units of analysis. As a sample unit within our work, we investigated separate newspaper articles, also short news, photos, etc. The unit of analysis was any source of information that was directly or indirectly the subject of surreptitious advertising. For example, in an article where covert advertising was not noticeable, but the article was close to an ad that matched the article thematically. According to our research, we have developed, identified and explained the categories among covert advertising that could be measurable. If the article had at least one of the following characteristics, we classified it as covert advertising:

- Positive statements of a selected product, company or service and its compliment;
- Citing the company or product in a very positive perspective in the title;
- Unambiguous connection in the text of the mentioned information with commercial business activities;
- The urgency of the text. The ad is a direct definite sales proposition;
- An article that did not include surreptitious advertising, but was next to the ad with which it was thematically related;
- Directing the attention of customers, and attracting them to the seller, manufacturer, brand, product, service or other features relevant to maintaining their interest;
- Press releases that promote a particular product, service, or company, and are published in the form of a journalistic article, but nowhere is it stated clearly that it is an advertisement;
- A logo appears in the text, which also represents covert advertising;
- The “artificial” nature of news, which is exaggerated or invented in the text, or has no realistic basis;
- Mentioning certain products (company, service) in relation to celebrities (politicians, sports people …);
- Editorial texts or columns next to photographs that are unambiguously neutral at first sight, but they mention the company and / or the prices of products or services;
- Focus on urgent sales of products or services (eg seasonal sales);
- Photographs next to articles that are unnecessary because the product or maturation is clearly visible on them.

Covert advertising does not include the:

- Editorial articles with an appropriate clear label (e.g. promotional material, PR message, paid notice, advertising services, advertised message, sponsored) which is clearly visible, and not colored or covered with anything else;
- Donations and sponsorships;
- Where the signs were clearly visible that they were not colored by or covering something else;
- Fashion reviews, despite highlighting certain brands or goods (see also Buari, 2001; Halmi, 1996; Tomažič & Jurišić 2009; Tomažič, Boras, Jurišić & Lesjak, 2014).

Based on the analyzes of some articles, we used an analytical approach; for each, we determined certain defined properties and transferred them to a data set, which enabled further quantitative analysis - the coding process. Our code list had the status of a questionnaire with predetermined possible answers. The results of the questionnaire were analyzed statistically, researched, compared, and, subsequently, compared with cross-tabulations using a data processing package for the Social Sciences, SPSS 16 (see also Balaban et al., 2016).

EMPIRICAL RESULTS

Our goal was to conduct an empirical research, with which we wanted to prepare a qualitative comparison between the three leading Slovenian newspapers (Delo, Večer, and Dnevnik) for parts of the years 2008 and 2012 in terms of articles that contained covert advertisements as typical examples of this medium's Corporate Social Irresponsibility. Each of the three newspapers (with supplements) was analyzed in one week, from January 3rd to January 9th of 2008, and from November 26th to December 1st of 2012. During the 6-day period of 2008, we found 84 texts with covert advertising, resulting in an average of 4.67 articles per newspaper, and 108 texts for the 6 days of 2012, an average of 6 articles per newspaper. We then compared the 2008 week with the 2012 week. The data from the research were processed using the SPSS 16 computer program (Statistics: $N=84, 108, \alpha=0.05$ $p<0.0005$, $p<\alpha$) and we came to the following conclusions:

- The amount of concealed covert advertising within the Slovenian print media for the 6 days of 2008 when compared to the 6 days of 2012 had increased, with an average of 4. 67 (N=84) articles per newspaper in 2008, to 6 (N=108) articles per newspaper in 2012. This tendency could be seen in Večer (with the number of covert texts having risen from 18 for 2008 to 40 for 2012) and Dnevnik (from 38 to 42), see Table 1 and Figure 1.

- Our research showed that 54.8% (N=84) of the texts identified as covert advertising for the 6 days of 2008 were edited within the newspapers themselves, and just 45.2% in the supplements. In the 6 days of 2012, the picture was totally different, namely, just 18.5% (N=108) were edited in the newspapers and 81.5% in the supplements. This tendency was, without any significant differences, a general feature for all the analyzed newspapers. Supplements attract greater interest from advertisers, as they provide more opportunities for hidden advertising and for editing promotional texts.

Table 1. Number and % of texts with covert advertising

Newspaper			Year		Total
			2008	2012	
	Delo	Count	28	26	54
		% within newspaper	51.9%	48.1%	100%
		% within year	33.3%	24.1%	28.1%
		% of Total	14.6%	13.5%	28.1%
	Večer	Count	18	40	58
		% within newspaper	31.0%	69.0%	100%
		% within year	21.4%	37%	30.2%
		% of Total	9.4%	20.8%	30.2%
	Dnevnik	Count	38	42	80
		% within newspaper	47.5%	52.5%	100%
		% within year	45.2%	38.9%	41.7%
		% of Total	19.8%	21.9%	41.7%
Total		Count	84	108	192
		% within newspaper	43.8%	56.3%	100%
		% within year	100%	100%	100%
		% of Total	43.8%	56.3%	100%

Therefore, we can conclude that, for the media, it is more desirable to gain profit than to practice CSR and earn a good reputation, see Table 2 and Figure 2.

Figure 1. Number of texts with covert advertising

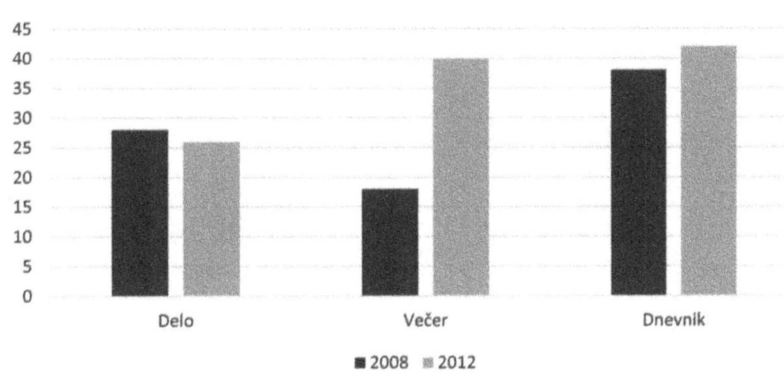

- Based on our research, most of the covert advertising texts appeared on Tuesdays and Wednesdays. It was obvious that this pattern was due to the fact that, on Tuesdays, a supplement for women was inserted in Delo and Večer, and on Wednesdays a supplement for women in Dnevnik, all of which are of special interest for advertisers. We noticed that 17 (42.5%) of the texts found as co-

Table 2. Place of publication

Place of Publication			Year		Total
			2008	2012	
	Newspaper	Count	46	20	66
		% within place	69.7%	30.3%	100%
		% within year	54.8%	18.5%	34.4%
		% of Total	24%	10.4%	34.4%
	Supplement	Count	38	88	126
		% within place	30.2%	69.8%	100%
		% within year	45.2%	81.5%	65.6%
		% of Total	19.8%	45.8%	65.6%
Total		Count	84	108	192
		% within place	43.8%	56.3%	100%
		% within year	100%	100%	100%
		% of Total	43.8%	56.3%	100%

vert advertising which were edited into the supplement on the Tuesday in 2008, had increased to 23 (57.5%) on the Tuesday in 2012. The most published texts with covert advertising were within the supplement on Wednesday, 22 (31.0%) in 2008, having increased to 49 (69.0%) in 2012. We can conclude once again that, for the media, it is more desirable to gain profit than to practice CSR and earn a good reputation, see Table 3 and Figure 3.

Figure 2. Place of publication

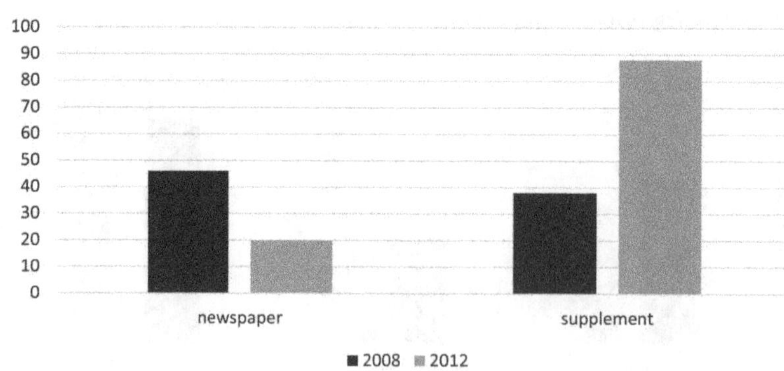

In the theoretical and empirical work, we used the method of comparison and the method of synthesis, and, based on the facts, we came to certain conclusions. We then formulated the key facts. Our research of the three more important Slovenian newspapers concluded that, on average, 4.72 covert articles appeared daily in 2008 regarding each daily newspaper, which had increased to an average of 6 in 2012. Moreover, it turned out that most of the hidden text texts were in the supplements, which increased from 45.2% in 2008 to 81.5% in 2012. The supplements are obviously more lucrative for advertisers,

Table 3. Covert advertising by day of the week

Day			Year		Total
			2008	2012	
	Monday	Count	12	9	21
		% within day	57.1%	42.9%	100%
		% within year	14.3%	8.3%	10.9%
		% of Total	6.3%	4.7%	10.9%
	Tuesday	Count	17	23	40
		% within day	42.5%	57.5%	100%
		% within year	20.2%	21.3%	20.8%
		% of Total	8.9%	12.0%	20.8%
	Wednesday	Count	22	49	71
		% within day	31.0%	69.0%	100%
		% within year	26.2%	45.4%	37.0%
		% of Total	11.5%	25.5%	37%
	Thursday	Count	7	5	12
		% within day	58.3%	41.7%	100%
		% within year	8.3%	4.6%	6.3%
		% of Total	3.6%	2.6%	6.3%
	Friday	Count	13	18	31
		% within day	41.9%	58.1%	100.0%
		% within year	15.5%	16.7%	16.1%
		% of Total	6.8%	9.4%	16.1%
	Saturday	Count	13	4	17
		% within day	76.5%	23.5%	100.0%
		% within year	15.5%	3.7%	8.9%
		% of Total	6.8%	2.1%	8.9%
Total		Count	84	108	192
		% within day	43.8%	56.3%	100%
		% within year	100%	100%	100%
		% of Total	43.8%	56.3%	100%

as they provide more opportunities for advertising hidden agendas and to gain a profit. Very similar pictures were shown by the data of increasing texts regarding covert advertising in female supplements on Tuesdays and Wednesdays, which are the most lucrative for advertisers, and were characteristic for all three newspapers.

Figure 3. Covert advertising by day of the week

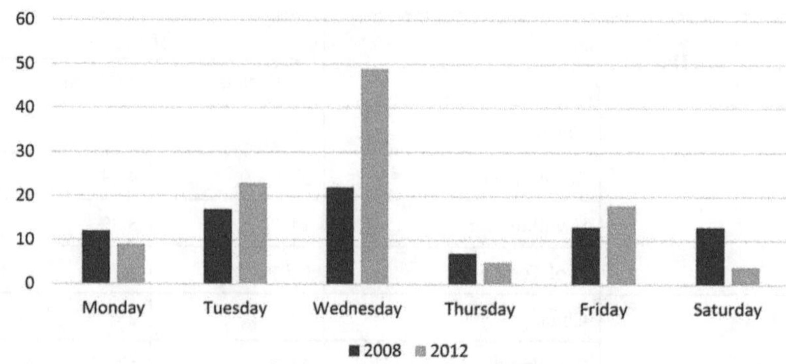

DISCUSSION AND CONCLUSION

Social responsibility is of great value for the long-term well-being of companies. At the same time, it creates the human side of the business, as it is an essential link with society in general, and with the community in which the company operates in particular. According to Dinu (2008), adherence to business ethics and standards is very important it requires time, money and human resources, and it depends on the ability to seize opportunities. Therefore, additional costs are required which could reduce, in particular, short-term profitability. According to Zentes et al. (2017), companies must meet different goals of equal importance: Be profitable and, at the same time, operate socially. When we talk about profit, we mean short-term, medium-term and long-term approaches that are not necessarily mutually compatible. Responsible retail companies often have higher costs, so Corporate Social Responsibility can reduce this short-term profit, but, in the long run, this measure will be satisfied.

Social responsibility and economic success are known mainly as opposites and as a negative correlation (Ruh et al. 2004). To compete in the economic field today, a company or corporation must master, among other things, the principles of marketing. The fundamental objective of any mass media industry is profit. Without it, the lifespan of any media organization would be a short one. In addition, an unprofitable procedure is questionable to attract the investment capital needed for expansion. On the other hand, making a profit allows an organization to invest in the talent and hardware essential for producing quality results. The critical factor regarding profitability is the ability of mass media industries to attract massive audiences in order to sell themselves to advertisers, media buyers or advertising agents. According to Suchaček, Seìa, Friedrich&Koutskˇ (2014) the possibility of monitoring whether the media are fulfilling their mission, ie. objective and impartial reporting, are, thus, eliminated. Namely, there are not so many individuals who would have enough time and energy to check the primary sources of information and check the validity of media information. Media companies have often lagged behind other industrial sectors in terms of the performance and inclusion of Corporate Social Responsibility. One indicator of this is that they have low scores on CSR indices. The potential of businesses is to take action based on supporting their community and their business goals. Therefore, social trust is of importance. According to Ziller (2017), the extent to which people rate their citizens as trustworthy depends on social trust. In the future, within our economic arena, we need to establish a new and constructive link between business and social responsibility, and between business and economic success. The media would not be an exception here.

However, in Eastern Europe, public administration is at an earlier stage of development. Namely, countries are still engaged in a rather basic restructuring. In addition, the exchange of best practices is fragmented and underdeveloped, and the internationalization of public administration has been proclaimed. The countries of Eastern Europe should, therefore, make greater efforts to have a more advanced and systematic understanding of European Standards to meet today's challenges in our globalized society (Kovac and Jukic, 2017), also, among others, with the concepts of business ethics, such as CSR and covert advertising in the context of public administration.

Based on our studies, we found that the number of texts with covert advertisements within the Slovenian press is concealed from year to year, which means that the media industry is more interested in making a profit than being corporate socially-responsible with an excellent, or even just good, reputation. However, obsessions with economic interests alone have recently caused basic problems regarding CSR. The authors of this paper believe that concentrating purely on making a profit should not be the only basic goal of a company. Furthermore, we agree that it is inappropriate to make a profit at the expense of neglecting customer satisfaction, the environment, and the employees.

According to Balaban et al. (2016), within the context of new tools and ways of communication, public administration, which is often characterized by rigidity and a conservative approach, will have to line up with the technological evolution and respond to the new society's needs.

Social and Managerial Recommendations

Based on this research, this article has presented the significant influences of those different factors that are necessary for forming and creating the fostering of profit and economic benefit, whilst, instead, acknowledging social responsibility. Within this framework, the media industry should not use covert advertising, because it damages their images and reputations. The public expects articles they can trust, but this is impossible in the form of covert advertising. The major problem is definitely the financial power of advertisers, and the media must not respond to pressures from them. According to Anghe, Grigore&Roşca (2011), consumers trust those companies who communicate openly, honestly and responsibly. Itis a moral frame that every organization should adopt. According to Halman&Van Ingen (2015), people are showing increasing freedom in forming their opinions, especially when we talk about moral and ethical issues. Furthermore, in the case of covert advertising, we can't talk about socially responsible advertising, therefore, advertisers and editors should prioritize the informative role of advertising, which will reflect in good ethical behavior and socially responsible treatment of consumers. Namely, ethical advertising has a crucial role in shaping a corporate socially responsible media industry, and not place profit as the single and most important goal. The media should be involved for the benefit of society as a whole, and not just do their job and only care about the consequences of it. We can conclude that the economic environment has a very strong influence on the media, and, consequently, on their ethical behavior.

LIMITATIONS AND FUTURE RESEARCH

We are aware of the possibility that our judgment will be, or could have been, wrong regarding certain examples, not only due to ambiguities when categorizing covert advertising, but also due to their questionable topics, causing us to rule them out as subjective assessments. We expected that the best results could be achieved by the method of participation (participatory monitoring would be best in the editorial office itself, or in the department marketing this medium), or by surveying the people directly involved in publishing the texts. But, we also knew that most of these people don't want to disclose information related to surreptitious advertising publicly.We are, therefore, aware of the possibility that our assessment of the evidence might be erroneous in some cases.

The study does not take into account the readers' perceptions, so we do not have enough information about the readers' interpretation of the articles that contain covert advertisements (see also Jakubowska, 2017).

We did not classify as being amongst covert advertising: Editorial contents marked with corresponding explicit labels. Furthermore, where the signs were clearly visible that they were not colored by or covering something else, fashion reviews, despite certain brands or goods being emphasized especially, and, finally, donations and sponsorships, because they are very complicated forms of advertising.

Another important limitation is associated with the fact that we investigated just covert advertising as examples of Corporate Social Irresponsibility.

This research was limited to the print medium only, and, as we considered just the texts of covert advertising in the press, it cannot be generalized regarding all media. Furthermore, the research was focused on just the Slovenian press.

For the future research the sample used for the empirical analysis could be upgraded, including media firms from different countries. The research could also be extended to any other media, for example social media, as a part of the new technologies.

REFERENCES

Anghel, L. D., Grigore, G. F., & Roşca, M. (2011). Cause-related marketing, part of Corporate Social Responsibility and its influence upon consumers' attitudes. *Amfiteatru Economic*, *13*, 29, 72–85.

Balaban, D. C. (2017). Online Communication and PR in Romanian Public Administration. The Case Study of Public Institutions From Transylvania. *Transylvanian Review of Administrative Sciences*, *47*(1), 20–34.

Bivins, T. (2004). *Mixed media*. Lawrence Erlbaum Associates.

Bonini, T., & Morello, G. (2014). No news from abroad: A comparative content analysis of news issues on Italian radio. *European Journal of Communication*, *29*(2), 218–229. doi:10.1177/0267323114520782

Brantner, C., & Herczeg, P. (2013). The life of a new generation: Content, values and mainstream media perception of transcultural ethnic media – An Austrian case. *Communications*, *38*(2), 211–235. doi:10.1515/commun-2013-0012

Buari, F. (2001). *Public relations or trust strategy*. Aspect Press.

Buckler, S. (2017). Corporate Social Responsibility, CSR, Sustainability, Ethics & Governance. In Imagined Communities Incorporated: Corporate Social Responsibility and Value Creation in a Globalised World. Springer International Publishing. doi:10.1007/978-3-319-35083-7_1

Conley, C., & Tosti-Kharas, J. (2014). Crowdsourcing content analysis for managerial research. *Management Decision, 52*(4), 675–688. doi:10.1108/MD-03-2012-0156

Costa, R., & Menichini, T. (2013). A multidimensional approach for CSR assessment: The importance of the stakeholder perception. *Expert Systems with Applications, 40*(1), 150–161. doi:10.1016/j.eswa.2012.07.028

Csigene-Nagypäl, N. (2014). Corporate Social Responsibility of Hungarian SMEs with good environmental practices. *Journal for East European Management Studies, 19*(3), 327–347. doi:10.5771/0949-6181-2014-3-327

Cymbala, A., & Owczarczuk, M. (2011). Variables importance in questionnaire data on advertising. *Expert Systems with Applications, 38*, 14218–14224. doi:10.1016/j.eswa.2011.04.234

Dinu, V. (2008, Feb. 23). The ethical dimension of business. *Amfiteatru Economic*, 7-8.

Dinu, V. (2011). Corporate Social Responsibility– opportunity for reconciliation between economic interests and social and environmental interests. *Amfiteatru Economic*, 13(29), 6–7.

Eteokleous, P. P., Leonidou, L. C., & Katsikeas, C. S. (2015). Corporate Social Responsibility in international marketing: Review, assessment, and future research. *International Marketing Review, 33*(4), 580–624. doi:10.1108/IMR-04-2014-0120

Fifka, M. S. (2013). Corporate responsibility reporting and its determinants in comparative perspective–a review of the empirical literature and a meta-analysis. *Business Strategy and the Environment, 22*(1), 1–35. doi:10.1002/bse.729

Gangone, A.-D., & Gănescu, M.-C. (2014). Corporate Social Responsibility in emerging and developing economies in Central and Eastern Europe – a measurement model from the stakeholder theory perspective. *Economic Research-Ekonomska Istraživanja, 27*(1), 539–558. doi:10.1080/1331677X.2014.967535

Gulyás, A. (2009). Corporate Social Responsibility in the British media industries – preliminary findings. *Media Culture & Society, 31*(4), 657–668. doi:10.1177/0163443709335281

Halman, L., & Van Ingen, E. (2015). Secularization and Changing Moral Views European Trends in Church Attendance and Views on Homosexuality, Divorce, Abortion, and Euthanasia. *European Sociological Review, 0*(5), 0, 1–12. doi:10.1093/esr/jcv064

Halmi, A. (1996). *Kvalitativna metodologija u društvenim znanostima*. Samobor: A.G. Matoš d.d.

Hartz, R., & Steger, T. (2010). Heroes, villains and 'honourable merchants': Narrative change in the German media discourse on corporate governance. *Organization, 17*(6), 767–785. doi:10.1177/1350508410372558

Howard, P. N. (2011). *Castells and the media*. Polity Press.

Ivancin, M. (2012). Business Ethics and the Return on Reputation. *Journal of Mass Media Ethics, 27*(2), 150–152. doi:10.1080/08900523.2012.684593

Jakubowska, H.(2017). Framing the Winter Olympic Games: A Content Analysisof Polish Newspapers Coverage of Female and Male Athletes. *Polish Sociological Review, 1*(197), 67-81.

Kovac, P., & Jukic, T. (2017). Declarations and Reality of Europeanized Public Administration in Eastern Europe: Journals Content Analysis in Slovenia and Croatia. *Transylvanian Review of Administrative Sciences*, *50*(3), 127–145. doi:10.24193/tras.2017.0008

Lee, S., & Jung, H. (2016). The effects of Corporate Social Responsibility on profitability The moderating roles of differentiation and outside investment. *Management Decision*, *54*(6), 1383–1406. doi:10.1108/MD-07-2015-0268

Leo'n-Soriano, R., Mun˜oz-Torres, M. J., & Chalmeta-Rosalen, R. (2010). Methodology for sustainability strategic planning and management. *Industrial Management & Data Systems*, *110*(2), 249–268. doi:10.1108/02635571011020331

Leszczynska, A. (2010). Manager's attitude toward environment. *Industrial Management & Data Systems*, *110*(8), 1234–1250. doi:10.1108/02635571011077852

Manduchi, A., & Picard, R. (2009). Circulations, Revenues, and Profits in a Newspaper Market with Fixed Advertising Costs. *Journal of Media Economics*, *22*(4), 211–238. doi:10.1080/08997760903375902

Miron, D., Petcu, M., & Sobolevschi, I. M. (2011). Corporate Social Responsibility and the sustainable competitive advantage. *Amfiteatru Economic*, *13*, 29, 162–179.

Patterson, P., & Wilkins, L. (1994). *Media ethics: issues and cases*. Brown & Benchmark, cop.

Peng, J., Zhang, G., Zhang, S., Dai, X., & Li, J. (2014). Effects of online advertising on automobile sales. *Management Decision*, *52*(5), 834–851. doi:10.1108/MD-10-2012-0753

Porter, M. E., & Kramer, M. R. (2006). Strategy & society. The link between competitive advantage and Corporate Social Responsibility. *Harvard Business Review*, (December), 5, 78–94. PMID:17183795

Ratković-Njegova, B., & Šiđanin, I. (2014). Public broadcasting crisis as management crisis: A case study of radio television of Vojvodina. *Journal for East European Management Studies*, *19*(3), 348–367. doi:10.5771/0949-6181-2014-3-348

Remisová, A., Lasáková, A., & Krzykala-Schaefer, R. (2013). Corporate Social Responsibility in European countries: The keystones of the concept and intercultural connotations. *Journal for East European Management Studies*, *18*(4), 512–543. doi:10.5771/0949-6181-2013-4-512

Rojek-Adamek, P. (2018). From Relational Space to Engagement—Designers' Ethics and Responsibility in the Light of Empirical Research. *Polish Sociological Review, 1*(201), 95-107.

Ruh, H. (2004). *Ethik im Management. Ethik und Erfolg verbunden sich*. Zuerich: Orell Fuessli Verlag AG.

Sandberg, H. (2011). Tiger talk and candy king: Marketing of unhealthy food and beverages to Swedish children. *Communications*, *36*(2), 217–244. doi:10.1515/comm.2011.011

Singer, J. B. (2013). The Ethical Implications of an Elite Press. *Journal of Mass Media Ethics*, *28*(3), 203–216. doi:10.1080/08900523.2013.802163

Sparks, C. (2013). Global media studies: Its development and dilemmas. *Media Culture & Society*, *35*(1), 121–131. doi:10.1177/0163443712464566

Spence, E., & Van Heekeren, B. (2004). *Advertising Ethics*. Pearson.

Sprott, D. E. (2008). The Policy, Consumer, and Ethical Dimensions of Covert Marketing. *Journal of Public Policy & Marketing*, *27*, 1, 4–6.

Stankevičienė, J., & Čepulytė, J. (2014). Sustainable value creation: Coherence of Corporate Social Responsibility and performance of socially responsible investment funds. *Economic Research-Ekonomska Istraživanja*, *27*(1), 882–898. doi:10.1080/1331677X.2014.976058

Sucháäek, J., Seìa, P., Friedrich, V., & Koutsk˘, J. (2014). Media portrayals of regions in the Czech Republic: Selected issues. *E+M. Ekonomie a Management*, *17*(4), 125–140. doi:10.15240/tul/001/2014-4-010

Tomažič, T. (2016). The Growth of Covert Advertising in the Context of Local Self-Government. *Lex localis - Journal of Local Self-Government, 14*(4), 739–758.

Tomažič, T., Boras, D., Jurišić, J., & Lesjak, D. (2014). Covert Advertising in the Context of Inadmissible Presentation of Information. *Industrial Management & Data Systems*, *114*(1), 107–122. doi:10.1108/IMDS-04-2013-0204

Tomažič, T., & Jurišić, J. (2009). Covert advertising in the context of media ethics on the example of Slovenian press. *Medijski dijalozi*, *2*(4), 57-67.

Tomkiewicz, J. (2017). The Social Consequences of Income Distribution—Are Increasing Company Profits a Threat to Economic and Social Stability? *Polish Sociological Review*, *2*(198), 203-218.

Van Reijmersdal, E. A., Tutaj, K., & Boerman, S. C. (2013). The effects of brand placement disclosures on scepticism and brand memory. *Communications*, *38*(2), 127–146. doi:10.1515/commun-2013-0008

Yoon, Y., Giirhan-Canli, Z., & Schwarz, N. (2006). The Effect of Corporate Social Responsibility (CSR) Activities on Companies with Bad Reputations. *Journal of Consumer Psychology*, *16*(4), 377–390. doi:10.120715327663jcp1604_9

Zentes, J. (2017). *Strategic Retail Management*. Springer Fachmedien. doi:10.1007/978-3-658-10183-1

Ziller, C. (2017). Equal Treatment Regulations and Ethnic Minority Social Trust. *European Sociological Review*, *0*(4), 0, 1–13. doi:10.1093/esr/jcx059

ADDITIONAL READING

Bachnik, K., & Nowacki, R. (2018). How to Build Consumer Trust: Socially Responsible or Controversial Advertising. *Sustainability*, *10*(7), 2173. doi:10.3390u10072173

Evans, N. J., & Park, D. (2015). Rethinking the Persuasion Knowledge Model: Schematic Antecedents and Associative Outcomes of Persuasion Knowledge Activation for Covert Advertising. *Journal of Current Issues and Research in Advertising*, *36*(2), 157–176. doi:10.1080/10641734.2015.1023873

Göbel, F., Meyer, A., Ramaseshan, B., & Bartsch, S. (2017). Consumer responses to covert advertising in social media. *Marketing Intelligence & Planning*, *35*(5), 578–593. doi:10.1108/MIP-11-2016-0212

Sabri, O. (2017). Does Viral Communication Context Increase the Harmfulness of Controversial Taboo Advertising? *Journal of Business Ethics, 141*(2), 235–247. doi:10.100710551-015-2751-9

Tomažič, T. (2016). *The growth of covert advertising in the context of local self-government.* Lex localis: revija za lokalno samoupravo, 14 (4), 739-758.

Tomažič, T., Boras, D., Jurišić, J., & Lesjak, D. (2014). Covert advertising as inadmissible presentation of information. *Industrial Management & Data Systems, 114*(1), 107–122. doi:10.1108/IMDS-04-2013-0204

Wang, Z., Deshpande, S., Waller, D. S., & Erdogan, B. Z. (2018). Religion and Perceptions of the Regulation of Controversial Advertising. *Journal of International Consumer Marketing, 30*(1), 29–44. doi: 10.1080/08961530.2017.1358681

Wojdynski, B. W., & Evans, N. J. (2020). The Covert Advertising Recognition and Effects (CARE) model: Processes of persuasion in native advertising and other masked formats. *International Journal of Advertising, 39*(1), 4–31. doi:10.1080/02650487.2019.1658438

KEY TERMS AND DEFINITIONS

Content Analysis: Content analysis is a research method used to analyze manifest or covert content in different media.

Corporate Social Responsibility: Corporate social responsibility is a long-term investment. This benefits both the company and the community as a whole. That's why we're talking about a smart investment. The company thus shows a degree of empathy towards stakeholders during the company's activity, which can be direct or indirect, financial or non-financial.

Covert Advertising: We refer to covert advertising in a specified media when the content, that is paid for and ordered, is presented as independent editorial content, even though it is an advertisment.

Print Media Industry: The print media industry is an industry related to news reporting, printing and distribution of news, mainly through newspapers and magazines.

Chapter 9
Marketing Ethics:
Digital Transformation and Privacy Ethics

Tansu Işıkay
Doğuş Üniversitesi, Turkey

ABSTRACT

The digital transformation caused significant changes in marketing activities. It has become effortless to collect, process, use, and transmit personal information of consumers. However, marketers' obtaining consumers' personal data in such an easy way has raised privacy ethics. This study examines privacy ethics, which has become an important subject in marketing due to digital transformation. For this, in the first part, the emergence of the ethics issue in marketing and the theories developed for making ethical decisions are mentioned. The second part focuses on the digital transformation of marketing and the matter of privacy ethics. It is also said who is responsible for protecting the privacy and how measures can be taken. This study also includes solutions, results, and suggestions for future research.

INTRODUCTION

Digital transformation is one of the most remarkable topics of today. The digital transformation process that started with the development of Web 2.0 technologies has affected all the functions of the businesses. One of the most affected business functions is marketing. The rapid development of technology and the widespread use of the Internet have enabled marketing activities to begin to be carried out on digital platforms. While traditional marketing methods allow consumers to review and purchase products or services only in stores, digital marketing has allowed consumers to research and purchase products or services without space and time limitations. Consumers can now supply even their most basic needs online. Accordingly, consumers' purchasing behavior and consumption habits have changed.

The transformation process in marketing has led traditional consumers to become online consumers. The number of consumers who prefer to use online channels is increasing day by day. This has enabled marketers to obtain information about consumers at a lower cost and more quickly. In other words, the development of digital marketing has made it easier to access information about consumers. Because consumers leave clues about themselves with every click on the internet. Online transactions such as social

DOI: 10.4018/978-1-7998-4117-3.ch009

media shares, emails, website visits and online purchases all cause consumers to leave digital cues about themselves. Marketers collect, process and use consumer information by following these digital cues.

The ability of online channels to access consumer information enables digital marketing activities to have many advantages over traditional marketing. First of all, marketers can use the information of the consumers (e-mail address, phone numbers, credit cards, website visits, product reviews, past buying behaviors, etc.) to analyze the requests, preferences and buying behaviors of the target audience. They can make strategic decisions for the target audience. In addition, they can measure and update how successful marketing practices are with digital metrics.

Although digital marketing has many advantages, it provides a very suitable environment for unethical behavior. The continuous collection, processing and use of consumer information, especially through digital marketing channels, causes issues regarding privacy ethics among businesses and consumers. Because of the internet and technological developments, consumers have lost control of keeping their personal information private. This situation, which occurs with the digital transformation of marketing, raises the issue of privacy ethics.

The fact that consumers clicks on the internet can be followed by marketers causes privacy concerns. Privacy concern refers to how worried consumers are about their personal information. These concerns differ for each user. Since the value that each user gives to their personal information is not the same, their level of privacy concern is not the same. For this reason, some consumers perceive lower privacy concerns than others, while others perceive higher. Consumers with lower privacy concerns are more likely to continue sharing information. Consumers with higher privacy concerns are expected to reduce or end information sharing. However, surprisingly, the expected situation may not happen. Because consumers need to share their personal information to perform online transactions. Otherwise, consumers can perform very few online transactions. Because consumers need to share their personal information in order to make transactions online. Otherwise, consumers can perform very few online transactions. Therefore, consumers continue to share information to utilize the benefits of digital marketing despite the possibility of unethical situations.

Researches are conducted on how to prevent unethical situations related to the privacy of consumers as a result of online marketing activities. Accordingly, all stakeholders who are parties to the relationship of change are considered responsible for the prevention of the issue of privacy ethics. These parties are government, businesses and consumers. Many countries around the world are making legal regulations to protect the privacy of consumers. One of the most important sides responsible for protecting consumer privacy is government. Because the regulations made by the government have the effect of deterring the violations that may be made regarding privacy. Another side responsible for the protection of consumer privacy is businesses. Businesses should request permission and state on how to use the information before collecting personal information about consumers. At the same time, they should publish their privacy and cookie policies. In addition, consumers should be able to access their own information, inquire about its accuracy and request it to be deleted at any time. There are also measures that consumers can take individually. Changing privacy settings, clearing cookies, using untrue information, and asking service providers to delete personal data are individual efforts.

Despite the many benefits that digital marketing provides, it has caused the emergence of ethical problems related to consumer privacy. For this reason, recently the subject of privacy has been frequently discussed in marketing. However, there are few studies examining privacy ethics issues. This study aims to contribute to this gap in the literature. This chapter begins by explaining the development process of marketing, the emergence of the concept of ethics in marketing and ethical theories. In digital marketing, it continues with consumer privacy ethics and solutions for the protection of privacy. In addition, it includes solutions and recommendations, future research directions ve conclusion.

BACKGROUND

Development Stages of Marketing and Marketing Ethics

In the process from the 1800s to the present day, marketing management orientation (MMO) have pass certain stages. These stages are; production orientation, product orientation, sales orientation, marketing orientation / customer-centric marketing and societal marketing orientation. The first stage, Production orientation, expresses the idea that a good product and service can sell itself (Mucuk, 2014, p. 8). At this stage, businesses focus on producing more and improving distribution channels. In other words, it wants to dominate the market with its low labor cost, high productivity and mass distribution strategies. However, Armstrong and Kotler (2009) stated that the understanding of production may cause marketing myopia. Marketing myopia refers to the problem that businesses focus on their products rather than their customers (Levitt, 1960). Therefore, this orientation will prevent businesses from focusing on consumer needs and establishing customer relationships. The second stage is Product orientation. It is based on the principle that consumers will prefer the best products in terms of quality, performance and innovation. According to this understanding, marketing strategy focuses on continuous improvement (Armstrong, Stewart, Denize, & Kotler, 2015, p. 11). The subsequent sales orientation refers to the period when it was not a problem to produce the goods, but to sell them. For this reason, more focus is on promotional activities (public relations, personal sales, advertisement etc.). Unethical methods such as deceptive advertising were frequently used in this period to increase sales. Sales orientation can be summarized as the typical way of thinking, "I sell whatever I produce as long as I know how to sell it." (Mucuk, 2014, p. 9). But realizing that unethical methods do not bring long-term profitability has led businesses to adopt marketing orientations, also known as customer-oriented marketing. Marketing orientation is based on identifying and satisfying the needs and wants of individual consumers and customers rather than mass market segments (Sheth, Sisodia, & Sharma, 2000, p. 56-57). The aim is to satisfy the customer and direct it to repurchase. Businesses that adopt this understanding can gain profitability and sustainable competitive advantage in the long run. The final stage of marketing management orientations is social marketing. Concerns about the rapid reduction of resources in the world and increasing environmental problems have led to the development of the concept of societal marketing (Prothero, 1990).

Events such as deceptive advertising, insistent sales and bribery encountered in the period of sales orientation brought the issue of marketing ethics to the agenda. It has begun to investigate how unethical situations can be prevented. Especially with the transition to the societal marketing period the subject of ethical behavior has gained importance in marketing ideas and practices. The increasing awareness of consumers on consumption, social responsibility awareness and sensitivity to protect the environment have shaped the marketing ethics. Marketing ethics has been researched by many researchers and various ethical models have been developed.

Ethics is a discipline that systematically examines right and wrong, good and bad, virtue and vice (Brinkmann, 2002). Marketing ethics is the application of ethical thinking to marketing strategy, tactics and decisions (Laczniak and Murphy, 2006). Although it is related to all the functions of an ethical business, the most ethical problem arises as a result of marketing practices. This made it necessary to analyze ethics in terms of marketing. Marketing ethics investigates what is good and bad in marketing. In addition to this it examines the moral challenges that marketing managers face during their marketing practice (Baker and Saren, 2010).

Ethical Models in Marketing

Various ethical theories have been developed to solve the ethical problems encountered in marketing. These theories help to understand how ethical and unethical behavior occurs. Cognitive Moral Development Model, The General Theory of Marketing Ethics, Contingency Model and Synthesis Model.

Cognitive Moral Development Model

The first study on ethical decision making in marketing was done by L. Kohlberg in 1969. In his study, L. Kohlberg shows the moral development stages of individuals in a theoretical framework. According to this framework, there are six moral development stages related to ethical decision making. The first is the stage where individuals do what is said for avoidance from punishment. The second stage is ego-centered. The individual makes a decision considering the reciprocity and exchange situation. Managers with this moral level display trickster and selfish behavior. The third stage refers to the individuals acting in line with the expectations of the society and avoiding conflict. The fourth stage speaks of the need to respect the law and authority. Individuals act in line with the expectations of the society for their own good (Rest, Turiel, & Kohlberg, 1969). The fifth is the moral stage, which is based on the contractual basis. At this stage, individuals consider the benefit of society when making decisions. The person avoids violations of the rights and will of other individuals. The sixth stage refers to universal ethical principles. Individuals direct their behavior in line with the ethical principles of their choice. These ethical principles are respect, equality of human rights and justice (Smith A. F., 1978).

L. Kohlberg states that not every individual has the same level of moral development. For this reason, he argues that the decisions taken by individuals during ethical dilemmas will be affected by the moral development levels of individuals. In other words, the ethical stage of the individual is effective on the ethical decision.

The individual moral development model has also been applied for organizations. The organizational moral development model has 5 stages: amoral, legalistic, responsive, emerging ethical and ethical. The model explains the balance between profit and ethics. The model is shown in figure 1.

Stage 1 shows organizational structures that do not give importance to ethics, but only focus on profit maximization. There are organizational structures that place more importance on ethical issues at every subsequent stage. Stage 5 refers to businesses that adopt ethical principles (Reidenbach & Robin, 1991).

The General Theory of Marketing Ethics

This theory was developed by Hunt and Vitell (1986) to explain the decision making process in situations that entail ethical evaluation in marketing. It is also called as H-V model in the literature. The model was reorganized in 1993 and 2006. The final version of the model appears in figure 2. The model tries to explain what factors affect the behavior in the presence of an ethical problem.

This model aims to determine how individuals perceive alternative ways necessary to solve these issues after encountering moral issues (Vermillion, Lassar, & Winsor, 2002). In the model, it is seen that

Figure 1. The model of organizational moral development
Source: (Reidenbach & Robin, 1991)

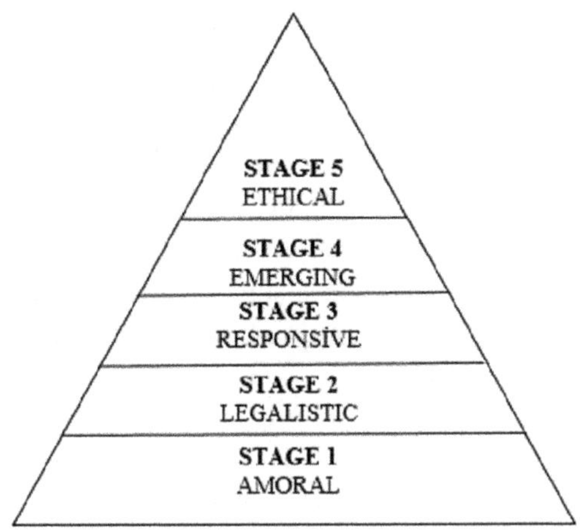

cultural environment, professional environment, industrial environment, organizational environment and personal characteristic variables have an indirect effect on teleological evaluations. However, these variables have a direct effect on perceived ethical problems, perceived alternatives, perceived consequences and deontological evaluations. Ethical evaluation emerges as a function of teleological and deontological evaluations. Theological evaluations decide whether it is morally right or wrong according to the result of a behavior. But deontological evaluations determine whether a behavior is morally correct, not ac-

Figure 2. The General Theory of Marketing Ethics
Source: (Hunt & Vitell, 2006)

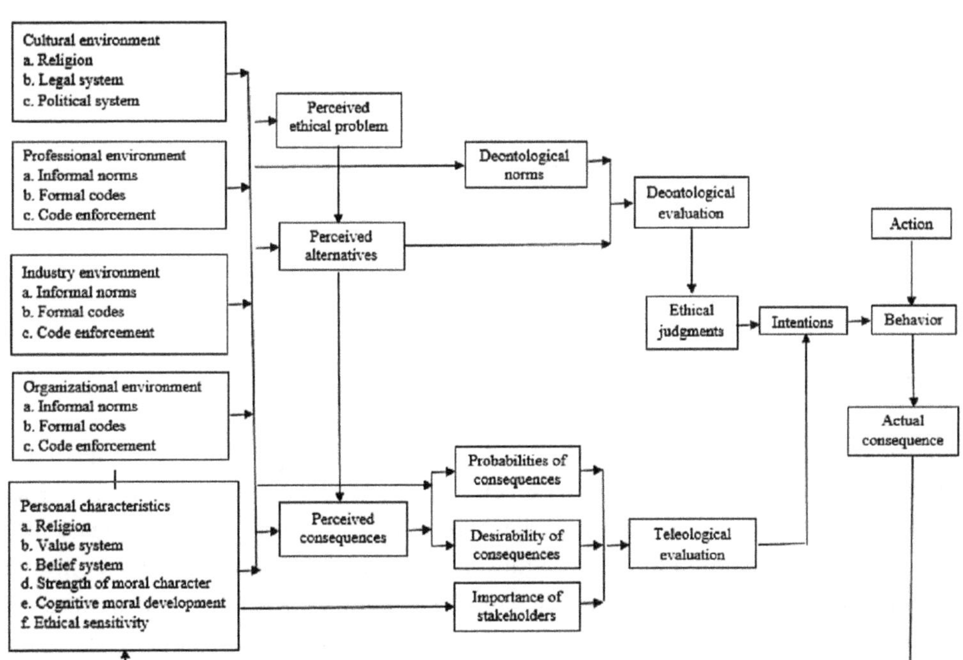

cording to the results, but according to its intention. In the model, it is seen that ethical and theological evaluations affect intention. Intention, action, control also affect behavior.

Contingency Model

The Contingency Model, proposed by Ferrell and Gresham in 1985, is another model developed to understand ethical decision making in marketing. The proposed model is multi-dimensional, process-oriented and probability type. The proposed model is multi-dimensional, process-oriented and probability type. As shown in Figure 3, the decision making process starts with the be awareness of the ethical problem or dilemma. In the model, personal and organizational variables seem to be two important factors that explain behavior (Ferrell & Gresham, 1985).

Figure 3. A Contingency Model of Desicion Making in a Marketing Ethics Organization
Source: (Ferrell and Gresham, 1985)

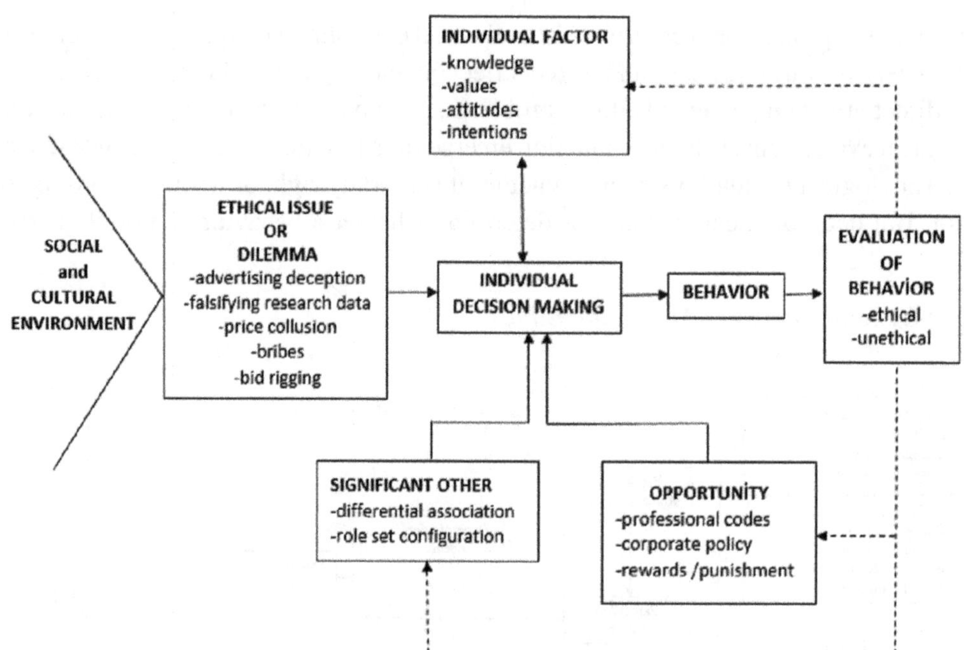

Synthesis Model

Developed in 1989 by Ferrell, Gresham and Fraedrich, the Synthesis Model combines various ethical decision making models with an integrated perspective. This model aims to explain the ethical decision making process of the enterprises more comprehensively. The Synthesis Model appears in Figure 4.

This model was created by combining the cognitive moral model of Kohlberg, the contingency model of Ferrell and Gresham, and the marketing ethics model of Hunt and Vitell. As with other ethical decision making models, the process begins with the awareness of the ethical problem. The second stage is

Figure 4. A Synthesis Integrated Model of Ethical Desicion Making in Business
Source: (Ferrell and Gresham, 1985)

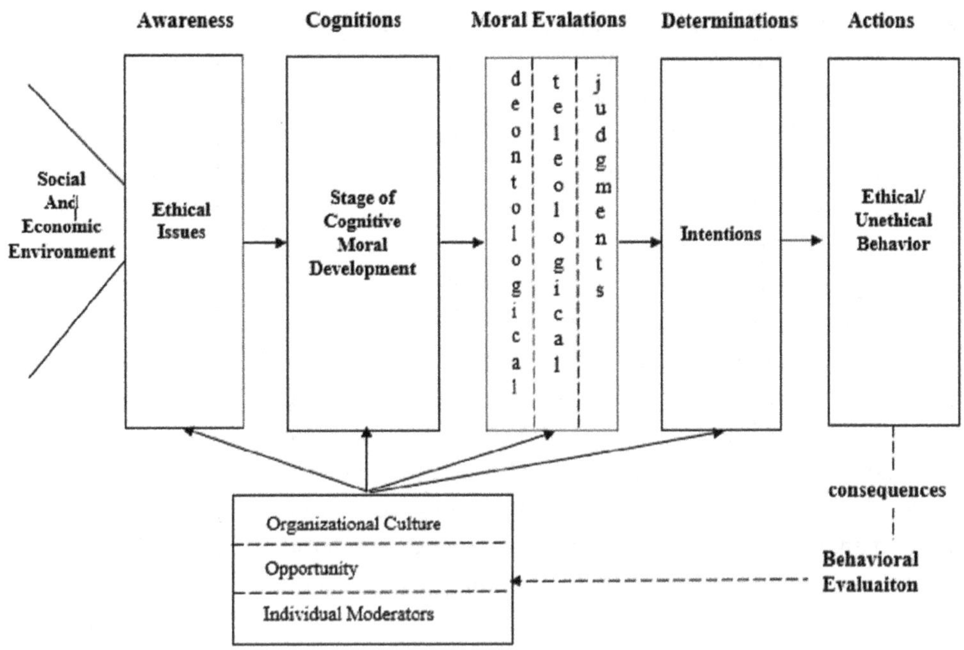

cognitions. This stage, based on Kohlberg's model, argues that each individual has a different level of moral development. It states that the moral development level of the individual directly affects ethical decision making (Jones, 1991). The third stage, according to the theory put forward by Hunt and Vitell, evaluates individuals morally through deontological approaches, teleological approaches and judgments. The fourth stage is determination. In the contingency model of Ferrell and Gresham, the effects of organizational culture, opportunity and individual factors on awareness, cognitions, moral evaluations, determination, action stages can be seen. The last stage of the model is action, that is, behavior (Ferrell, Gresham, & Fraedrich, 1989).

Digital Transformation of Marketing and Privacy Ethics

The developments in the internet and information technologies in the 1970s initiated the digital transformation process of marketing. This transformation started as a result of the developments in web 2.0 technologies. Many applications were developed during the Web 2.0 era, and computers, smartphones, etc. technological devices have become an indispensable part of human life. Developments in Web 3.0 technologies enabled human mind and machine learning to be done together. Cloud systems, artificial intelligence and augmented reality other technologies are developments in web 4.0 era. These developments in web technologies have radically changed marketing. Marketers moved their marketing activities from traditional to digital to gain competitive advantage.

Digital marketing can be explained as making traditional marketing strategies and decisions through digital channels, but it has significant advantages compared to traditional marketing (Bulunmaz, 2016). While traditional marketing allows one-way and non-interaction a communication process, digital marketing allows two-ways and an interactive communication process. Since the communication process has a dynamic structure, users can transfer information to the content of the websites and share the content with other users (O'Reilly, 2007). However, compared to traditional marketing, digital marketing has a lower cost, faster and updated structure (Wind & Mahajan, 2001). In addition, digital marketing has many metrics that allow the measurement and evaluation of marketing activities for the target audience. These metrics are: Margins and Profits, Product and Portfolio Management, Customer Profitability, Sales Force and Channel Management, Pricing Strategy, Promotion, Advertising Media and Web Metric, Marketing and Finance (Farris, Bendle, Pfeifer, & Reibstein, 2010). With the help of these metrics, marketing managers can monitor campaign processes, check whether targets are met and reorganize campaigns. In addition, Chaffey and Smith (2012) stated that digital marketing has significant advantages such as identifying, anticipating and satisfying compared to traditional marketing.

Another advantage of digital transformation of marketing is the increased efficiency of the marketing information system. PBS is the process of converting the obtained data into useful information for marketing managers and the internet provides more data for PBS (Marangoz, 2018). In order to carry out marketing activities, it is necessary to determine the target market, to create a marketing mix suitable for this target market and to update these decisions in accordance with the changing environmental conditions. Marketing managers constantly need information from both the internal and external environment to make the right marketing decisions. The Internet has provided a unique opportunity to access, collect and process this needed information into the database.

Figure 5. Framework of a Marketing Information System
Source: (Li, McLeod Jr, & Rogers, 2001)

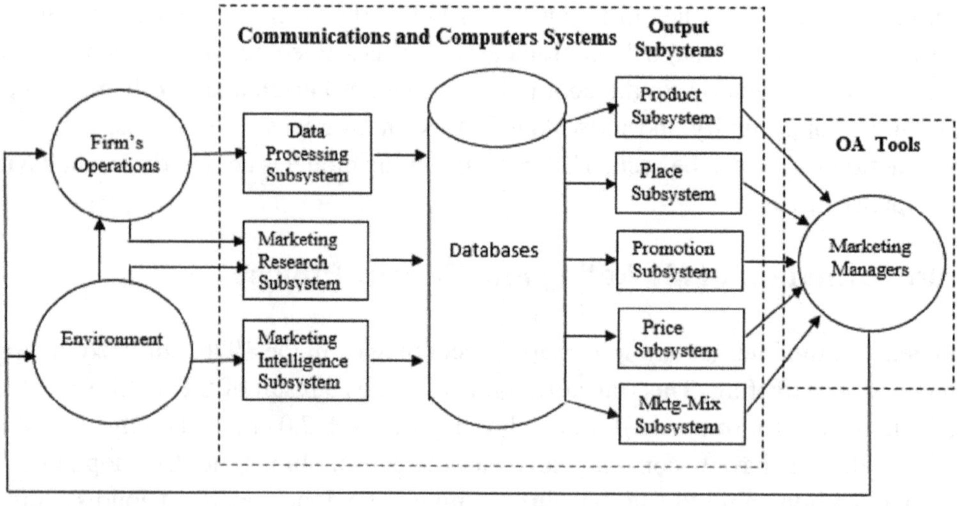

According to the Marketing Information System (PBS), there are three sub-input systems: Data processing, Marketing Research, Marketing Intelligence. Data processing subsystem is the internal accounting information system of the business. This system allows managers to easily follow their marketing activities. Marketing research subsystem is a system that collects data on certain topics to assist management's decision making process. Marketing intelligence subsystem provides data on external issues such as consumers, competitors, sales and government agencies. These data are then stored in databases. The data in the database is transformed into information through five output subsystems and sent to marketing managers. Marketing-Mix subsystem expresses the ability of the marketing mix to be integrated into the marketing strategy (Li, McLeod Jr, & Rogers, 2001; Venture, 2003). In summary, the digital transformation of marketing has simplified the marketers' process of collecting, processing and evaluating consumer data.

Chaffey (2016) has defined marketing as the management process responsible for profitably defining, predicting and meeting customer needs. This definition highlights the importance of accessing information about consumers. Consumer information is easily obtained through marketing activities carried out through digital channels. Because consumers need to share their personal information in order to make transactions online. For example, in order to make a purchase online, a consumer must first be a member of the website, complete the order steps and record their contact information (e-mail address, phone number, invoice and delivery address) in the system. It should also share credit card information as the last step in the purchase. This information, which is provided by consumers, is continuously collected, processed and used by marketers. Consumer information may be transmitted to third parties or shared with group companies in order to perform technical and logistics functions, or used for marketing applications of businesses with similar products / services. This situation leads to the emergence of ethical issues / dilemmas between marketers and consumers.

The issue of ethics in marketing started to be discussed in the 1990s. Ethical issues that emerged during traditional marketing methods were investigated (Murphy, 2002). Ethical issues related to the marketing mix (ethical issues related to product decisions, ethical issues related to pricing, public relations and ethical issues, ethical issues related to distribution, and ethical issues related to promotion) were frequently examined. The development of the internet and the start of marketing activities in digital environments led to the discussion of new ethical issues. Because the digital transformation of marketing has made the personal information of consumers easily accessible, collectable, processable, useable and transmittable. With the help of this personal information, marketers can produce in-depth insights about consumers' socio-demographic characteristics, lifestyles, interests, product preferences, brand likes and personal goals. However, when this information is not used carefully, it may create ethical issues for consumers. Misuse of online users' personal data can lead to unethical situations such as fraud and cyber attacks.

Richard O. Mason (1986) stated long ago that the information age would lead to various ethical issues. The growth of information technologies, the development of the internet and the digital transformation of marketing provided access to consumer data, the ability to store and transmit this data. However, this situation has led to the emergence of ethical issues regarding the privacy of individuals. Similarly, Milberg et al. (1995) mentioned the new ethical issues caused by the internet and technology in their study and stated that the privacy of personal information will be one of the most important ethical issues of the future. In fact, collecting personal information of consumers by marketers and tracking consumer behavior is nothing new. Because marketers have long been collecting and analyzing personal data of consumers to decide on the choice of the target market, marketing mix strategies and tactics. So, why privacy is an ethical issue or dilemma that is often discussed today. Caudill and Murphy (2000) explain this situation with the assumption of anonymity. According to the anonymity assumption, consumers are not concerned about privacy due to personal information collected through traditional marketing methods. Because consumers disclose their personal information under their control and publicly. However, continuous monitoring of consumers through digital channels, collecting their information (name, surname, phone number, e-mail address, credit card, etc.), storing this information indefinitely and selling it to other businesses without permission causes privacy issuess related to privacy (Bloom, Milne, & Adler, 1994).

Privacy means the right of an individual to keep confidential information they think is personal or sensitive (Rusk, 2014). According to another definition, privacy refers to the individual's ability to control the conditions in which the personal information is acquired and used (Culnan & Armstrong, 1999). The concept of Information privacy, which emerges with the development of the Internet, refers to when, how and to what extent the information collected about individuals, groups or institutions are transmitted to others (Malhotra, Kim, & Agarwal, 2004). In short, the fact that digital marketing activities allow easy access to personal information has raised privacy concerns. Privacy concerns indicate how concerned a user is about their privacy online (Hart ve Dinev, 2006). According to another definition, privacy concern is the desire to keep personal information away from others (O'Neil, 2001). At the same time, the values that individuals give to their personal information are different, so their privacy concerns are not the same. In other words, consumers have different levels of privacy concern (Cespedes & Smith, 1993; Caudill & Murphy, 2000; Malhotra, Kim, & Agarwal, 2004). Therefore, some consumers may perceive lower (higher) privacy concerns than others. Consumers with relatively lower privacy concerns are more likely to continue sharing information on digital media. Consumers with high privacy concerns are more likely to behave towards protecting their privacy. But surprisingly, consumers with high privacy concerns do not end their behavior of sharing information. Because consumers' information sharing improves and personalizes services, simplifies and accelerates online transactions. For this reason, consumers do not disable services that potentially threaten their privacy, although they have concerns about their personal information. This discrepancy situation is called the privacy paradox. The privacy paradox means that consumers continue to disclose or share information, even though they are concerned about their personal information.

Lutz and Strathoff tried to explain the privacy paradox with the rational choice approach. Accordingly, online users or consumers allow the collection of their personal data in exchange for benefits such as personalized service, discounts, coupons, promotions. On the other hand, consumers share their personal data with businesses and take on costs such as phishing, stalking, and cyberbullying that they cannot foresee. However, the benefits of sharing personal information online for the individual are generally more than their costs. However, evaluating the privacy paradox only in terms of rational choice approach causes to overlook the emotional aspects of behavior. In order to understand the privacy paradox better, emotional aspects of behavior should also be taken into consideration. Trust is a psychological factor that can positively affect the individual's intentions and behaviors. For this reason, trust is very important in understanding the privacy paradox (Lutz & Strathoff, 2014).

In the literature, the issue of privacy of consumers has often been studied along with the issue of trust. Consumers who think a business is reliable tend to share their personal information more. In this context, trust has a positive effect on consumers' intention to share information. But concerns about privacy have a negative impact on consumers' intention to share information. This is an important obstacle to the growth of e-commerce companies. Trust is an important predictor, which both reduces privacy concerns and increases the intention to use e-commerce sites (Gefen, Karahanna, & Straub, 2003). For this reason, privacy concerns, trust, intention, and behavioral relationships are frequently studied together in the literature (Malhotra, Kim, & Agarwal, 2004; Wu, Huang, Yen, & Popova, 2012; Bansal, Zahedi, & Gefen, 2016).

Another variable used in the literature to understand the issue of privacy is data breach experience. Experiencing situations where personal data is breached increases concerns about consumer privacy.

During the US elections in 2016, the fact that the data analytics company named Cambridge Analytica influenced the public by using the personal data of 50 million Facebook users without permission caused a major crisis. During this process, a large number of users supported the sharing made to delete Facebook. This privacy violation issue has caused Facebook's stocks to rapidly lose value within a week. Facebook founder Mark Zuckerberg apologized for the violation of users' personal data and stated that they took precautions (Cadwalladr & Graham-Harrison, 2018). However, news about the violation of user data frequently continued to be brought to the agenda in the media. Therefore, users believe that their private data may be exposed to potential abuse (Awad & Krishnan, 2006; Li, Wang, Li, & Che, 2016).

Anic et al. (2019) stated that there are studies in the literature that examine the issues of privacy of e-commerce businesses, but the conceptual frameworks developed for privacy concerns are insufficient. For this reason, they proposed and experimentally tested a model that shows the relationships between the antecedent and the consequents of privacy concerns. Accordingly, they claimed that online privacy concern was a function of privacy awareness, previos online experience, desire for information control, government online privacy regulation. However, research findings show that age, gender, and privacy awareness variables have little effect on online privacy concern. Perceived control and government regulations have a reducing effect on privacy concerns. Another finding is that privacy concerns will reduce consumers' online fabricating, withholding and sharing of personal information. However, online privacy concern directly and indirectly (through attitudes) negatively affects online purchasing behavior.

In the literature, the importance of government regulation and self-regulation of businesses is mentioned in the solution of problems related to privacy ethics. In other words, consumers consider businesses and government regulators responsible for the protection of their personal information. Because they have acquired responsibilities as the main parties of the exchange activities in marketing. Government regulations are critical to protecting consumer privacy. For this reason, governments are doing a lot of legal work to regulate the issues related to the privacy of consumers or internet users. The first efforts to protect the privacy of consumers began in the USA in 1973. Similarly, the European Union has made various legal arrangements to protect the personal data of consumers. In 2016, the law on protection of personal data was adopted in our country.

Miltgen and Smith (2015) developed a model to examine how individuals' knowledge levels of government regulations affect the decision-making process. This model explains attitudes and behavior relationships between government regulations and privacy concerns with the help of some variables. According to this research, perceived privacy regulatory protection increases as the level of knowledge consumers have about government regulations increases. Privacy regulatory protection and trust variables have a reducing effect on privcay risk concern. However, as the risk of privacy decreases, consumer protection behavior and regulatory preferences also decrease. It also concluded that the rewards perceived by consumers will lead to weakening of protection behavior and regulatory preferences (Miltgen & Smith, 2015).

It is not only businesses and the state that are responsible for protecting consumer privacy. Consumers are responsible for the protection of their personal information as another side of the exchange relationship experienced in marketing. Consumers can also take various measures to protect their personal information. These measures include changing privacy settings, tracking personal information withheld by businesses online, using false information, cleaning cookies to prevent tracking of online transactions, and requesting deletion of personal data from service providers. Consumers can also take various preventions to protect their personal information. These preventions include changing privacy settings, tracking personal information withheld by businesses online, using false information, cleaning cookies to prevent tracking of online transactions, and requesting deletion of personal data from service providers. Consumers should also have the ability to take this kind of prevention in order to prevent problems they may encounter regarding privacy ethics. Because it is possible that the consumers who have advanced internet use skills will increase their efforts to protect their privacy (Büchi, Just, & Latzer, 2016).

Researchers have made several studies to understand the issue of privacy, one of the important ethical issues of our time. In these studies, concerns about privacy were investigated for various antecedents and consequences. Especially, the fact that digital marketing activities accelerated the spreading of personal information caused privacy problems. Westin (1968), in his study *Privacy And Freedom*, evaluated the conflicts between the privacy of electronic devices and the perceived surveillance created by the society. It also noted the growing importance of government regulations on the protection of the privacy of individuals in the community (Westin, 1968).

One of the first studies investigating information privacy was done by Smith and others. In the study of Smith and others in 1996, they examined the issue of information privacy as a multi-dimensional structure. The first of these dimensions is the collection of personal information. In particular, the advances in information technologies and the widespread use of the Internet cause concerns in the public as it creates the perception that consumers are gathering too much personal data. İkincisi, internal unauthorized secondary use of personal information. This dimension expresses the concern that the personal information collected by a business to be used for a specific purpose is used again for a different purpose without permission. Third, external unauthorized secondary use of personal information. The concern is that personal information collected by a business for a specific purpose is shared with another business without permission and reused for a different purpose. Fourth, errors in personal information. Concerns about whether businesses have the ability to prevent mistakes that can be made as deliberate or indeliberate. Fifth, improper access to personal information. Consumers' concerns about whether personal data can be viewed or accessed by unauthorized people (Smith, Milberg, & Burke, 1996).

Malhotra et al. (2004) also worked to evaluate the privacy of internet users. They examined information privacy in terms of three factors (collection, awareness and control) in terms of social contract theory. According to this theory, privacy concerns are reduced as perceiving the existence of a fair situation between businesses and consumers. For this reason, they stated that consumers should be informed about the purpose of using personal data accumulated by businesses and consumers should have control over their data (Malhotra, Kim, & Agarwal, 2004).

Martin and Murphy (2017) have created a holistic framework for the understanding of data privacy in marketing by synthesizing consumer, organization, ethical and legal perspectives together. This holistic framework theoretically presents the antecedents and consequences that lead to data privacy in marketing. In the model, consumer antecedents (privacy concern, consumer vulnerability, individual differences, organizational privacy) are linked to consumer outcomes variables (purchase intention, disclosure willingness, click-through, falsifying information, negative WOM, switching). However, privacy enhancing (organizational trust, personalization value, consumer control, data transparency) has a reducing effect on consumer outcomes. Consumer Outcomes are also effective on organizational outcomes (firm performance, brand equity, customer relationships, privacy self –regulation, privacy as strategy). In addition, the privacy failure (data breach, identity theft, hyper -competitive actions) variables have a negative effect on both consumer outcomes and organizational outcomes.

SOLUTIONS AND RECOMMENDATIONS

The issue of how to prevent unethical issues that may occur as a result of marketing activities directly concerns the government, businesses and consumers. Because they have an acquired responsibility as the parties to the relationship of change in marketing. The frequent emergence of ethical issues related to consumer privacy in recent years has provided increased government regulations. Many countries make personal data protection laws to protect the privacy of consumers. In addition, the rapidly changing technological environment causes unexpected ethical issues. Therefore, in order to prevent unethical situations regarding consumer privacy, the changing technological environment should be constantly monitored. A system that regulates, updates and controls itself should be created. In other words, it is necessary to have a proactive perspective on the protection of privacy. Businesses must also comply with ethical standards set by the government. Before collecting information from consumers, they should ask for permission and provide information on how and where the information is used. Consumers should be able to access their personal information whenever they want and check its accuracy. It should also be able to request correction and deletion of personal data. Consumers should also be careful to share the information they think is sensitive and adjust their privacy settings.

FUTURE RESEARCH DIRECTIONS

Marketers access consumers' personal information through devices such as computers, tablets and smartphones. From these technological devices, smart phones allow to reach more personal information about consumers. Because devices such as computers and tablets can have more than one user, which negatively affects marketing decisions. For this reason, marketers organize various campaigns to direct consumers to mobile. Since the information coming from mobile usage gives access to more personal information than other smart devices (computer, tablet etc.), more precise information is obtained about consumer requests, preferences and purchasing behavior. The information that marketers obtain is not limited to e-mail addresses, phone numbers, credit cards, website visits, product reviews, past buying behavior.

Mobile usage offers consumers / users a more efficient and effective use through applications. Because applications are software developed for specific purposes, they simplify and speed up their online transactions. Therefore, consumers download a large number of mobile applications. However, consumers must accept certain permission requests before they can use mobile apps after downloading. When consumers accept these permission requests, they actually accept that businesses have access to their personal information. Thus, consumers who accept mobile application permissions share the control of many tools such as media, file, camera on their phones. This allows businesses and marketers to closely monitor consumers. But it increases the likelihood of unethical situations regarding privacy.

Consumers continue to use technological devices to take advantage of the ease and benefits provided by digital channels. However, the use of smartphones can pose serious threats to the privacy of consumers. Future research may examine consumers' use of mobile applications in terms of privacy ethics.

CONCLUSION

The digital transformation of marketing has led to the emergence of ethical issues related to consumer privacy. The fact that this transformation allowed the collection, processing, use and transmission of personal information about consumers raised privacy issues. In fact, collecting consumer data by marketers and withholding it in the marketing information system is nothing new. Since the personal information obtained through traditional marketing methods is made in line with the consumers' own controls and permissions, no privacy ethics issues have been experienced. Since the personal information obtained through traditional marketing methods are made in line with the consumers' own controls and permissions, privacy issues have not been caused. However, in digital marketing methods, the collection of personal information without the knowledge and permission of consumers has raised the issue of privacy ethics. Digital marketing has prepared an environment suitable for unethical situations related to the privacy of consumers. Today, it has made the issue of consumer privacy one of the frequently discussed ethical issues in marketing. In the near future, technological developments are likely to cause larger unethical issues regarding privacy. Therefore, the issue of consumer privacy attracts a lot of attention. In recent years, the number of studies on privacy ethics has been increasing. However, more research is needed to ensure that privacy is understood in marketing.

REFERENCES

Anic, I.-D., Škare, V., & Milaković, I. K. (2019). The determinants and effects of online privacy concerns in the context of ecommerce. *Electronic Commerce Research and Applications*, *36*, 1–10. doi:10.1016/j.elerap.2019.100868

Armstrong, G., Stewart, A., Denize, S., & Kotler, P. (2015). *Principles of Marketing*. Pearson.

Awad, N. F., & Krishnan, M. S. (2006). The Personalization Privacy Paradox: An Empirical Evaluation of Information Transparency and the Willingness to Be Profiled Online for Personalization. *Management Information Systems Quarterly*, *30*(1), 13–28. doi:10.2307/25148715

Ay, C., Kartal, B., & Nardalı, S. (2010). *Pazarlamada Etik Yaklaşımlar*. Beta.

Bansal, G., Zahedi, F. M., & Gefen, D. (2016). Do context and personality matter? Trust and privacy concerns in disclosing private information online. *Information & Management*, *53*(1), 1–21. doi:10.1016/j.im.2015.08.001

Bloom, P. N., Milne, G. R., & Adler, R. (1994). Avoiding Misuse of New Information Technologies: Legal and Societal Considerations. *Journal of Marketing*, *58*(1), 98–110. doi:10.1177/002224299405800108

Bowie, N. E., & Jamal, K. (2006). Privacy Rights on the Internet: Self-Regulation or Government Regulation? *Business Ethics Quarterly*, *16*(3), 323–342. doi:10.5840/beq200616340

Brinkmann, J. (2002). Business and marketing Ethics: Concepts, Approaches and Typologies. *Journal of Business Ethics*, *41*(1/2), 159–177. doi:10.1023/A:1021318710382

Büchi, M., Just, N., & Latzer, M. (2016). Caring Is Not Enough: The Importance of Internet Skills for Online Privacy Protection. *Information Communication and Society*, 1261–1278.

Bulunmaz, B. (2016). Gelişen Teknolojiyle Birlikte Değişen Pazarlama Yöntemleri ve Dijital Pazarlama. *TRT Akademi Dergisi*, *1*(2), 347–365.

Cadwalladr, C., & Graham-Harrison, E. (2018). *Revealed: 50 million Facebook profiles harvested for Cambridge Analytica in major data breach*. https://www.theguardian.com/news/2018/mar/17/all

Caudill, E. M., & Murphy, P. E. (2000). Consumer Online Privacy: Legal and Ethical Issues. *Journal of Public Policy & Marketing, 19*(1), 7-19.

Cespedes, F. V., & Smith, J. H. (1993). Database Marketing: New Rules for Policy and Practice. *Sloan Management Review*, *34*(4), 7–22.

Chaffey, D., & Ellis-Chadwick, F. (2016). *Digital Marketing Strategy ımplementation and Practice*. Pearson Education.

Chaffey, D., & Smith, P. (2012). *Emarketing Excellence, Planning and optimising your digital marketing*. Taylor & Francis.

Culnan, M. J. (2000). Protecting Privacy Online: Is Self-Regulation Working? *Journal of Public Policy & Marketing*, *19*(1), 20–26. doi:10.1509/jppm.19.1.20.16944

Culnan, M. J., & Armstrong, P. K. (1999). Information privacy concerns, procedural fairness, and impersonal trust: An empirical investigation. *Organization Science*, *10*(1), 104–115. doi:10.1287/orsc.10.1.104

Farris, P. F., Bendle, N. T., Pfeifer, P. E., & Reibstein, D. J. (2010). *Marketing Metrics:The definitive guide to measuring marketing performance*. Pearson Education.

Ferrell, O., & Gresham, L. (1985). A Contingency Framework for Understanding Ethical Decision Making in Marketing. *Understanding Ethical Decision Making in Marketing*, *49*, 87–96.

Ferrell, O. C., Gresham, L. G., & Fraedrich, J. (1989). *A Synthesis of Ethical Decision Models for Marketing*. Academic Press.

Gefen, D., Karahanna, E., & Straub, D. W. (2003). Trust and TAM in Online Shopping: An Integrated Model. *Management Information Systems Quarterly*, *27*(1), 51–90. doi:10.2307/30036519

Godin, S. (1999). *Permission Marketing: Turning Strangers Into Friends And Friends Into Customers*. Simon&Schuster.

Hart, P., & Dinev, T. (2006). An Extended Privacy Calculus Model for E-Commerce Transactions. *Information Systems Research*, 61–80.

HoofnagleC. J. (2001). Privacy Practices Below the Lowest Common Denominator: The Federal Trade Commission's Initial Application of Unfair and Deceptive Trade Practices Authority to Protect Consumer Privacy (1997-2000). https://papers.ssrn.com/sol3/papers.cfn?abstractid=507582

Hunt, S., & Vitell, S. (1986). The General Theory of Marketing Ethics. *Journal ofMacromarketing*, 1–12.

Jones, T. M. (1991). Ethical Decision Making by Individuals in Organizations: An Issue-Contingent Model. *Academy of Management Review*, *16*(2), 366–395. doi:10.5465/amr.1991.4278958

Krafft, M., Arden, C. M., & Verhoef, P. C. (2017). Permission Marketing and Privacy Concerns— Why Do Customers (Not) Grant Permissions? *Journal of Interactive Marketing*, *39*, 39–54. doi:10.1016/j.intmar.2017.03.001

Levitt, T. (1960). Marketing Myopia. *Harvard Business Review*, 45–56. PMID:15252891

Li, E. Y., McLeod, R. Jr, & Rogers, J. C. (2001). Marketing Information Systems in Fortune 500 Companies: A Longitudinal Analysis of 1980, 1990 and 2000. *Information & Management*, *38*(5), 307–322. doi:10.1016/S0378-7206(00)00073-2

Li, K., Wang, X., Li, K., & Che, J. (2016). Information privacy disclosure on social network sitess: An empirical investigation from social exchange perspective. *Nankai Business Review International*, *7*(3), 282–300. doi:10.1108/NBRI-02-2015-0005

Lutz, C., & Strathoff, P. (2014). Privacy Concerns and Online Behavior – Not so Paradoxical After All?Viewing the Privacy Paradox through different theoretical lenses. *SSRN Electronic Journal*, 81-99. doi:10.2139srn.2425132

Malhotra, N. K., Kim, S. S., & Agarwal, J. (2004). *Internet Users' Information Privacy Concerns (IUIPC): The Construct, the Scale, and a Causal Model.* Academic Press.

Marangoz, M. (2018). İnternette Pazarlama. *BETA Bulletin of Experimental Treatments for AIDS.*

Martin, K. D., & Murphy, P. E. (2017). The role of data privacy in marketing. *Journal of the Academy of Marketing Science*, *45*(2), 135–155. doi:10.100711747-016-0495-4

Mason, R. O. (1986). Four Ethical Issues of the Information Age. Management Information Systems Research Center, 5-12.

Milberg, S. J., Burke, S. J., Smith, H. J., & Kallman, E. A. (1995). Values, Personal Information, Privacy and Regulatory Approaches. *Communications of the ACM*, *38*(12), 65–74. doi:10.1145/219663.219683

Miltgen, C. L., & Smith, H. J. (2015). Makale adı: Exploring information privacy regulation, risks, trust, and behavior. *Information & Management*, *52*(6), 741–759. doi:10.1016/j.im.2015.06.006

Mucuk, İ. (2014). *Pazarlama İlkeleri*. Türkmen.

Murphy, P. B. (2002). *Marketing ethics at the millennium: Review, reflections, and recommendations. In The Blackwell Guide to Business.* Blackwell.

Nunan, D., & Domenico, M. D. (2012). Market research and the ethics of big data. *International Journal of Market Research*, *55*(4), 505–520. doi:10.2501/IJMR-2013-015

O'Neil, D. (2001). Analysis of Internet Users' Level of Online Privacy Concerns. *Social Science Computer Review*, *19*(1), 17–31. doi:10.1177/089443930101900103

O'Reilly, T. (2007). What Is Web 2.0: Design Patterns and Business Models for the Next Generation of Software. *Munich Personal RePEc Archive*, 17-37. https://mpra.ub.uni-muenchen.de/4578/

Prothero, A. (1990). Green Consumerism and the Societal Marketing Concept: Marketing Strategies for the 1990s. *Journal of Marketing Management*, *6*(2), 91. doi:10.1080/0267257X.1990.9964119

Reidenbach, R. E., & Robin, D. P. (1991). A Conceptual Model of Corporate A Conceptual Model of Corporate. *Journal of Business Ethics*, 273–284. doi:10.1007/BF00382966

Rest, J., Turiel, E., & Kohlberg, L. (1969). Level of moral development as a determinant of preference and comprehension of moral judgments made by others. *Journal of Personality, 37*(2), 225-252.

Rusk, J.-D. (2014). Trust and Decision Making in the Privacy Paradox. *Proceedings of the Southern Association for Information Systems Conference*, 1-6.

Sheth, J. N., Sisodia, R. S., & Sharma, A. (2000). The Antecedents and Consequences of Customer-Centric Marketing. *Journal of the Academy of Marketing Science*, *28*(1), 55–66. doi:10.1177/0092070300281006

Smith, A. F. (1978). Lawrence Kohlberg's cognitive stage theory of the development of moral judgment. *New Directions for Student Services*, 53-67.

Smith, J., Milberg, S. J., & Burke, S. J. (1996). Information Privacy: Measuring Individuals' Concerns about Organizational Practices. *Management Information Systems Research Center*, *20*(2), 167–196. doi:10.2307/249477

Venture, K. (2003). Pazarlama Araştırmaları Kapsamında Yaşanan Teknoloji-Tabanlı Değişim. *Ege Üniversitesi İktisadi ve İdari Bilimler Fakültesi*, 78-89.

Vermillion, L., Lassar, W., & Winsor, R. (2002). The Hunt-Vitell general theory of marketing ethics: Can it enhance our understanding of principal-agent relationships in channels of distribution? *Journal of Business Ethics*, *41*(3), 267–285. doi:10.1023/A:1021284922440

Vitell, S., & Hunt, S. (2006). The General Theory of Marketing Ethics: A Revision and Three Questions. *Journal of Macromarketing*, *26*(2), 1–11.

Westin, A. F. (1968). Privacy And Freedom. *Washington and Lee Law Review*, *25*(1), 166–170.

Wind, J., & Mahajan, V. (2001). *Digital MarketingGlobal Strategies from the World's Leading Experts*. Wiley.

Wu, K.-W., Huang, S. Y., Yen, D. C., & Popova, I. (2012). The effect of online privacy policy on consumer privacy concern and trust. *Computers in Human Behavior*, *28*(3), 889–897. doi:10.1016/j.chb.2011.12.008

ADDITIONAL READING

Bhatia, V. (2020). Drivers and barriers of permission-based marketing. *Journal of Research in Interactive Marketing*, *14*(1), 51–70. doi:10.1108/JRIM-07-2018-0088

Bornschein, R., Schmidt, L., & Maier, E. (2020). The Effect of Consumers' Perceived Power and Risk in Digital Information Privacy: The Example of Cookie Notices. *Journal of Public Policy & Marketing*, *39*(2), 135–154. doi:10.1177/0743915620902143

Foxman, E. R., & Kilcoyne, P. (1993). Information Technology, Marketing Practice, and Consumer Privacy: Ethical Issues. *Journal of Public Policy & Marketing*, *12*(1), 106–119. doi:10.1177/074391569501200111

Hunter, G. L., & Taylor, S. A. (2018). The relationship between preference for privacy and social media usage. *Journal of Consumer Marketing*, 43–54.

Martin, K. D., Borah, A., & Palmatier, R. W. (2017). Data Privacy: Effects on Customer and Firm Performance. *Journal of Marketing*, *81*(1), 36–58. doi:10.1509/jm.15.0497

Milne, G. R. (2000). Privacy and Ethical Issues in Database/Interactive Marketing and Public Policy: A Research Framework and Overview of the Special Issue. *Journal of Public Policy & Marketing*, *19*(1), 1–6. doi:10.1509/jppm.19.1.1.16934

Phelps, J. E., D'Souza, G., & Nowak, G. J. (2001). Antecedents and consequences of consumer privacy concerns: An empirical investigation. *Journal of Interactive Marketing*, *15*(4), 2–17. doi:10.1002/dir.1019

Xu, H., Luo, X. R., Carroll, J. M., & Rosson, M. B. (2011). The personalization privacy paradox: An exploratory study of decision making process for location-aware marketing. *Journal of Research in Interactive Marketing*, 42–52.

KEY TERMS AND DEFINITIONS

Digital Marketing: It is all of the marketing activities carried out with technological devices for the promotion, awareness and sales of products/services.

Marketing Ethics: It means that marketing activities are realized within the framework of moral and ethical values and by considering the interests of the society.

Personal Data: It refers to all kinds of information (name, phone number, picture, image, sound records, shopping habits, purchase background, etc.) about a person.

Privacy: It is the special informations that an individual wants to keep secret from others.

Privacy Concern: It refers to the worry that the personal data of individuals may be used by others in an unpredictable way.

Privacy Paradox: It refers to individuals' continuing to share their personal information despite potential privacy issues that may arise.

Privacy Protection: Precautions that can be taken against privacy issues.

Trust Towards Websites: It is the sense of confidence that individuals have in websites (e-commerce sites, social networking sites, video sharing sites, etc.).

Chapter 10
Social Media Short Video-Sharing TikTok Application and Ethics:
Data Privacy and Addiction Issues

Kevser Zeynep Meral
https://orcid.org/0000-0002-2514-8744
İstanbul Bahçeşehir University, Turkey

ABSTRACT

With 3.8 billion users, social media created ethical problems as well. The Cambridge Analytica scandal has been a serious issue with data security lately. Contents not in compliance with general moral rules is another important violation of ethics. TikTok application, the fastest-rising short video-sharing website, is examined. As a result of the literature review, it is observed that TikTok application also had ethical violations issues like lack of private data safety, not sufficient precautionary system barriers for the young generation, and addiction risk. Furthermore, considering that the young people can interact with malicious users through the fake accounts and the risk of sharing their exceptional videos to have more viewers, it is suggested that the sanctions should be arranged as a deterrent in violation of the rules. Parents and young users must be educated about the risks and ethical violations of social media.

INTRODUCTION

With the number of social media users increasing to 3.8 billion, and users spending more than one third of their time in social media created ethical problems as well. In particular, the lack of security in keeping private data of social media users, furthermore selling users private date has been a major issue with Cambridge Analytica scandal lately. With the expansion of social media, the companies using artificial intelligence, started using tailored advertisements which ended up exposure of users to excessive advertisements was another ethical violation. Contents not in compliance with general moral rules is among the most important ethical violations which causes problems especially for the young

DOI: 10.4018/978-1-7998-4117-3.ch010

generation. In this study, TikTok application, the fastest rising short video-sharing website of social media is examined. TikTok, a short video-sharing platform with 800 million active users, although still behind WhatsApp, Facebook, WeChat and Instagram, and other social platforms will be examined with literature review. Although has been expanding tremendously, there are few researches about TikTok Short Video-Sharing application.

According to Digital 2019 report, social media is mostly used between 18-35 age group and least used above 55 age group, which shows that social media is used mostly by young age group. In Turkey, according to a research conducted by Ministry of Youth and Sports (Bulut, 2013), one in three young person spends minimum 3 hours in social media and 86 percent of young population enters social media minimum once per day, 72 percent of them enter social media minimum a few times per day. Social media user numbers according to the networks are Facebook 2 billion 271 million, Youtube 1 billion 900 million, Instagram 1 billion, TikTok 500 million, Twitter 326 million, Linkedin 303 million, Pinterest 250 million.

SOCIAL MEDIA CONCEPTUAL BACKGROUND

Social media generally covers the social network sites like Facebook, blogs created by users like blogspot, video-sharing sites like TikTok, Youtube, collabarative sites like Wiki where contributors share knowledge, business networking sites like Linkedin, business communities like Amazon and academic social networks ResearchGate or Academia (Emirza et al.2012). Social media includes interactive on-line communication channels where communities are connected with each other and developed by the users participation (Kalafatoğlu, 2011). Social media web sitesi are internet-based application in the cyber world providing users exchangeble contents with web based technologic infrastructure (Kaplan and Haenlein, 2010). In other words social media is created by blogs, discussion boards of companies, email, forums blogs consisting extensive and bilateral interaction based on online networks (Mangold and Faulds, 2009).

According to Mayfield (2008), social media covers communities of common interests like photography etc which enables them sharing comments, information and communicate with other people, other web sites or resources. Safko (2009) defines social media as web based application where people or communities share their behaviours, ideas, relationships and activities by words, photos, sounds and videos. The difference of social media from the traditional media is that it enables comments or additions to the posts. Safko in his research found that more than 70 percent of 600 participants were not familiar with social media however the majority could define it. Social media apart from being a communication platform, it is a mass communication network with campaigns and an important role even in radical social changes issues like Arab Spring (Meral and Meral, 2020). Schreck and Keim (2013), state that social media are all type of media formats where user groups create, share and develop knowledge via web network connections.

According to a social media specialist, digital strategy manager, Michael Fruchter (Fruchter, 2009), who worked in Dairy Queen social media campaign which can be considered as a case study in social media. Dairy Queen company is an ice cream company established in 1940, and is a leader in soft service and fast food restaurant industry with local and international shops more than in twenty countries, is a globally wellknown social brand. With social media when everything changed from traditional advertising to social media in 2008, Dairy Queen have changed as well by entering social media world.

They started to negotiate with their customers, with a new technology which connected them to their customers not by only giving a free five dollar coupon at the same time by using facebook fan page to like their page, by treating them a five dollar free give away social media enables them to remind and share their childhood/family days. Social media reduced their costs, and enabled customers to share their opinions as well.

Evans (2008) stated that the social media is that the users integrate on the common point, learn and share information. Today, social media is expanded as covering all social processes and at the same all social relationships, for this reason social media is along with contribution to individuals cultural, social, art related, economic, political aspects, it is at the same time contributes to corporate activities as well, by enabling cheaper and quicker delivery of messages and news to their target audiences.

Baruah (2012), added that social media with its extended users numbers and usage area, provides a virtual environment, which changed the communication systems, styles and changed the quality of the social processes as well. A communication platform provides vital and professional changes and opportunities. In this context, social media is an integral part of corporates and their strategies and applications (Peters, et al. 2013). Social media with its increased user number, global expansion contributes to production process as well with the shared contents, application, information at the same time (Kaplan and Haenlein, 2010). Barabasi (2003) argues that all members are being a member of a worldwide social net and nobody will be left out of this group. We do not know everybody in the world however it is possible that two different person in same social web group has a relation between each other.

Solis (2010), claims that the social media with the culture and lifestyle improvements, keeps the social renaissance live in other words social media is a tool representing the live ecosystem. Social media by communication and collaboration between relatives, friends and active people, keeps the people together by posting stories of human and with affection, words, pictures, videos, experience, observation, news, understanding, opportunity and privileges. Social media at the same is seen as a process of converting individuals from content consumers to content sharers, in other words it is seen as democratization of knowledge.

SOCIAL MEDIA AND ETHICS

The main ethical problems of social media are given as follows.

Privacy of Personal Life

Privacy of personal life is the general living space of the individual includes public living spaces that everyone can share with everyone. It includes the parts of life you share with certain people and qualities required for the private living space. The right to privacy in the "human dignity" environment is violated in various ways on social media (Araslı, 1979)

Failure to Secure Personal Data

Personal data means any information that shows the personal, professional and family characteristics of the individual and that is suitable for distinguishing the individual from other individuals and revealing their qualities (Şen, 2014). Protection of personal data covers the collection, storage and use of personal data and all stages of the data processing process consisting of data matching. Information shared in social media can create a database of personal information. Increasing experience in social media environments is the most important development that reveals the need to ensure the security of personal data (Binark and Bayraktutan 2013).

Failure to Show the Original Source of the Content and Copyright / Patent Rights

Although "copy-paste" is primarily seen in internet journalism and journalism, it can be seen in all areas with the increase in dynamic social media content, types and numbers. Undoubtedly, a site broadcasting on the internet and social media may publish information on another site in order to increase its circulation. However, the publication of the source producing the original production from which the information in question is not cited is the most common form of plagiarism in social media. Failure to show the original source of the content raises the issue of copyright / patent rights. (Castells 2009).

Dissemination of the Content Produced Before Accuracy

In terms of processing information on social media, the accuracy of both news and information must be checked, in other words, editorial control is required to produce content that is both publicly available and distinctive from advertising content or promotional content. (Deuze and Dimoudi 2002).

News and Advertising Interwine in Social Media

The main sources of income for websites and online news applications are advertising revenues, just like television broadcasting. However, with the characteristics of the social media, it is possible to reach the specific individuals or groups whose specific features can be defined, in order to attract the attention of users to certain advertisements, it is observed that the advertisements are mostly presented as if they were news. (Binark and Bayraktutan 2013).

Digital Surveillance and Data Mining

Digitization and surveillance as a result of modernity and has addressed two different meanings of surveillance and surveillance within modernism: First, surveillance is encrypted knowledge. It is about the coding and counting of individuals by turning them into objects. The information collected is not just a simple collection of information, as this information collected includes a certain classification and differentiation process, it is a qualified information storage process (Giddens, 2008). Companies not only collect personal information of users during shopping, but also develop marketing strategies according to these users by monitoring which sites users enter, what they like on social networks, what topics they watch, what words they use in their e-mail. Companies do not only use this collected user data for themselves, but also sell it to other companies without the user's knowledge and consent (Arslantaş et al. 2012).

Exposure of the User to Intensive Advertising / Misleading Labeling and Titling to Attract User

In social media, the user is constantly exposed to advertising content and has to watch the advertisement in order to reach the information or to reach the page. The main sources of income for commercial websites are advertisements. In order to achieve this goal, mostly intentional and sensational tagging and capping are used in social media. Thus, social media users surfing the internet open the link by wondering about the sensational title or tag it encounters and display a large number of ads on their computer while learning what the content is on that page (Binark and Bayraktutan, 2013).

Hate Speech

The widespread use of hate speech in circulation prepares the ground for hate crimes. It is stated that there is a link between OSCE ODIHR (Democratic Institutions and Human Rights Unit), hate crime and hate speech. He underlines that hate-based violence often occurs in hate speech, although it is difficult to prove that it is a direct link. ODIHR emphasizes that organized hate groups express their opinions through various publications, popular music, the Internet and shows, and that these actions are aimed at normalizing and legitimating hate speech and hate crimes (ODIHR 2009: 53). Hate speech types are generally classified under six headings (Binark and Çomu 2012):

1. Political hate speech, targeting a political thought and its followers.
2. Hate speech against women, starting from sexist expressions, it can be discussed in more advanced dimensions. They are often not revealed in the form of open hate. However, it is seen as criticism of what women do or do not.
3. Hate speech against foreigners and immigrants, can target foreigners, immigrants and / or ethnic groups.
4. Hate speech caused by sexual identity, hate speech targeting people with different sexual identities and sexual identity orientation.
5. Hate speech of faith and sect origin, can be produced for people or groups with different religions and sects.
6. Hate speech for disabled and various diseases, it is produced for people with physical or mental disabilities or people with certain diseases.

Creating and Disseminating Content Against General Morality

Mavnacıoğlu 2009: 64), content against general morality is among the most important ethical violations that both children and young people can negatively affect their physical, psychological and sexual development and that are incompatible with general moral rules.

Anonymity and Trol (Spam Blog)

Anonymity can also be called a user with a fake ID. It is possible for users to create a fake identity, open an account on social media, freely exchange ideas, enter discussions, and support various movements or organizations. Fake identity users, other users' personality etc. violating their rights also appears to be violating the ethical rule.

Troll tries to sow seeds on the Internet to delight people or start a dispute; In online communities, such as forums, blogs, chat rooms, they are people who intentionally provoke readers to send emotional responses or distribute the topic of a topic (tr.wikipedia.org, 2020). This issue is directly related to user-derived content production. The troll phenomenon is the same as sending spam blog. The fact that internet users create content and share it with other users in order to make money instead of expressing themselves, sharing their professional experience, knowledge and opinions is one of the behaviors that contradict the purpose of social media. It is seen that blogs called "spam blogs" have been created solely for the purpose of advertising. The fact that users make monetization as a primary goal renders the content created ethically problematic (Mavnacıoğlu 2009).

Language

This new language created with social media, which is almost impossible to follow due to its dependence on technological progress, has emerged in a pluralistic but uncontrolled and unpredictable medium as individuals become both a source of production and consumption. Any vehicle itself is not good or bad. The attitude of the human element determines the function of the article. For example, it is not television or the internet that makes people victim; bad television or bad internet use (Küçükyılmaz, 2013).

Suggestions Against Ethical Violations in Social Media

The 10 Golden Rules on Computer Ethics define the ethics rules of internet rules (Computer Ethics, 2020). The major golden rules is "Definitely, don't use the computer to harm others." and the second rule is "Definitely, don't write a program without thinking about its social content."

According to a common view, the internet has negative effects on people's social lives. In an environment different from the real world, "in the virtual world, human relations are changing. In most cases, people who interact with each other like an internet address, an e-mail address, etc. On the other hand, the internet also removes many features that may be important in the real world, such as distance between people, age, gender, race, culture, etc. Usage such as on-site shopping, on-site banking, even working from home without going to work etc. are factors that can affect the social life of people (Erol, 2009).

Responsibilities fall to all users, content providers and regulatory boards against ethical violations. Unethical behaviors occur among individuals, even if individuals do not realize, because it works socially with the logic of social media, chat and gossip. Content providers should implement the following practices to prevent unethical behavior:

- A more effective identity and data verification system should be applied for content creators, with the filter system, fake profiles should be prevented.
- In line with data privacy, restrictions should be placed on the copying of data in common sharing.
- In addition to the user contract, the binding ethics contract should be implemented.

Regulatory boards should implement the following practices to prevent unethical behavior (Mavnacıoğlu 2009).

Furthermore,

- Taking into account the unique nature of social media, legal and ethical limits regarding content sharing and creation should be determined.
- Collaboration with international content providers on combating unethical and harmful content.
- As a result, all users should not be punished for unethical content, only the user who performs that unethical behavior should be held responsible. Regardless, the main responsibility lies with the user. Audit or measures to be taken by the content provider or regulatory boards will not be more effective than the self-control of the user. The important thing is that social media users become aware of the dangers and risks that await them and their children in these environments. In order to provide self-control against ethical violations in the social media environments of the user, media literacy implemented for traditional media should be implemented for the new media and social media within the scope (Binark ve Bayraktutan 2013).

For ethical rules, a text is prepared by the Turkish Informatics Foundation under the name Internet ethics "Internet protocol" in Turkey. The text of Internet protocols of TBV is as follows: respect for others, formal care, care about content, other topics (Türkiye Bilişim Vakfı, 2009).

SHORT VIDEO-SHARING PLATFORMS AND 'TIK TOK' APPLICATION

Short video platforms are very popular among young generation including Tiktok (Patrick, 2018). TikTok, the most popular short video platform is defined as that their mission is to bring joy and inspire creativity, and be the leader for short mobile videos (Tiktok.com, 2020). The TikTok previously known as Musical.ly, is a video-sharing application. The Chinese ByteDance company has bought Musical.ly in 2017 and renamed as TikTok. It has increased its user number incredibly in 2019 (Statista.com, 2020).

TikTok's name is Douyin in China and very popular. 57 percent of Douying users are mobile internet users as of 2019, with nearly 470 millions user in 2020. TikTok i.e. Douying in China is used to share daily life of Chinese users. Worldwide video-sharing users via TikTok Android devices are more than 167 million monthly. Via iOS devices nearly 8,2 million Daily users. users In other words, TikTok is one of the most downloaded entertainment apps after WhatsApp and Facebook Messeger in 2019. India TikTok users downloaded 18.7 million times, in December 2019. The short 15-second video-sharing platform TikTok is developed by Bytedance company located in China. TikTok's firs time installment numbers were 738 million times in 2019 which is more than 655 million times first time downloads in 2018.

According to Priori Data (2020) it has been already downloaded 41 million times in US only in ten months of 2019, and with India total downloads reached to 190 million. The growth rate has been incredible in 2019, by country Romania has increased 1,508 percent in December, 2019 compared to November 2019. The downloaded numbers can be seen in the following figure where India is number one with 190,6 million downloads, followed with USA with 41 million downloads and Turkey ranked as number 3 with 23.2 million downloads, followed with Russia 19.9 million and Pakistan with 19.5 million downloads as of end of October, 2019.

TikTok U.S. statistics as of January 2020 can be summarized as;

- 46.9 percent (Android) app owners accessed Daily to TikTok, 78.4 percent of them were monthly active users.

- User ratio, by frequency, 27.5 percent of TikTok were heavy users, using the app more than 2/3 of a month, 36 percent of (Android owners) were light users of photo sharing.
- 56.5 percent of active user were female, 43.5 percent male (on the Android platform).
- 37.2 million users in 2019, increased by 21.9 percent and increased to 45.4 million users in 2020.

Furthermore, Priori Data, reports growth rate of TikTok as France is the number one as being the fastest-growing TikTok on downloading video-sharing app with 120 percent increased compared to November 2019, in December, 2019, with 790 thousand times download of TikTok app via Google Play. USA had 4,6 million downloads of TikTok video-sharing in December 2019 via iOS user. The most remarkable increase about TikTok statistics is first time installations per quarter between 2016-2019, of the short 15-second video-sharing application TikTok is the most downloaded application in the World following WhatsApp and Facebook Messenger in 2019. TikTok short video-sharing application was downloaded 219 million times first time in the last three months of 2019, increased its downloading 13 percent more than its 655 million first time installation in 2018.

TikTok was the third in the World with over 1.5 billion downloads in 2019 (Priori Data, 2020). Being one of the fastest-growing social media apps TikTok is very popular especially with young users in U.S by engagement as well, in September 2019, Facebook was used 769,16 minutes decreased 26 percent compared to 2017 where Instagram increased 6 percent during the same period, however TikTok was the winner by increasing 1.533 percent during the same time. The reasons of this incredible increase are first of all downloading is free, easy to use, upon downloading you can start using the short video-sharing application immediately (Beutel, 2020).

When it comes to revenues, Priori Data reports that TikTok app 75 percent of revenues totalled upto 2,2 million dollars in US from iOS and 53 percent of revenues totalling 294 thousand dollars from Google Play in December 2019. TikTok users spent175 million U.S. dollars in October 2019. The TikTok App generated 18.2 million U.S. dollars from users spending in October 2019.

TikTok short video-sharing is available globally, in 34 language and widely used not only in Asia, it is widely Europe and U.S. The application has a pop-up alert to remind the user if the user did not enter the app for 24 hours. With easy access, the short videos, which can be personalized, with rich content, music and interactive structure, (PR Newswire, 2018), users numbers increased and hit 500 million users all over the world with 300 million in China in 2018 (Jon, 2018).

REASONS OF USING TIKTOK AND SOCIAL MEDIA

There is very limited research about TikTok. De Veirman et al, (2019) says even no studies yet focused on TikTok.

TikTok has changed video-sharing in internet which was created by a limited number of producers to an unlimited creators (Khan, 2016). However this change has made it difficult to differentiate the difference between the consumers and producers which enabled consumers to use networking sites for their own motives, Cooke and Buckley, (2008). If a created video is shared and becomes viral, it is considered as successful in online environment (Alhabash and McAlister, 2012), and fan size of online video is considered like social capital in offline (Feroz Khan and Vong, 2012).

Researches show that the majority of the people in social media are passive users, that is they just read, follow and browse (Preece, Nonnecke, Andrews, 2004). These passive followers are defined as users who don't post any message in online websites (Takahashi, Fujimoto and N. Yamasaki, 2003) they are like consumers who show consumption behavior (Khan, 2017). According to Shao (2009), user generated media cover participating users by liking, adding, sharing, posting, commenting and producing by creating and sharing created content and at the same time cover passive consuming users who browse, watch, read and view the contents.

The research about reasons of using social media, focused on social and psychologic factors (Chen, 2011), (Correa, Hinsley, and Zuniga, 1977). Some researchers focused on personality traits effects about using social media (Butt and Phillips, 2008), (Seidman, 2013), (Tuten, Bosnjak, 2001), (Wang, 2014) and (Khan, 2017). Other researchs, about user generated media have focused and identified audience reasons of using social media via social network websites (Smock et al, 2011), about social media (Whiting and Williams, 2016), Youtube (Chiang, Hsiao, 2015) and Instagram (Eunji and L. Jung-ah, 2015).

Results show different reasons for using social media like Youtube using reasons were that the user's self-expression affected their sharing behaviour (Chiang and Hsiao, 2015). Caplan, (2003), argues that users fulfill their communication and interaction needs via social media, wherelse, Sung, Lee, Kim, and Choi, (2016) proposed that social media postings enable users to remember their special occasions and they can go back to their memories via archived posts. Other researchers suggest that via creating their own videos users are self-expressing themselves, (Gibbs, Ellison, Heino, 2006), (Kramer, Chung, 2011), (Papacharissi, 2002). Other users create and share videos to show themselves, to draw attention of others and to have connections supporting them (Dominick, 1999), (Smith, 1998), (Walker, 2000). Whiting and Williams (2016), claim that users want to runaway from their daily pressures via using social media.

Omar and Dequan (2020) have argued that TikTok users like other social media users create and share short video with the same reasons like other social media users, their findings show that they use TikTok to run away from daily pressures, for social interaction, for archiving purposes and to express themselves.

SHORT VIDEO-SHARING (TIKTOK) AND DATA PRIVACY

With technology and globalization, children in digital era, not knowing how life was without internet and smart technology, childhood is changing as well, where it is described as innocence, dependency and participation (Beutell, 2020). Pew Research Center (2018) reports that 95 percent of the children participating the survey between the age group of 13-17 had a smart phone or were able to access smart phone and 45 percent of them were constantly online, enabling information very easy to access. Although mobile phones were available among young population, and children users accessing smart phones increased, there is very few research about their effect on young age group and their social development (Tahir, Husin, 2017). Prout, reported that childhood concept of today is now disappearing because the boundary between childhood and adulthood is blurring due to TV and internet's easy access to information. Children are exposed to adult information very easily (Prout, 2005). Children and social media literature can not keep update with popular social media applications expolisions and the changing environment with the ongoing changing environment of online social media (Grand-Clement, 2017).

The 'digital media literacy' concept has emerged with internet (Untila Kaplan, 2019), especially with personal data sharing. Data privacy is very important especially in social media. While using social media platforms, private data is available for the app and users are more concerned about data privacy especially after Cambridge Analytica scandal. Facebook users are more careful with their posts and they intent to use network less, in other words behavior of Facebook users have changed, after private data treatment of Facebook, known as Cambridge Analytica scandal in 2018 (Statistista.com, 2019), in a survey in April 2018, only 20 percent of the participants declared that they would use the app as in the past. The scandal was about Facebook's enabling fake news spreading in 2016 election and that Facebook users all over the world data were used without millions of users' consent. Statista.com (2020) also stated that TikTok App owner (Chinese ByteDance) is also under investigation in U.S. for users' data collection and cencorship suspection.

In 2019, TikTok has paid 5.7 million fine for collecting children's data under 13 (Cnn.com, 2019). TikTok app owner company Bytedance has not paid fines only in U.S. has paid fines for pornographic content and fake advertisements, in China as well. India also banned (lifted after improving safety) TikTok from its stored in April, 2019 for illicit content pornography encouragement, with a warning from the court about exposing children to predators, TikTok had to withdraw 6 million videos from its platform. Indonesia blocked TikTok app. in July, 2018, for pornographic content (Livemint.com, 2019).

However, with all these lawsuits, fines paid, TikTok is increases users among teenagers, TikTok reported that they have exceeded 1 billion downloads in February 2019 (Yeh, 2019).

SHORT VIDEO-SHARING AND ADDICTION RISK

Kumar (2019), reported that with social media revolution, the young age group of millennial generation, used social media to express themselves and to get in touch with their age group, set up theigr profiles, in social media, different than themselves, by creating a 'virtual identity' of themselves, of where the boundary might not be very clear between reality and fantasy (Turkle, 1995). He drew attention to TikTok, where they might be attracted to ego-enhancing activities (Wang et al, 2012) claimed that the sudden increase in users number of TikTok also might be of young age group's wish to imitate the celebrity's lifestyle, and not theigr true personalities, in their selfies (Back et al. 2010). Just to draw more attention, Barry et al (2017) argued that highly insecure personalities are more likely to share sexualized photographs, to contruct themselves a a desirable person. Additionally, psychological–neurobiological models (Brand et al. 2016) reported that addiction to social media network sites with their interactive structure, provokes the addicted adolescent's desire to be viewed by larger audience to share very sensitive, private short videos. The initial starting point entering the platform where the adolescent escapes from real world's problems, can become focus of the society's reflections. He concluded that social media is like an ocean where the individuals meet problems starting with solution of psychologic problems to addiction to social network platforms. The users of TikTok spend average of 52 minutes per day and this covers like approximately 200 short videos (Smolentceva, 2019).

Although short video-sharing application has become very popular, with increasing very high percent in users number, the excessive using the platform is once of the main concerns of the public. Despite the negative results of excessive using results in spending quite a lot of time by using the application might be accepted, for example, 22 percent of the current 150 million daily active current users who access to the platform average 4,7 times per day, spend time more than one hour per day, (Biznext, 2018).

Using internet and video game playing are among addiction behaviours (Young, 1999). The excessive internet using may cause many problems, i.e. addiction indicators like including time management problems, concentration and it might even lead to less learning time (Hong et al., 2014), (Gao et al., 2017). Facebook addiction which is an Internet addiction type, short video application can be another Internet addiction type because it is also like the Internet addiction which causes the person not being able to control using Internet which creats psychological, work, school difficulties by causing functional impairment and distress (Burnay et al., 2015). Furtlermore, internet addiction have negative effects like mental health problems, learning difficulties, low performance and poor interpersonal relations (Ho et al., 2017). Actually it is easy to get lost in short form videos because TikTok enables users to use special effect filters, video editing tools to enrich their videos and funny stickers. TikTok, enabling users to create their short video contents very easily, at the same time by using AI (Artificial Intelligence) to catch user's interest. TikTok make correct predictions by analyzing users' behavior (Knowledge, 2019). As a result, social media may be used to fill people's psychological gaps my end up with social media addiction (Blackwell et al., 2017; Gao et al., 2017).

Actually attachment is caring and emotional attachment between members of the online web site members (Ren et al., 2012). Furthermore, studies show that addiction to social media is related with people's personal characteristics (Blackwell et al., 2017; Hong et al., 2014).

In their research (Zhang et al. 2019) about short form video applications excess using might be addictive or not they found that social interaction anxiety and social isolation affect interpersonal attachment positively. At the same time, personalization and entertain have a positive relationship with addiction to social site. Furthermore, interpersonal and website attachment have positive effect on short video application addiction.

Kim et al. (2015) stated that attachment to a group of buying site influences loyalty to the site as well. In another study the following year Kim et al. (2016) found that site attachment had a positive effect on group attachment and relationship between others and in return has affects altruism for each other. On the other hand VanMeter et al. (2015) found that social media attachment has an effect on consumers defending and supporting actions. Choi and Seo (2015) have shown that attachment has related with Internet and smartphone addiction.

Short form video-sharing apps have at the same time a social aspect like Facebook, Instagram and Twitter, in other words, these apps are socio-technical information sites. Mc Cord et al (2014) found that social media using is affected from social factors namely anxiety of social interaction and social isolation. Ho et al. (2017) have confirmed that social media by providing a safe environment and by enabling users to communicate with peers helped to reduce social isolation. Hence, Valkenburg and Peter (2007) stated that social interaction anxiety weakens behavior in online dating site.

Ahn and Shin (2013) in the contrary stated that social media via video or communication does not affect social isolation. Teo (2002) on the other hand argued that individual differences are related with different online activities, like demographic and personality characteristics are mentioned as addiction risk factors. Hence, Foregger (2008) found that gender difference that is women have tendency to become heavier social media users than men.

FUTURE RESEARCH DIRECTIONS

Short video-sharing app TikTok is a highly downloaded, with very high expansion in social media however as it is comparatively new in social media, there are really very few researchs related with TikTok, and the researches are usually are usually about social media users' behaviours, advertisements etc. Future researchers are recommended to research about the reasons of the young generations preferring short video-sharing, what precautions must be taken to avoid adolescents to be addicts of short video applications and social media, and training aspects of users, and parents to be aware of the potential risks and dangers of the social media ocean.

CONCLUSION

Short video-sharing application namely TikTok has been spread all over the world, the growth rate surprised everybody. The short videos sharing, accompanied with music and other technical user friendly features provided by the app attracted users and it has been the highest ranked downloaded app in 2019. Furthermore, because it was provided free with easy access no registration required, just entering the main page and starting to watch the short videos shared was another reason for the expansion. However as it started to expand globally governments started to take precautions against the application. One of the main reasons for precautions was the private data security or safety of the data of the users especially after the Cambridge analytica scandal it was one of the priorities of all parties, as users have lost trust about their data to be used without consent by the apps. Another concern was the adolescents, using the app, and addiction to the social media especially to the short video-sharing app was another problem. The videos are fun, short and it was viral sharing enjoying moments via social media. However especially with the adolescents, who wanted to be more and more watched by the users, sometimes shared very intimate personal videos to attract users attention. Furthermore there were other risks like predators who were ready to abuse the users. Anybody under 18 is considered as a child according to the United Nations and children are to be protected by all means. The apps themselves must not leave it to users themselves to use the app appropriately because it is known that children under age of 18 use social media by fake age declaration to be able to use the apps. Governments, social media users, families everybody must be aware of the risk of abuse, addiction to social media and contents not appropriate for children must be avoided, precautions must be taken to protect them.

REFERENCES

Ahn, D., & Shin, D. H. (2013). Is the social use of media for seeking connectedness or for avoiding social isolation? Mechanisms underlying media use and subjective well-being. *Computers in Human Behavior*, *29*(6), 2453–2462. doi:10.1016/j.chb.2012.12.022

Araslı, O. (1979). *Özel yaşamın gizliliği hakkı ve TC Anayasasında düzenlenişi*. Yayınlanmamış Doçentlik Tezi.

Arslantaş, S., Binark, M., Dikmen, E. Ş., & Fidaner, I. B. (2012). *Türkiye'de Dijital Gözetim: T.C. Kimlik Numarasından E-Kimlik Kartlarına Yurttaşın Sayısal Bedenlenişi*. https://ekitap.alternatifbilisim.org/turkiyede-dijitalgozetim.html

Back, M. D., Stopfer, J. M., Vazire, S., Gaddis, S., Schmukle, S. C., Egloff, B., & Gosling, S. D. (2010). Facebook profiles reflect actual personality, not self-idealization. *Psychological Science*, *21*(3), 372–374. doi:10.1177/0956797609360756 PMID:20424071

Barabasi, A. L. (2003). *Linked: How Everything Is Connected To Everything Else And What It Means*. Penguin Group.

Barry, C. T., Doucette, H., Loflin, D. C., Rivera-Hudson, N., & Herrington, L. L. (2017). "Let me take a selfie": Associations between self-photography, narcissism, and self-esteem. *Psychology of Popular Media Culture*, *6*(1), 48–60. doi:10.1037/ppm0000089

Baruah, T. D. (2012). Effectiveness of Social Media as a tool of communication and its potential for technology enabled connections: A micro-level study. *International Journal of Scientific and Research Publications*, *2*(5), 1–10.

Baym, N. K., Zhang, Y. B., & Lin, M. C. (2004). Social interactions across media: Interpersonal communication on the internet, telephone and face-to-face. *New Media & Society*, *6*(3), 299–318. doi:10.1177/1461444804041438

Beutell, J. M. (2020). *Children's rights and social media: An analysis of TikTok's Terms of Service through the lens of a young user*. Academic Press.

Binark, M., & Bayraktutan, G. (2013). *Ayın karanlık yüzü: yeni medya ve etik*. Kalkedon.

Binark, M., & Çomu, T. (2012). *Sosyal Medyanın Nefret Söylemi İçin Kullanılması İfade Özgürlüğü Değildir!* Retrieved from https://yenimedya.wordpress.com/2012/01/20/sosyalmedyanin-nefret-soylemi-icin- kullanilmasi-ifade-ozgurlugu-degildir/

Blackwell, D., Leaman, C., Tramposch, R., Osborne, C., & Liss, M. (2017). Extraversion, neuroticism, attachment style and fear of missing out as predictors of social media use and addiction. *Personality and Individual Differences, 116*, 69–72. doi:10.1016/j.paid.2017.04.039

Brand, M., Young, K. S., Laier, C., Wölfling, K., & Potenza, M. N. (2016). Integrating psychological and neurobiological considerations regarding the development and maintenance of specific internet-use disorders: An Interaction of Person-Affect-Cognition-Execution (I-PACE) model. *Neuroscience and Biobehavioral Reviews*, *71*, 252–266. doi:10.1016/j.neubiorev.2016.08.033 PMID:27590829

Burnay, J., Billieux, J., Blairy, S., & Larøi, F. (2015). Which psychological factors influence Internet addiction? Evidence through an integrative model. *Computers in Human Behavior*, *43*, 28–34. doi:10.1016/j.chb.2014.10.039

Butt, S., & Phillips, J. G. (2008). Personality and self reported mobile phone use. *Computers in Human Behavior*, *24*(2), 346–360. doi:10.1016/j.chb.2007.01.019

Caplan, S. E. (2003). Preference for online social interaction: A theory of problematic Internet use and psychosocial well-being. *Communication Research*, *30*(6), 625–648. doi:10.1177/0093650203257842

Castells, M. (2004). *The network society A cross-cultural perspective*. Edward Elgar.

Changing childhood in a globalizing world. (2005). InProut, A. (Ed.), *The future of childhood*. Routledge. Retrieved from https://learn.illinois.edu/pluginfile.php/5030953/mod_resource/content/1/Prout%2C%20A.%20%28Ed.%29%20%282005%29.%20Changing%20childhood%20in%20a%20globalizing%20world.pdf

Chen, G. M. (2011). Tweet this: A uses and gratifications perspective on how active Twitter use gratifies a need to connect with others. *Computers in Human Behavior*, *27*(2), 755–762. doi:10.1016/j.chb.2010.10.023

Chiang, H. S., & Hsiao, K. L. (2015). YouTube stickiness: The needs, personal, and environmental perspective. *Internet Research, 25*(1), 85–106. doi:10.1108/IntR-11-2013-0236

Choi, Y. Y., & Seo, Y. S. (2015). The relationship between insecure adult attachment and smartphone addiction: the mediation effect of impulsiveness moderated by social support. *Korean J Psychol Couns Psychother, 27*(3), 749e72.

Cnn.com. (2019). Retrieved from https://edition.cnn.com/2019/02/28/tech/tiktok-ftc-fine-children/index.html

Computer Ethics. (2020). Retrieved from https://www.cmpe.boun.edu.tr/~say/c150/intro/lit10.html

Cooke, M., & Buckley, N. (2008). Web 2.0, social networks and the future of market research. *International Journal of Market Research, 50*(2), 267–292. doi:10.1177/147078530805000208

Correa, T., Hinsley, A. W., & De Zuniga, H. G. (2010). Who interacts on the Web?: The intersection of users' personality and social media use. *Computers in Human Behavior, 26*(2), 247–253. doi:10.1016/j.chb.2009.09.003

De Veirman, M., Hudders, L., & Nelson, M. R. (2019). What Is Influencer Marketing and How Does It Target Children? A Review and Direction for Future Research. *Frontiers in Psychology, 10*, 2685. doi:10.3389/fpsyg.2019.02685 PMID:31849783

Deuze, M., & Dimoudi, C. (2002). Online journalists in the Netherlands: Towards a profile of a new profession. *Journalism, 3*(1), 85–100. doi:10.1177/146488490200300103

Dominick, J. R. (1999). Who do you think you are? Personal home pages and self-presentation on the World Wide Web. *Journalism & Mass Communication Quarterly, 76*(4), 646–658. doi:10.1177/107769909907600403

Emirza, E., İştahlı, S. B., & İştahlı, Y. H. (2012). Sosyal Medyanın Çok Katlı Pazarlamada Kullanımı ve İş Tatmini Üzerindeki Etkileri. *Yönetim ve Ekonomi Araştırmaları Dergisi, 10*(18), 20–32.

Erol, G. (2009). Medya ve Etik: İnternet Haberciliğinde Nesnellik, Etik İlkelerın İçeriği, Uygulamadaki Sorunlar Ve Televizyon Haberciliği İle Karşılaştırılması. *Fırat Üniversitesi Medya ve Etik Sempozyumu*, 7-9.

Evans, D. (2008). Social media marketing: An hour a day. Academic Press.

Foregger, S. K. (2008). *Uses and gratifications of Facebook. com.* Michigan State University. Department of Communication.

Gao, W., Liu, Z., & Li, J. (2017). How does social presence influence SNS addiction? A belongingness theory perspective. *Computers in Human Behavior, 77*, 347–355. doi:10.1016/j.chb.2017.09.002

Gibbs, J. L., Ellison, N. B., & Heino, R. D. (2006). Self-presentation in online personals: The role of anticipated future interaction, self-disclosure, and perceived success in Internet dating. *Communication Research, 33*(2), 152–177. doi:10.1177/0093650205285368

Giddens, A. (2008). *Ulus Devlet ve Şiddet.* Kalkedon Yayınları.

Grand-Clement, S. (2017). *Digital learning: Education and skills in the digital age.* RAND Europe. Retrieved from https://www.rand.org/content/dam/rand/pubs/conf_proceedings/CF300/CF369/RAND_CF369.pdf

Hong, F. Y., Huang, D. H., Lin, H. Y., & Chiu, S. L. (2014). Analysis of the psychological traits, Facebook usage, and Facebook addiction model of Taiwanese university students. *Telematics and Informatics, 31*(4), 597–606. doi:10.1016/j.tele.2014.01.001

Jon, R. (2018). *China's ByteDance leapfrogs Uber to becomes world's most valuable startup.* Retrieved from https://techcrunch.com/2018/10/26/chinas-bytedance-leapfrogs-uber-to-becomesworlds-most-valuable-startup/

Kalafatoğlu, Y. (2011). *Sosyal Medya ve E-Pazarlama İlişkisi.* Academic Press.

Kaplan, A. M., & Haenlein, M. (2010). Users of the world, unite! The challenges and opportunities of Social Media. *Business Horizons, 53*(1), 59–68. doi:10.1016/j.bushor.2009.09.003

Khan, G. F., & Vong, S. (2014). Virality over YouTube: An empirical analysis. *Internet Research.*

Khan, M. L. (2017). Social media engagement: What motivates user participation and consumption on YouTube? *Computers in Human Behavior, 66*, 236–247. doi:10.1016/j.chb.2016.09.024

Kim, M. J., Lee, C. K., & Preis, M. W. (2016). Seniors' loyalty to social network sites: Effects of social capital and attachment. *International Journal of Information Management, 36*(6), 1020–1032. doi:10.1016/j.ijinfomgt.2016.04.019

Kim, Y., Wang, Y., & Oh, J. (2016). Digital media use and social engagement: How social media and smartphone use influence social activities of college students. *Cyberpsychology, Behavior, and Social Networking, 19*(4), 264–269. doi:10.1089/cyber.2015.0408 PMID:26991638

Knowledge, V. I. (2019). *The TikTok Strategy: Using AI Platforms to Take Over the World.* Academic Press.

Kramer, A., & Chung, C. (2011, July). Dimensions of self-expression in Facebook status updates. In *Proceedings of the International AAAI Conference on Web and Social Media* (*Vol. 5*, No. 1). AAAI.

Küçükyılmaz M. (2013, Feb. 16). Sosyal Medyada Henüz Reşit Olamadık. *Star Gazetesi.*

Kumar, V. D., & Prabha, M. S. (2019). Getting glued to TikTok®. Undermining the psychology behind widespread inclination toward dub-mashed videos. *Arch Ment Health, 20*(2), 76–77. doi:10.4103/AMH.AMH_7_19

Livemint.com. (2019). Retrieved from https://www.livemint.com/industry/media/the-chinese-tech-guru-behind-tiktok-s-success-11574883537442.html

Mangold, W. G., & Faulds, D. J. (2009). Social media: The new hybrid element of the promotion mix. *Business Horizons, 52*(4), 357–365. doi:10.1016/j.bushor.2009.03.002

Mavnacıoğlu, K. (2009). İnternette Kullanıcıların Oluşturduğu ve Dağıttığı İçeriklerin Etik Açıdan İncelenmesi: Sosyal Medya Örnekleri. *Fırat Üniversitesi İletişim Fakültesi Medya ve Etik Semp*, 7-9.

Mayfield, A. (2008). *What is social media*. Academic Press.

Meral, K. Z., & Meral, Y. (2021). The Role of Social Media in Arab Spring. *e-Journal of New Media, 5*(1), 26-34.

ODIHR. (2009). *Hate Crime Laws A Practical Guide*. Retrieved from http://www.osce.org/odihr/36426?download=true

Omar, B., & Dequan, W. (2020). *Watch, share or create: The influence of personality traitsand user motivation on TikTok mobile video usage*. Academic Press.

Papacharissi, Z. (2002). The presentation of self in virtual life: Characteristics of personal home pages. *Journalism & Mass Communication Quarterly, 79*(3), 643–660. doi:10.1177/107769900207900307

Patrick, M. S. (2018). *TikTok video app has become a petri dish for youth culture*. Retrieved September 20, 2018, from https://www.japantimes.co.jp/news/2018/08/25/national/media-national/tiktok-video-appbecome-petri-dish-youth-culture/#.XIda5ohKjid

Peters, K., Chen, Y., Kaplan, A. M., Ognibeni, B., & Pauwels, K. (2013). Social media metrics—A framework and guidelines for managing social media. *Journal of Interactive Marketing, 27*(4), 281–298. doi:10.1016/j.intmar.2013.09.007

Preece, J., Nonnecke, B., & Andrews, D. (2004). The top five reasons for lurking: Improving community experiences for everyone. *Computers in Human Behavior, 20*(2), 201–223. doi:10.1016/j.chb.2003.10.015

Priori Data. (2020). Retrieved from https://prioridata.com/publishers/tiktok-pte-ltd-1322881000/profile?publisher_id=1322881000&from=1&free=true&paid=true&free_plus=true&paid_plus=true&downloads=true&metrics_country=GG&time_period=last_cal_month

Priori Data. (2020a). *Leading Android apps in the Google play store worldwide in November 2019, by number of downloads (in millions)*. Statista. Retrieved from www.statista.com/ statistics/693944/leading-android-apps-world wide-by-downloads/

Priori Data. (2020b). *Leading iphone apps in the Apple app store worldwide in November 2019, by number (in millions)*. Statista. Retrieved from www.statista.com/statistics/695791/ top-iphone-apps-worldwide-by-number-ofdownloads/)

PRNewswire.com. (2018). Retrieved from https://www.prnewswire.com/news-releases/tiktok-captures-the-worlds-most-creative-videos-in-year-in-rewind-compilation-300767154.html

Ren, Y., Harper, F. M., Drenner, S., Terveen, L., Kiesler, S., Riedl, J., & Kraut, R. E. (2012). Building member attachment in online communities: Applying theories of group identity and interpersonal bonds. *Management Information Systems Quarterly, 36*(3), 841–864. doi:10.2307/41703483

Safko, L. (2010). *The social media bible: tactics, tools, and strategies for business success*. John Wiley & Sons.

Schreck, T., & Keim, D. (2012). Visual analysis of social media data. *Computer, 46*(5), 68–75. doi:10.1109/MC.2012.430

Seidman, G. (2013). The need to belong: Desire for interpersonal attachments as a fundamental human motivation. *Personality and Individual Differences*, *54*(3), 402–407. doi:10.1016/j.paid.2012.10.009

Şen, E. (2014). *Kişisel Verilerin Kaydedilmesi*. Retrieved from http://www.haber7.com/yazarlar/ prof-dr-ersan-sen/1235168-kisisel-verilerin-kaydedilmesi

Shao, G. (2009). Understanding the appeal of user-generated media: A uses and gratification perspective. *Internet Research*, *19*(1), 7–25. doi:10.1108/10662240910927795

Smith, M. J. (1998). *E-merging strategies of identity: The rhetorical construction of self in personal web sites* (Doctoral dissertation). Ohio University.

Smock, A. D., Ellison, N. B., Lampe, C., & Wohn, D. Y. (2011). Facebook as a toolkit: A uses and gratification approach to unbundling feature use. *Computers in Human Behavior*, *27*(6), 2322–2329. doi:10.1016/j.chb.2011.07.011

Smolentceva, N. (2019). *TikTok: World's most successful video app faces security concerns*. Retrieved from https://www.dw.com/en/tiktok-worlds-most-successful-video-app-faces-security-concerns/a-48063869

Solis, B. (2008). *The essential guide to social media.* http://www. scribd. com/doc/3283966/The-Essential-Guide-to-Social-Media

Statista.com. (2019). Retrieved from https://www.statista.com/chart/16939/tiktok-app-downloads/

Statista.com. (2019). Retrieved from https://www.statista.com/statistics/248074/most-popular-us-social-networking-apps-ranked-by-audience/

Sung, Y., Lee, J. A., Kim, E., & Choi, S. M. (2016). Why we post selfies: Understanding motivations for posting pictures of oneself. *Personality and Individual Differences*, *97*, 260–265. doi:10.1016/j.paid.2016.03.032

Tahir, T. B. M., & Husin, M. H. B. (2017). Online social media and risks: An exploration into existing children practice. In *2017 International Conference on Electrical Engineering and Informatics (ICELTICs)* (pp. 195-200). IEEE. 10.1109/ICELTICS.2017.8253250

Takahashi, M., Fujimoto, M., & Yamasaki, N. (2003, November). The active lurker: influence of an in-house online community on its outside environment. In *Proceedings of the 2003 international ACM SIGGROUP conference on Supporting group work* (pp. 1-10). 10.1145/958160.958162

Teo, T. S. (2002). Attitudes toward online shopping and the Internet. *Behaviour & Information Technology*, *21*(4), 259–271. doi:10.1080/0144929021000018342

TikTok.com. (2020). Retrieved from https://www.tiktok.com/about?lang=en

Türkiye Bilişim Vakfı. (2009). Retrieved from http:// www.tbv.org.tr/default.asp?ID=30)

Turkle, S. (1995). *Life on the Screen: Identity in the Age of the Internet*. Simon and Schuster.

Tuten, T. L., & Bosnjak, M. (2001). Understanding differences in web usage: The role of need for cognition and the five factor model of personality. *Social Behavior and Personality*, *29*(4), 391–398. doi:10.2224bp.2001.29.4.391

ulut, M. (Ed.). (2013). *Gençlik ve sosyal medya araştırma raporu.* TC Gençlik ve Spor Bakanlığı.

Untila Kaplan, O. (2019). *Media literacy in a changing process in Turkey. 11th International Conference on Electronics, Computers and Artificial Intelligence (ECAI)*, Pitesti, Romania. 10.1109/ECAI46879.2019.9042089

Valkenburg, P. M., & Peter, J. (2007). Who visits online dating sites? Exploring some characteristics of online daters. *Cyberpsychology & Behavior*, *10*(6), 849–852. doi:10.1089/cpb.2007.9941 PMID:18085977

VanMeter, R. A., Grisaffe, D. B., & Chonko, L. B. (2015). Of "likes" and "pins": The effects of consumers' attachment to social media. *Journal of Interactive Marketing*, *32*, 70–88. doi:10.1016/j.intmar.2015.09.001

Walker, K. (2000). "It's difficult to hide it": The presentation of self on Internet home pages. *Qualitative Sociology*, *23*(1), 99–120. doi:10.1023/A:1005407717409

Wang, J. L., Jackson, L. A., Zhang, D. J., & Su, Z. Q. (2012). The relationships among big five personality factors, self-esteem, narcissism, and sensation-seeking to Chinese university student's uses of Social Networking Sites(SNSs). *Computers in Human Behavior*, *28*(6), 2313–2319. doi:10.1016/j.chb.2012.07.001

Wang, T. L. (2014). The usage behaviors, motivations and gratifications of using User-Generated Media: The case study of Taiwan's YouTube. *Advances in Journalism and Communication*, *2*(04), 137–150. doi:10.4236/ajc.2014.24015

Whiting, A., & Williams, D. (2013). Why people use social media: A uses and gratifications approach. *Qualitative Market Research*, *16*(4), 362–369. doi:10.1108/QMR-06-2013-0041

Wikipedia.com. (2020). Retrieved from https://tr.wikipedia.org/wiki/Trol

Yeh, O. (2019). TikTok surpasses one billion installs on the app store and google play. *Sensor Tower*. Retrieved from https://sensortower.com/blog/tiktok-downloads-one-billion

Young, K. S. (1999). Internet addiction: Symptoms, evaluation and treatment. *Innov. Clin. Pract.*, *17*(17), 351–352.

Zhang, X., Wu, Y., & Liu, S. (2019). Exploring short-form video application addiction: Socio-technical and attachment perspectives. *Telematics and Informatics*, *42*, 101243. doi:10.1016/j.tele.2019.101243

ADDITIONAL READING

Duradoni, M., Innocenti, F., & Guazzini, A. (2020). Well-Being and Social Media: A Systematic Review of Bergen Addiction Scales. *Future Internet, 12*(2), 24. doi:10.3390/fi12020024

Taşkıran, N. Ö., & Yilmaz, R. (2015). *Handbook of research on effective advertising strategies in the social media age*. Business Science Reference. doi:10.4018/978-1-4666-8125-5

KEY TERMS AND DEFINITIONS

Data Breach: When the information is stolen or used without consent of the system's owner, the data stolen may cover confidential information like credit cards or passwords.

Hate Speech: Hate speech usually prepares ground for hate crimes and hate speech can be against political groups/followers, women, foreigners/immigrants, different sexual identities, different religions, and disabled/various diseases.

Social Media: Social media covers media, blogs, company discussion boards, e-mail, forums, blogs which enables online interaction among users via online web sites.

Social Media Addiction: Internet addiction covers social media addiction, where excessive use of using social media affecting daily lives negatively, like little or less quality sleeping, exaggerated use of social media, not being able to control and limit the use of social media in internet.

TikTok: TikTok, the most popular short video platform is defined in their web site as that their mission is to bring joy and inspire creativity, and be the leader for short mobile videos. The TikTok previously known as Musical.ly, is a video-sharing application. The Chinese ByteDance company has bought Musical.ly in 2017 and renamed as TikTok. It has increased its user number incredibly in 2019.

Troll/Troller: Usually a fake account created in social media used for posts aiming to quarrel or to upset people in online social media.

Chapter 11
Social Media Ethics and Children in the Digital Era:
Social Media Risks and Precautions

Kevser Zeynep Meral
https://orcid.org/0000-0002-2514-8744
Bahçeşehir University, Turkey

ABSTRACT

Social media users have increased to 3.5 billion including children users. Social media affects children as well. Although children's rights are protected by United Nations Convention, and advertising addressing children is not ethical, the embedded, integrated advertising in social media is more difficult for children due to their lack of advertising literacy because it is much easier to differentiate advertising in traditional methods compared to digital advertising via social media, where social media integrated advertising is used. This study aims to clarify social media and ethics concepts and how social media advertisements affect children in the digital era.

INTRODUCTION

Ethics comes from the Greek word "ethos" which means character, tradition and habit. Ethics deals mainly two basic subjects. The first one consists of personal characteristics about how to be a good person, and the second consists of social and strict rules that determine the people's right or wrong behaviours (Shaw, 1999). Aristotle's best known work in the field of ethics is "Ethics for Nikomakhos", he has converted words such as "good, virtue, freedom, happiness" into concepts (Aristotle, 1984). Ethic is a philosophy discipline that examines concepts and questions human behavior and moral behavior like good, bad, duty, necessity, permission, happiness, pleasure, benefit, will, responsibility, freedom, justice, equality, obligation, fair, unfair, etc. Ethic concerns everyone in the community who, as an individual with a sense of responsibility and consciousness, wishes to shape the current forms of communication, action and behavior and lead to the good together with other individuals. Ethics states how to behave in order for

DOI: 10.4018/978-1-7998-4117-3.ch011

an action to be accepted morally and aims to reach constant and permanent conclusions. Ethics has no goals such as moralization, transformation into ideology, or revealing a world view.

Pieper, explains the purpose of ethics together with a number of subordinates, like illumination of human practice in terms of moral quality, to present ethical arguments that can develop a critical, morally consciousness and to enter the basic processes, moral action is not an arbitrary action that one can take if he wishes, or give up if he does not; on the contrary, to be able to show that it represents the expression of an indispensable feature of human existence, that is, to teach to love human being. Ethics is often considered as a moral philosophy in the ethical literature. It is through ethics to convey the meaning of moral action to the individual. On the one hand, while defining and analyzing moral behavior models and attitudes, on the other hand, it reveals criticism that will enable the necessity of acting ethically, at the same time, it opens the concept of moral action and enables people to grasp this concept with their mind (Pieper, 1991).

Social media is an integral part of our lives and it affects everything. Especially children being affected by social media is a major concern for parents in digital era. This chapter aims to describe the effects of social media advertisements on children via literature review method. Why ethics in social media is very important and how the children can be saved as especially in social media, where embedded commercial advertisements can not be differentiated by the children.

SOCIAL MEDIA MARKETING ETHICS

Marketing ethics is a systematic work for standard marketing decisions, behavior and how these decisions and behavior are applied to institutions (Laczniak & Murphy, 1993). It has two different fields normative and value focused philosophy and positive and descriptive usually ampiric social sciences. Both dimensions are necessary for ethic marketing application to improve ethic marketing practice. The normative and descriptive difference can be defined as positive ethics which based on observation and data and usually explains what happened on moral based situations. Normative ethics focuses on a specific standard where can be applied and reasoning and at the same time describing the reasons why to continue with such an ideal. At the same time normative ethics, like positivism methods, when applied to marketing issues, it is a part of written analytic process, and it is not related to like sayings "customers are always right" etc. (Smith, 2001).

Current marketing groups have scientific discussions covering ethics including normative ethics, like what should normative ethics be and positive ethics like what can be done or what could positive ethics be), consumer ethics like which morel rules could guide the consumers and virtue ethic like what is ethics approach (Murphy et al., 2007).

Internet marketing has two-way communication comparing to one-way communication traditional marketing therefore it has different ethic and privacy issues (Malhotra et al., 2004). Researches related about marketing based on private characteristics tried to find out how they had ethical decisions (Singhapakdi et al., 1996). For example, Hunt and Vittell (2006), claim that people reply differently to ethical questions or situations depending on their ethic sensitivity.

Tuten and Solomon has defined social media marketing as an organization to use social media technologies, channels and softwares to communicate, to present and to exchange offers for their stake holders (Tuten and Solomon, 2017).

Social media is celebrated for enabling users to express themselves and their via the content the people themselves create (van Dijck, 2009), however a lot of people express their anxieties about their privacy to be used by third parties (Hargittai, Marwick, 2019). Even with the social media data open to the public people might have privacy expectation (Gruzd et al., 2018). When the organizations want to use the rich data from social media and internet extensively serious privacy and ethical issues are raised (Ward, 2018). Furthermore, Sparks and Hunt (1998), found that when the marketing experts take a role in a decision-making process, most of them failed to on ethical issues, and this makes it worse when they have to use social media data. While understanding marketing professionals decision-making process, to provide information about current social media marketing will not contribute to how marketing ethics should be, as Malhotra and Miller (1998) mentioned it is inevitable to remember that consumer is an important part of marketing process, however marketing research about ethical conflicts or dilemmas view are also very important and more efforts should be diverted towards targeting efforts to the consumer's ethical dilemma than focusing only through business side. For this reason, marketers' must understand consumer's view about social media data using in marketing.

Barget et al (2016) claimed that fragmentation arised in social media marketing discipline and invited for more research to understand consumers' participation in favour of consumers and organizations benefit. Kapoor et al., (2018) in their literature review about social media, have found that social media has been adapted and widely used as a marketing tool. Social media is used as a communication method to introduce products and services by the private sector and is generally used for information sharing and to motivate users' participation in public sector (Gruzd et al., 2018). Marketers, including customers, they use marketing strategies to have satisfactory long-term mutual relationship with key parties (Kang and Kim, 2017).

Social media related researches about social media show that advertising (Alalwan, 2018), perceived practicableness (Chang et al., 2015) and relevance (Jung, 2017), efficiency of social media (Kapoor et al, 2018). In other words, according to Alalwan (2018) customers who find social media advertising useful and more advantageous, are willing to buy the targeted products given in social media ads.

Consumers use social media to create information, to share experience via tweets, online group posts, with friends and their company (Bailey et al., 2018). Organizations strategies in social media and with social media tools are more attractive (Misirlis and Vlachopoulou, 2018). Using social media changes people's learning style (Gruzd et al., 2016), life style (Quan-Haase and Young, 2010), communication style (van Dijck, 2012) and changes business styles fundamentally.

Although it is very important to listen to social media, organizations try to understand what the customers and public say about their products and services (Lee, 2018), the consumers might not be very happy with this application (Dumois et al., 2018). Furthermore, with the Facebook data violation case etc. has increased people's private data sharing regulations anxieties and their awareness (Cochrane, 2018), which is one of the main issues of social media. Although marketers had to deal with different ethic issues, with adopting and using social media widely arised new difficult issues for marketing ethics application (Laczniak and Murphy, 2006). Likewise, Uzun (2007) in Turkey, explains the effect of convergence towards ethical issues in social media as communication technologies developing rapidly in recent years also caused new ethical problems by expanding the existing ethical problem areas. The emerging communication technologies also concern the press, publishing, advertising and public relations sectors. On the other hand, speed of social media is another issue in social media ethics. Bayraktutan (2012) referred to this issue in his study. The concept of interaction, another feature of social media, diverts the discussion of ethics in social media from the discussion of ethics in traditional media. Ethics in social media; speed is about interaction, intercultural and increasing the size of the interaction. Everything is developing fast on social media. Making quick updates and recycling is the most important feature of social media. Ethical decision making requires time to decide what is right and what is wrong. In this context, speed in social media causes the ethical decision-making process to be violated from time to time (Bayraktutan, 2012).

Ethics and Children Advertisements

Clow and Baack listed the criticisms and concerns towards advertisements as follows (Clow, Baack, 2012):

- Advertisements show people many reasons to purchase products and services,
- Advertisements emphasize on materialism frequently, so to say, luxury consumption is encouraged, and it is highlighted that social status and respect is gained via some products people own,
- Advertisements increase the costs of products and services,
- Advertisements sustain stereotyped judgements about men and women.
- Advertisements demonstrate products like alcohol and smoking cigarettes etc. as appealing.
- Violent elements are frequently included in advertisements,
- Advertisements targeting children are not ethical.

Children might repeat the songs and slogans from advertisements again and again throughout the day, remembering some of the advertisements completely. They might find their parents after the ad and say "I want the same". They even insist on the advertised product or the package which has the Barbie printed on rather than the one their parent has picked out. When children understand the intention of advertisements is persuasion, they begin to believe less in advertising messages. This applies to general advertising as well as to individual advertising messages. According to a study conducted in Belgium, 500 children between the ages of 7-9 were asked the question "Do you believe in advertisements" and there were great differences in the analysis of the results depending on whether the children were informed about the persuasion intention or not. While only 8% of the children who were aware of the persuasive power of the advertisement stated that they always believe in the advertisements, 66% of them responded they believed advertisements sometimes and 26% of them responded they never believe in ads. The rates of the children who did not know about the persuasive purpose of the advertisement were as follows: 19%, 71% and 10%. In addition to these results, children who have learned about persuasive nature of the advertisements also know that what is said in the advertisements does not all reflect the truth (Kapferer, 1992).

Advertisements that do not directly target children are concerned with a social structure in which the child participates economically, rather than aiming to advance or relocate the child socially, culturally or educationally. In this way, the child is directed towards a certain form of behavior and consumption. For advertisers, a child can have various roles (Öcel, 2002):

- Children are assets that can make products for adults more adorable and is seen as a bridge to reach adults.
- Children are assets that can direct the consumption of adults towards children, and ultimately causes adult spending to be child-centered.
- The children are in a position to direct the consumption for children.

- The child is also seen as an asset that can direct consumption for children and adults and direct future forms of production and consumption with individual spending authority.

The cognitive effect of advertising is the ability of children to grasp the advertisement. The cognitive effect of advertising; defining the difference between advertisements, recognizing the source of the message, perceiving the presence of the message as the target audience, detecting the informative message in the advertisement, detecting the persuasive message in the advertisement, and putting symbolic meanings on the characters in the advertisement (Rossiter, 1979)

The effect of advertisement on attitude is to examine whether there is a change related to the age of children on understanding the advertisements. Research shows that children's attitudes towards advertisements vary by age (Rossiter, 1979). Negative attitudes towards advertisements emerge as children grow older. Children begin to understand the deceptive aspect of advertising (Hite and Eck, 1987). As the ages of the children get older, the rate of believing in the advertisements decreases with the separation of advertising and programs is established. The average age of children who are capable of making this distinction in Turkey are children who are nine years old (Elden and Ulukök, 2006).

Advertisers use various unethical ways in advertisements targeting children. These ways (Schlegelmilch, 1998):

- They manipulate the pure feelings and thoughts of children by making people think that they will be more competent and more while using the product.
- They make claims that appear unreasonable in advertisements to be included in the advertisements, which can also be important information.
- They may display the products used differently from their normal and real environments or settings.
- They can make deceptive or misleading claims about the performance of the products.

Since a young child normally does not have enough financial power to buy a product, he shows his buying behavior by pressing his family (Hite and Eck, 1987). As children grow older, their pressure to purchase a product they see in advertisements decreases (Rossiter, 1979).

Another important effect of television advertisement is that for the vast majority, television advertising affects parents' authority over their children, because children are exposed to messages targeting children in television advertisements (Treise et al., 1994).

CHILDREN AND SOCIAL MEDIA

Since 1970's many researches have been done about children's literacy of classical television advertisements, whether the children can recognize and understand the convincing messages and whether they can cope with them (Wright et al., 2005). Classical advertisements types like television ads first of all, focus on product's quality and specifications and try to convince via factual or proposing messages (Moore and Rideout, 2007). Today it is claimed that one of three internet users is a child (Livingstone, Carr and Byrne, 2015) and younger ages are increasing more and more online (UNICEF, 2017). When children grow up, they become online players and social network players although these networks are designed for adults and young teenagers not for children (Stoilova, Livingstone and Nandagiri, 2019). Online playing apps, e-learning platforms etc. enable commercial, private and sometimes public organizations to collect private, personel information data of children (Lupton, Williamson, 2017). Thousands of mobile application used by children collect data and share with third parties, children's locations' are followed up and used to privatized ads addressing to children or without children affecting children's behavior or parents' knowledge (Reyes et al., 2018). Hence, Amazon was sued in US, for recording children with smart speakers while playing with the devices without their or their parent's consent, for violation of Privacy Act (Kelion, 2019). Whereelse, 80 percent of parents in US allow their children under age of 11 to use YouTube where the children are subject to watch ads before a video in YouTube (Pew Research Center, 2018). Ryan's World, with embedded branded toys ads with more 35,18 million children audience was the highest viewed YouTube site (Statistica.com, 2020), and according to the statistics, digital advertising expenses for children was 1,2 billion US dollars (Statistica.com, 2020).

Hudders et al., (2016) found that advertisements replaced in the media content and integrate more and more, by involving consumers with the commercial content (like ad plays) and motivate them by interacting and by collecting personal data, they adopt the content according to the consumer's preferences, they use the new media options cleverly. Furthermore, by different ways like repeating the same messages, connecting with positive aspects like humour and by embedding the content in the title they reinforced their messages (Rozendaal et al., 2011). Day by day increasing the enjoyable and charming contents of which the brands and products are integrated the new formats are trying the covered persuasion method (Nairn and Fine, 2008). To focus on covert and hidden advertisement to persuade children for example such advertisements affecting children without even they are aware and using their collected private data for marketing without their consent is more important than the new ethical problems (Nebenzahl and Jaffe, 1998).

Profile Marketing and Children

Profile marketing can be defined as advertisers using shared private data of users like age, favorite hobbies, brands, musics, books etc so that advertisers can merge their marketing campaigns according to private consumers favorites is called profile marketing (Villiard and Moreno 2012). Children are not qualified to understand hidden, advertising tactics like profile marketing methods (Rozendaal et al., 2011). It is not fair and not appropriate to use such advertising targeting children (Nairn and Fine 2008). Related researches like van Reijmersdal et al., (2017) has investigated effects of profile marketing on children and found that targeting the product had more positive in brand attitude and purchasing intent.

Influencers Marketing and Children

Vloggers, give ideas about their favorite brands they use in daily life and even directly advice which product they must use or what products they must not use to their followers (Meral, Özbay, 2020).

De Jans (2019) drew attention to the brands usually adding children influencers i.e. Stella and Blaise, The McClure Twins, AnnieLeblanc veya EvanTubeHD to their marketing strategy because of their trustworthiness and access children followers.

De Veirman remarks that majority of the researches about influencers marketing are that how the influencers affect the purchasing behavior of their followers and about mainly the adults (De Veirman et al., 2017) and that research on how little children influencer affecting children are still limited and despite the precautions on the children influencers are increasing, they are used more widely in children's media (De Jans et al., 2019). Their youtube channels also earn advertisement ads via them because of their followers and although legally they are obliged to reveal these advertised based endorsements of the products, they do not declare their advertised/commercial side of their posts (De Veirman et al., 2017). Although it looks as if they are ordinary children, some of them for example Ryan earned 22 million dollars in 2018 (Robehmed and Berg, 2018). Actually children are main targets of advertisors because they affect their parents' purchasing decisions furthermore they are at the same time future adult consumers (Calvert, 2008).

Hence, Hudders et al, (2016), claim that, branded contents or advertisements are fully entegrated to media of children's media. Children's literacy about advertising is not sufficient yet. Their skills, cognitive ability, emotional and morally are not developed under age of 12 and they are still immature. Strong literacy about advertisement requires critical thinking about advertisement and to convince subconscious persuasion. Hence it is not surprising various studies about children under age of 12 findings showing that the children can not cope with embedded advertisements. The new media, changing advertising methods are affecting children, many new developments affect their way of coping with the ads.

In addition to Hudders et al, other researchers have drawn attention to the same issue as children under age of 12, (Valkenburg and Cantor 2001) also emphasized that especially children younger than 12 is special target for advertisers, because they can not differentiate advertisement and media content, they can not process ads like adults, although only children above 7 years can read ad names and slogans. Children are affected by persuasive ads messages more than adults because they are not able to critise advertisements yet (Kunkel et al., 2004). Hence, Folkvord et al (2019) claim that even parents may not understand the YouTube content is actually an advertisement. As the advertisement and fun border is blurred and as content of advertisement is entegration into different media platforms, it makes it more difficult fort he children to differentiate between commercial and fun media content (Waiguny, Nelson ve Terlutter 2014). Furthermore, as the contemporary ads are more interactive and more fun therefore children are exposed to the commercial message for a longer time (van Reijmersdal et al., 2010).

Besides these researches, Blades et al., (2014), state that children do not only ads or embedded ads in printed media or television only, when they enter to YouTube meet with ads messages of vloggers, read sponsored articles in web sites, play with advertisement plays with tablet and see the targeted ads in Facebook and communicate with online advertisement via advertisement plays, social media advertisement or banner advertisements and comes closer to the advertisement like in their tablets or smart phones advertisements.

On the other hand, in their research, Clarke and Svanaes (2012) investigated academic researches covering 17-18 years old adolescents. They have found that none of the abstracts of the researches focused on youth' and adolescents' focus on their social, cognitive and emotional development. They have presented two literature review, first on digital advertisement in 2012 and the second on food and beverage related digital advertisement in 2014 and as a result they found that all efforts were on investigating digital advertisement to children along with other aspects however as digital technologies non-stop progressed with innovations, researches requiring time for investigation process could not keep up with the developments. They suggested not only digital food advertisements and that researches must search other aspects of digital marketing like digital marketing's effect on children's social life.

Carlson and Clarke (2014) also draw attention to food advertisement and effects. These findings show that there are food advertisements especially targeting children. Galbraith-Emami ve Lobstein (2013) also reviewed especially researches about food advertisements on children. Folkvord et al., (2016) presented a model of how food marketing affects children's eating habits. Most of them focus on children in digital platforms or food advertisements (Clarke and Svanaes 2012, 2014).

SOCIAL MEDIA RISKS AND CHILDREN

UNICEF classifies online risks into three groups, as content risk, contact and conduct risk. Content risks are the risks of which the children are exposed to content which are not for children, like pornographic or violence content. Furthermore, advertisements with contents of racism, discrimination or hate or recommendation of unhealthy products or danger, and suicidal or anorexic web sites affecting children with self-harm. Second group of risk classified by UNICEF for children is that the contact risk, risky communication with strangers which enables an adult to contact the child for sexual abuse or by persuading him to be in danger for himself. The last group of the classification is the conduct risk, that is the children are persuaded to post/share writing or sexual images about themselves or other children like friends of which they might have created (UNICEF, 2017). According to the report, when the risk is converted to harm, it might give extreme harm to children which yields to eating disorders, pro-suicide behavior, cyberbullying, sexual abuse of the child (Slavtcheva-Petkova, Nash and Bulger, 2015.

According to UNODC study, child abuse including children kidnapped by strangers or sexual abuse is not new however the format is changed with information and communication technology (ICT) which enables instant communication via internet with each other, ICT has facilitated and increased access to children abuse and increased criminal profit and increased harm. ICT has increased risk for children, children are under more risk for harm and more children are under risk of sexual abuse including child sexual abuse material like live sex abuse broadcasting and self-generated content, including sexual abuse of children via travel and tourism. The study has found that children trafficking for sexual exploitation, in other words sex travel and tourism costs are decreased with ICT (UNODC, 2015) because now they can have access and extend their networks via mobile phones, tablets by advertising and a modern-day slavery is created via digital world (U.S. Department of State, 2013)

Extreme harm the cyberbullying risk can even end up with committing suicide like the Canadian adolescent, Todd case where she committed suicide at the age of 15 when she was blackmailed by a stranger after being persuaded to strip in web cam and her photo was shared with her friends, after making a youtube video explaining her situation, she committed suicide which started the government to take precautions (CBS news, 2017).

EU report has found found that 92% of 57.335 child sexual abuse material belonged to five countries, ranking as Netherlands, US, Canada, France and Russia in 2016 (European Cybercrime Centre, 2015). Internet Watch Foundation (IWF) report pointed out that 53% of child abuse victims were 10 years old or younger in 2016 and which was 69% in 2015, however 11-15 old children images have increased from 30% to 45% in 2016, due to self-produced content share increase in 2016 (Internet Watch Foundation, 2017).

According to the UNICEF Report by Keeley and Little, in 2017, it was found that mainly the girls, children from poor income families, children who are out of school, children with disabilities and who have mental health problems and children of marginalized groups who are vulnerable, who are more likely subject to online risks are. Furthermore, digital access without guidance and not being aware of the risks increase online risk for children. According to this report, a child abuse victim was from Philippines, who reported that strangers from abroad like USA, Australia, UK and China, instructed them to get naked and earned US$3.00 per show, and was kidnapped by one of American abusers, was saved by the police, when she ran away from her abuser. Another example was from El Salvador, a 14 year old girl, was sexually abused not by an outsider but by her ex-boyfriend who took her nude photos and opened a social media account and posted her nude photos and shared it with her friends and family after their breakup and he was not punished (Keeley and Little, 2017).

SOCIAL MEDIA PRECAUTIONS FOR CHILDREN SAFETY

Being a child, a child might consider a potential risk different than an adult, for example where an adult might consider a child meeting a stranger in internet very dangerous however a child might consider this an opportunity to meet a new friend, and a teen might even think to have a romantic relationship (Mascheroni, Cuman, 2014). Children's online privacy is also important, they see overprotective parents, who spy on them, businesses and goverments as well (Third, 2014).

In a study for children 13-17 age group, they claimed that, they themselves can handle these risks, They also said they felt equipped to handle these risks and that they can take these risks to be online connectedness (Samuels, et al., 2013).

In another study for children of 15–17 age group, percent who knew how to activate and change their privacy settings has increased from 68 per cent to 98 per cent in Argentina, the Philippines, Serbia and South Africa. However, for younger children, the rate was significantly lower, ranging from 5 per cent to 40 per cent (Bryne, et al., 2016).

European Union

European Union set out their internet strategy in 2012 (European Union, 2012) and set new standards in 2019 (European Advertising Standards, 2019). As children started using internet and mobile before they learn how to read or write, therefore safe internet is vital and that interactive, creative and educational online content is necessary. They motivated and promoted creative and educational online content, aiming positive online experiences for young children. By empowering, teaching digital literacy and online safety in European Union schools, tried to increase awareness. They aimed to create a safer internet via age specific settings, classified contents, parental controls. EU Commission and Member States came together to work with mobile operators, manufacturers, social network providers to find solutions to combat child online sexual abuse and exploitation and published the strategy document on their website.

A report funded by EU (Mascheroni, Cuman, 2014), suggested the precautions advised for children safety as follows:

- Content classification, age-appropriate privacy settings, easy reporting mechanisms, and saety applications which motivates parents and children must be encouraged on mobile and services.
- Governments and non governmental organizations, schools are advised to make policies to make internet safer for children.
- Group communication via mobile devices yielded to aggressive communication and peer victimization like excluding group members from WhatsApp groups. Cyber-bullying prevention compaigns including using mobile devices and mobile applications responsibly.
- Children must be aware of using their personal devices because they might have unauthorized Access to their photos etc.
- Campaigns to make children aware of not sharing their phone numbers via social network sitesi.
- Private photos shared might be ruined once the relationship is over and 'revenge sexting' should be avoided.
- Children must be taught how to deal with social pressure and overdependence, to accept not to reply immediately to messages, to learn tips and tricks for self-management and to support each other.

- The children must be encouraged to get help from parents, teachers, peers. Parents must be communicating with their children about their online experiences and they must find mediating strategies with their children.
- Schools must prevent bullying via social network sites (SNS) because the communication is mainly between school friends and as bullying is age-specific, messages must be created for different age groups.
- Parents must be educated for potential risks and be aware of parental controls which are already designed for smart phones. These apps must be easy to use for parents who may not be very quick learners as their children. These parentel control mechanisms must be parametric, flexible, adoptable to children's specific needs, like their age etc.
- Schools might consider whether their regulations might be amended to involve mobile technology in class learning activities as although it is forbidden unfortunately students might be busy with their mobiles during class. Teachers might discourage copy/paste and teach to avoid plagiarism.
- Teachers and parents must be educated to use digital technologies.

UNICEF

UNICEF has started a global program to combat online child sexual abuse and exploitation under the program WePROTECT. The achievements under the program can be taken as a sample of precautions to combat with children sex abuse and exploitation since 2013 as follows (United Nations Children's Fund, Unicef, 2015).

- Governments commitment to end child sex abuse (77 countries),
- Children, teachers, parents were given education about to mitigate online risks for children (60 thousand in 12 countries),
- ICT representatives have participated for their role and responsibilities about online child protection (23 countries),
- Judges, law enforcement officers and prosecutors improved their capacity to investigate and prosecute online child sexual exploitation (more than one thousand in 14 countries),
- Specialized units and prosecutors' offices national capacity are increased to investigate and prosecute online children sex abuse and exploitation (9 countries).
- Cybercrime Investigation Unit and National Prosecutor's Office have found out two networks producing online child sexual abuse materials and demolished. (Guatemala).
- New established unit for online crimes against children, had a specialized training and increased children's awareness of online abuse and exploitation risks by visiting schools informed the students how to protect themselves and how to report (21 cases were dealt in Jordan).

Europol (Europol's European Cybercrime Centre, 2015). has drawn attention to digital currencies and encriptions help predators and hide online and offline child sexual abuse and exploitation, and that facial recognition and location finder software also increases risks for children who can identify and locate potential vulnerable children victims via these applications (Internet Organised Crime Threat Assessment-IOCTA, 2016) along it might also help law enforcement to rescue victims. Microsoft has donated PhotoDNA technology to International Centre for Missing and Exploited Children, the technology can be used by INTERPOL's International Child Sexual Exploitation's database integration. This program by generating a unique image (fingerprint) created from a photo/or video can be used online anywhere, and identifies approximately seven child and enabled 3,800 offenders arrested, 7,800 victims from 50 countries in 2015 (Interpol, 2015).

SOLUTIONS AND RECOMMENDATIONS

To be in compliance with international accepted ethical rules especially for the children who are exposed to embedded or located ads in social media is very important because the children especially under age of 12 can not even differentiate commercial and fun from each other. Therefore children need to be protected against potential negative effects of advertising. Although regulations are set by worldwide institutions like United Nations, EU and self-regulatory institutions like European Advertisement Standards (EASA, 2019) to protect children from advertising effects, including influencer marketing, the precautions are not sufficient yet.

As social media is so active, dynamic and innovative that it is nearly impossible to follow up/to control everything posted including embedded advertisements in children media, therefore to rely on rules for the most precious, special group, the children, who are the future would be very optimistic. Therefore, elated all parties must be aware of the potential affects of these advertisements in children media must be responsible, apply ethical rules, parents must be alert and follow up their children's social media. The advertisers, while making these advertisements must know ethical rules and the legislation about children and must avoid making ads affecting children negatively.

FUTURE RESEARCH DIRECTIONS

Social media ethic rules applications in different countries and different cultures can be investigated. Furthermore, how to improve controlling methods especially embedded ads, if they can be followed and differentiated via system controls can be investigated as well. Furthermore, as new tools in children media like TikTok and especially negative effects of advertisements in children media, how children are affected are recommended for future researchers to investigate.

CONCLUSION

Advertisements affecting children are not accepted ethical. Children are not capable of differentiating commercial in children's social media. It is not fair for children to be exposed to such embedded commercial ads. Advertisers, companies acting in compliance with advertising ethics is very important. Especially profile marketing, influencer marketing by using children's favorites must be avoided. Although there are rules and regulations set by international institutions and local governmental bodies, to protect children, the precautions are not enough to protect children yet. As setting rules and regulations are not enough to protect children, parents, teachers, families, must be educated about embedded advertisements targeting children and about negative effects of social media advertising affecting children. Furthermore, to avoid advertisements affecting children, technical innovation, software solutions like how to differentiate commercial and which will automatically block such ads for children users must be used to protect children in social media.

REFERENCES

Alalwan, A. A. (2018). Investigating the impact of social media advertising features on customer purchase intention. *International Journal of Information Management*, *42*, 65–77. doi:10.1016/j.ijinfomgt.2018.06.001

Aristotle. (1984). The Complete Works of Aristotle (vols. 1-2, J. Barnes, Ed.). Princeton University Press.

Bailey, M., Christensen, J. E., Danis, C. M., Ellis, J. B., Erickson, T. D., Farrell, R. G., & Kellogg, W. A. (2014). *U.S. Patent No. 8,903,847*. Washington, DC: U.S. Patent and Trademark Office.

Bayraktutan, G., Binark, M., Çomu, T., Doğu, B., İslamoğlu, G., & Telli Aydemir, A. (2012). *Social media in political communication and the use of twitter in the 2011 general elections in Turkey*. Academic Press.

Calvert, S. L. (2008). Children as consumers: Advertising and marketing. *The Future of Children*, ▪▪▪, 205–234.

CBS News. (2017). *Dutch man charged in Amanda Todd case allegedly targeted 2nd Canadian child.* https://www.cbc.ca/news/canada/british-columbia/amanda-todd-aydin-coban-1.3951334

Chang, Y. T., Yu, H., & Lu, H. P. (2015). Persuasive messages, popularity cohesion, and message diffusion in social media marketing. *Journal of Business Research*, *68*(4), 777–782.

Clow, K. E., & Baack, D. (2012). *Integrated Advertising, Promotion and Marketing Communication.* Pearson.

Cochrane, U. K. (2018). *Facebook video genom Twitter*. Academic Press.

De Jans, S., Van de Sompel, D., Hudders, L., & Cauberghe, V. (2019). Advertising targeting young children: An overview of 10 years of research (2006–2016). *International Journal of Advertising*, *38*(2), 173–206. doi:10.1080/02650487.2017.1411056

European Advertising Standards (EASE). (2019). *Children*. Retrieved from: https://www.easa-alliance.org/issues/children

European Union. (2012). *European Strategy for a Better Internet for Children.* Retrieved from https://ec.europa.eu/digital-single-market/en/european-strategy-better-internet-children

Europol's European Cybercrime Centre. (2015, Oct.). Virtual Global Taskforce Child Sexual Exploitation Environmental Scan 2015. EC3-Europol, 12.

Folkvord, F., Bevelander, K. E., Rozendaal, E., & Hermans, R. (2019). Children's bonding with popular YouTube vloggers and their attitudes toward brand and product endorsements in vlogs: An explorative study. *Young Consumers*, *20*(2). Advance online publication. doi:10.1108/YC-12-2018-0896

Galbraith-Emami, S., & Lobstein, T. (2013). The impact of initiatives to limit the advertising of food and beverage products to children: A systematic review. *Obesity Reviews, 14*(12), 960–974. doi:10.1111/obr.12060 PMID:23845093

Gruzd, A., Haythornthwaite, C., Paulin, D., Gilbert, S., & Del Valle, M. E. (2018). Uses and gratifications factors for social media use in teaching: Instructors' perspectives. *New Media & Society, 20*(2), 475–494. doi:10.1177/1461444816662933

Gruzd, A., Jacobson, J., Wellman, B., & Mai, P. (2016). *Understanding communities in an age of social media: the good, the bad, and the complicated*. Academic Press.

Hite, R. E., & Eck, R. (1987). Advertising to Children-Attitudes of Business vs Consumers. *Journal of Advertising Research, 27*(5), 40–53.

Hudders, L., Cauberghe, V., & Panic, K. (2016). How advertising literacy training affect children's responses to television commercials versus advergames. *International Journal of Advertising, 35*(6), 909–931. doi:10.1080/02650487.2015.1090045

Hunt, S. D., & Vitell, S. J. (2006). The general theory of marketing ethics: A revision and three questions. *Journal of Macromarketing, 26*(2), 143–153. doi:10.1177/0276146706290923

Internet Organised Crime Threat Assessment (IOCTA). (2016). Report. Author.

Internet Watch Foundation. (2017). *Annual Report 2016*. IWF.

Interpol. (2015). *Crimes against Children*. Fact Sheet, Interpol. Retrieved from https://www.interpol.int/en/News-and-Events/News/2015/Online-child-sex-abuser-forums-targeted-in-INTERPOL-coordinated-operation

Jung, A. R. (2017). The influence of perceived ad relevance on social media advertising: An empirical examination of a mediating role of privacy concern. *Computers in Human Behavior, 70*, 303–309. doi:10.1016/j.chb.2017.01.008

Kang, J. Y. M., & Kim, J. (2017). Online customer relationship marketing tactics through social media and perceived customer retention orientation of the green retailer. *Journal of Fashion Marketing and Management, 21*(3), 298–316. doi:10.1108/JFMM-08-2016-0071

Kapferer, J. N. (1992). *Strategic Brand Management*. Kogan Page.

Kapoor, K. K., Tamilmani, K., Rana, N. P., Patil, P., Dwivedi, Y. K., & Nerur, S. (2018). Advances in social media research: Past, present and future. *Information Systems Frontiers, 20*(3), 531–558. doi:10.100710796-017-9810-y

Keeley, B., & Little, C. (2017). *The State of the Worlds Children 2017: Children in a Digital World*. UNICEF. Retrieved from: https://files.eric.ed.gov/fulltext/ED590013.pdf

Kelion, L. (2019). *Amazon sued over Alexa child recordings*. Retrieved from: https://www.bbc.com/news/technology-48623914

Kunkel, D., Wilcox, B. L., Cantor, J., Palmer, E., Linn, S., & Dowrick, P. (2004). Report of the APA task force on advertising and children. Washington, DC: American Psychological Association.

Laczniak, G., & Murphy, P. (1993). *Ethical Marketing Decisions: the Higher Road*. Allen and Bacon.

Laczniak, G. R., & Murphy, P. E. (2006). Normative perspectives for ethical and socially responsible marketing. *Journal of Macromarketing*, *26*(2), 154–177. doi:10.1177/0276146706290924

Lee, I. (2018). Social media analytics for enterprises: Typology, methods, and processes. *Business Horizons*, *61*(2), 199–210. doi:10.1016/j.bushor.2017.11.002

Livingstone, S., Carr, J., & Byrne, J. (2015). *One in three: Internet governance and children's rights*. Global Commission on Internet Governance Paper Series.

Lupton, D., & Williamson, B. (2017). The datafied child: The dataveillance of children and implications for their rights. *New Media & Society*, *19*(5), 780–794. doi:10.1177/1461444816686328

Malhotra, N. K., Kim, S. S., & Agarwal, J. (2004). Internet users' information privacy concerns (IUIPC): The construct, the scale, and a causal model. *Information Systems Research*, *15*(4), 336–355. doi:10.1287/isre.1040.0032

Malhotra, N. K., & Miller, G. L. (1998). An integrated model for ethical decisions in marketing research. *Journal of Business Ethics*, *17*(3), 263–280. doi:10.1023/A:1005711311426

Marwick, A., & Hargittai, E. (2019). Nothing to hide, nothing to lose? Incentives and disincentives to sharing information with institutions online. *Information Communication and Society*, *22*(12), 1697–1713. doi:10.1080/1369118X.2018.1450432

Mascheroni, G., & Cuman, A. (2014). *Net Children Go Mobile: Final report*. Deliverables D6.4 and D5.2. Milano: EDucatt. Retrieved from http://netchildrengomobile.eu/reports/

Meral, Y., & Özbay, D. E. (2020). Electronic Trading, Electronic Advertising, and Social Media Literacy: Using Local Turkish Influencers in Social Media for International Trade Products Marketing. In N. Taskiran (Ed.), *Handbook of Research on Multidisciplinary Approaches to Literacy in the Digital Age* (pp. 224–261). IGI Global. doi:10.4018/978-1-7998-1534-1.ch012

Misirlis, N., & Vlachopoulou, M. (2018). Social media metrics and analytics in marketing–S3M: A mapping literature review. *International Journal of Information Management*, *38*(1), 270–276. doi:10.1016/j.ijinfomgt.2017.10.005

Moore, E. S., & Rideout, V. J. (2007). The online marketing of food to children: Is it just fun and games? *Journal of Public Policy & Marketing*, *26*(2), 202–220. doi:10.1509/jppm.26.2.202

Murphy, P. E., Laczniak, G. R., & Wood, G. (2007). An ethical basis for relationship marketing: A virtue ethics perspective. *European Journal of Marketing*, *41*(1/2), 37–57. doi:10.1108/03090560710718102

Nairn, A., & Fine, C. (2008). Who's messing with my mind? The implications of dual-process models for the ethics of advertising to children. *International Journal of Advertising*, *27*(3), 447–470. doi:10.2501/S0265048708080062

Nebenzhal, I. D., & Jaffe, E. D. (1998). Ethical dimensions of advertising executions. *Journal of Business Ethics*, *17*(7), 805–815. doi:10.1023/A:1005850812845

Özel. Kurguları, & Algıları. (n.d.). *Bildiriler Kitabı*, 397. Pew Research Center. Retrieved from: http://gradelevelreading.net/wp-content/uploads/2018/10/Role-of-Media-in-Supporting-Parent-Success-Pre-Readings-Combined.pdf

Pieper, J. (1991). *Philosophie*. Kontemplation, Weisheit.

Quan-Haase, A., & Young, A. L. (2010). Uses and gratifications of social media: A comparison of Facebook and instant messaging. *Bulletin of Science, Technology & Society*, *30*(5), 350–361. doi:10.1177/0270467610380009

Reyes, I., Wijesekera, P., Reardon, J., On, A. E. B., Razaghpanah, A., Vallina-Rodriguez, N., & Egelman, S. (2018). "Won't somebody think of the children?" examining COPPA compliance at scale. *Proceedings on Privacy Enhancing Technologies*, *2018*(3), 63–83. doi:10.1515/popets-2018-0021

Robehmed, N., & Berg, M. (2018). Highest-paid YouTube stars 2018: Markiplier, Jake Paul, PewDiePie and more. *Forbes.* Retrieved from https://www. forbes. com/sites/natalierobehmed/2018/12/03/highest-paid-youtube-stars-2018-markiplierjake-paul-pewdiepie-and-more

Rossiter, J. R., & Percy, L. (1979). *Visual communication in advertising*. Columbia University, Graduate School of Business.

Rozendaal, E., Buijzen, M., & Valkenburg, P. (2011). Children's understanding of advertisers' persuasive tactics. *International Journal of Advertising*, *30*(2), 329–350. doi:10.2501/IJA-30-2-329-350

Samuels, C., Brown, Q., Leoschut, L., Jantjies, J., & Burton, P. (2013). *Connected dot com: Young people's navigation of online risks. Social media, ICT's and online safety*. Centre for Justice and Crime Prevention and UNICEF.

Schlegelmilch, B. B. (1998). *Marketing ethics: An international perspective*. Cengage Learning EMEA.

Shaw, W. (1999). *Contemporary ethics: Taking account of utilitarianism*. Academic Press.

Singhapakdi, A., Vitell, S. J., & Kraft, K. L. (1996). Moral intensity and ethical decision-making of marketing professionals. *Journal of Business Research*, *36*(3), 245–255. doi:10.1016/0148-2963(95)00155-7

Slavtcheva-Petkova, V., Nash, V. J., & Bulger, M. (2015). Evidence on the Extent of Harms Experienced by Children as a Result of Online Risks: Implications for policy and research. *Information Communication and Society*, *18*(1), 48–62. doi:10.1080/1369118X.2014.934387

Smith, N. C. (2001). Ethical guidelines for marketing practice: A reply to Gaski and some observations on the role of normative marketing ethics. *Journal of Business Ethics*, *32*(1), 3–18. doi:10.1023/A:1010685524206

Statistica.com. (2020a). *Most viewed YouTube channel owners of all time as of January 2020, by views (in billions)*. Retrieved from: https://www.statista.com/statistics/373753/most-viewed-youtubers-all-time/

Statistica.com. (2020b). *Spending on advertising to children worldwide from 2012 to 2021, by format (in billion U.S. dollars).* Retrieved from: https://www.statista.com/statistics/750865/kids-advertising-spending-worldwide/

Stoilova, M., Livingstone, S., & Nandagiri, R. (2019). *Children's data and privacy online: growing up in a digital age: research findings*. Academic Press.

Third, A., Bellerose, D., Dawkins, U., Keltie, E., & Pihl, K. (2014). *Children's rights in the digital age: A download from children around the world*. Academic Press.

Treise, D., Weigold, M. F., Conna, J., & Garrison, H. (1994). Ethics in advertising: Ideological correlates of consumer perceptions. *Journal of Advertising*, *23*(3), 59–69. doi:10.1080/00913367.1994.10673450

Tuten, T. L., & Solomon, M. R. (2017). Social media marketing. *Sage (Atlanta, Ga.)*.

UNICEF. (2017). *Risks and opportunities related to children's online practices: Ghana country report*. Author.

United Nations. (1948). *United Nations' Universal Declaration of Human Rights of 1948 Articles 18 and 19 of the Universal Declaration of Human Rights in 2.2.2.*

United Nations Children's Fund. (2015). *UNICEF Global Programme to Build Capacity to Tackle Online Sexual Exploitation*. Final Report to the Government of the United Kingdom of Great Britain and Northern Ireland, Reporting Period: 24 May 2015–30 March 2016, UNICEF Child Protection Section, Programme Division, New York.

U.S. Department of State. (2013). Trafficking in Persons Report. U.S. Department of State.

Uzun, R. (2007). *İletişim Etiği: Sorunlar ve Sorumluluklar*. Gazi Üniversitesi İletişim Fakültesi.

Valkenburg, P. M., & Cantor, J. (2001). The development of a child into a consumer. *Journal of Applied Developmental Psychology*, *22*(1), 61–72. doi:10.1016/S0193-3973(00)00066-6

Van Dijck, J. (2009). Users like you? Theorizing agency in user-generated content. *Media Culture & Society*, *31*(1), 41–58. doi:10.1177/0163443708098245

Van Dijck, J. (2012). Facebook as a tool for producing sociality and connectivity. *Television & New Media*, *13*(2), 160–176. doi:10.1177/1527476411415291

Van Reijmersdal, E. A., Jansz, J., Peters, O., & Van Noort, G. (2010). The effects of interactive brand placements in online games on children's cognitive, affective, and conative brand responses. *Computers in Human Behavior*, *26*(6), 1787–1794. doi:10.1016/j.chb.2010.07.006

Van Reijmersdal, E. A., Rozendaal, E., Smink, N., Van Noort, G., & Buijzen, M. (2017). Processes and effects of targeted online advertising among children. *International Journal of Advertising*, *36*(3), 396–414. doi:10.1080/02650487.2016.1196904

Villiard, H., & Moreno, M. A. (2012). Fitness on facebook: Advertisements generated in response to profile content. *Cyberpsychology, Behavior, and Social Networking*, *15*(10), 564–568. doi:10.1089/cyber.2011.0642 PMID:22963337

Waiguny, M. K., Nelson, M. R., & Terlutter, R. (2014). The relationship of persuasion knowledge, identification of commercial intent and persuasion outcomes in advergames—The role of media context and presence. *Journal of Consumer Policy*, *37*(2), 257–277. doi:10.100710603-013-9227-z

Wright, P., Friestad, M., & Boush, D. (2005). The development of marketplace persuasion knowledge in children, adolescents and young adults. *Journal of Public Policy & Marketing*, *24*(2), 222–233. doi:10.1509/jppm.2005.24.2.222

ADDITIONAL READING

Akseki, P., & Baş, Ş. (2019). *Korkunç Güzel İnternet-Çocuktan Al Bilgiyi*. Genç Timaş Yayınevi.

Caesar, L. (2015). *Cyberbullying: The Ultimate Guide for How to Protect You and Your Children From A Cyber Bully (Online Bullying, Online Reputation, Bullying Cure, eBully, Cyber Stalking, Bullying Free, Abuse)*. Paperback.

Tuncer, N. (2000). Çocuk ve internet kullanımı. *Türk Kütüphaneciliği, 14*(2), 205–212.

KEY TERMS AND DEFINITIONS

Influencer: Influencers unlike Vlogger or Youtuber, produces content and reaches a wide audience. Influencer is the person who can change and affect the behavior of people in new media.

Influencers Marketing: With Influencers, a new marketing concept of "Influencer marketing" has emerged with Influencers.

Information and Communication Technology (ICT): It is similar to Information Technology (IT), but mainly focuses on communication via wireless network, cell phones, internet and other ways of instant communication websites like video conferencing (zoom), Facebook, Instagram, WhatsApp which enable online real time, instant communication.

Social Media: The online network where users or businesses publish their own content is defined as social media.

Vlogger: Abbreviation for the word "Video-logger". Vloggers are the people who record, produce, and create videos. Vloggers create mostly conceptual content. Vlogger is the video version of "blogger."

YouTuber: YouTube is the world's largest video sharing website. Youtubers, who share their videos on YouTube with their followers earn revenue from YouTube and through many other promotions and advertisements such as sponsored content, promotion of brands, promotion of products and applications.

Section 3

Business Ethics as a Changing Phenomenon in Its Historical Depth

Chapter 12
Business Ethics:
The Changing Face of Working Life

Neslihan Yilmaz
https://orcid.org/0000-0003-3302-8598
Istanbul Medipol University, Turkey

ABSTRACT

Ethics is about the behaviors that people enact on a social scale and that have consequences affecting others and the thinking processes that shape them. Business life is a part of social life and it has to develop principles compatible with value judgments in society. Increasing professional corruption is a situation where there is a complexity of values within the society and utilitarianism overcomes everything over time, and an ethical process is required. This can only be achieved through effective leadership.

INTRODUCTION

The concept of business ethics has recently become a concept that all organizations are closely interested in. The effect of globalization forces organizations to operate within the framework of ethical principles. Creating ethical codes of organizations in this respect is among the most common methods. The most important factor that pushes organizations towards such a method is that businesses operating in compliance with business ethics principles believe that they will be successful in the long term. Ethical breakdowns of businesses that do not comply with business ethics reveal the necessity of business ethics. The result of this has revealed an ethical collapse that has been exposed to business scandals in America and Europe. Interesting and known of all accounting scandals from the past; ancient and middle ages, 17th-18th. centuries can be listed as 19th century (Jones, 2011: 115), 20th century and 21st century scandals (Yardımcıoğlu& Ada, pp.44-52).

The business community has clearly understood the necessity and importance of business ethics through ethical collapses. In addition, the ethical collapse in the businesses has been examined in terms of its causes and consequences.

DOI: 10.4018/978-1-7998-4117-3.ch012

The delay of legal arrangements in parallel with technological and social developments in business life, the lack of knowledge of laws, rules and methods, material and moral dissatisfaction cause professional corruption and weaken professional ties.

We have separated generations while studying work ethics. We examined the effects of generational differences on business life and the relationships between business ethics. When we examined the effects, we evaluated them both as employers and employees. Without an ethical climate, we cannot talk about success.

ETHICS AND ETHICAL VALUES

Ethics, Ethical Principles, and Ethical Systems

Ethics is about the behaviors, actions, thinking processes that the ethical person performs on a social scale and that affects others. Within the framework of this related formation, the concrete emergence of ethics occurs in three forms:

1. Ethics as a main area of philosophy addresses the issue of what good is and what is good in the context of behavior and aims to formulate the ideal behavior of man.
2. A set of rules for ethics in social life, what people should do and avoid in their relationships with each other; briefly it is general morality.

Professional ethics, in which the rules to be followed by those who are active in a certain field are determined and the desired behaviors in the practices specific to that field are examined; It is newer than the first two and in a way is a synthesis of them.

Ethics refers to a historical, social and dynamic process based on various factors of societies including history, social, political, economic and cultural factors. The principles and rules created within the framework of ethics are not created by law but by social reactions (Kutlu,2015, pp.2).

Business life is a part of social life and it must develop principles that are compatible with value judgments in society. It has a very broad meaning as it covers the activities carried out by all professions. The "business life" that is spread in such a wide area cannot be considered devoid of ethical principles. Although it is possible to increase the number of these principles, we can list the ethical principles that should exist in business life as follows.

1. Obey the laws. The first basic principle of social responsibility and managerial morality is to believe in the benefit of the law and obey its power.
2. Say the truth. Telling the truth is very effective in gaining the trust of interest groups.
3. Respect people. The philosophy of treating people with respect has a long history in ethics studies.
4. Ste up the rule of "doing everyone a favor". The idea of "doing a favor to others so that others can do a favor to you" is a measure to evaluate the ethical dimensions of business activities.
5. First of all, do not harm anything. This principle is seen as the basic idea of ethical thinking and easily adapts to the job.
6. *Adopt participation in practice, say no to fatherly management. This principle aims to learn what their needs are, rather than making decisions about what is good for interest groups.*

7. Always act when you are responsible. Managers should have the responsibility to take action when they have the capacity and resources to do something.
8. *Employ the right people. This feature is seen as an important part of the business's own qualities.*
9. *Set standards rather than rules. It is not possible to write a transaction code that will contain every result, there will be always be something missing. An individual prone to erroneous and unethical behavior will, in any case, refrain from considering the code. Therefore, time should not be spent on extremely strict and comprehensive regulations. Instead, clear standards should be put in place. These are essential for high-level performance.*
10. *Do not isolate yourself as management. If executives act in a dream world far from the realities of the world, they may not be able to analyze their market and lose track of their competitors. They may also not be able to see what's going on in their own businesses.*
11. *Easily establish a relationship and interact with different internal and external interest groups. The purpose of this principle is that the business sees the benefit of its interest groups as part of its own benefit, that is, to consider and take into account the interests of others as much as their own.*
12. *Always be an ethical example. Managers should be aware of the signals they send to those around them. This understanding will motivate employees and spread the notion that their success is hidden in the details. Mark Twain says below on the subject.*

"Always do the right thing, this will be a surprise for some people. However, doing the right thing will trigger others over time" (Kutlu,2015, pp.90).

The approaches that are based on the development of ethical principles are the principle of equity, human rights, utilitarianism, and individuality. Fairness emphasizes that all decisions are consistent, impartial and factual. Human rights focus on the existence, integrity and basic human rights of individuals. Utilitarianism undertakes the decision that will be good for everyone. Individuality advocates that the main goal of the individual is to increase his personal earnings in the long run. Different perspectives represent different ethical systems. Each of the ethical systems alone is not enough to solve the ethical problems and dilemmas encountered in all situations.

Instead of a single ethical system, it is necessary to develop effective thinking skills that will enable individual behaviors to be directed towards the correct and the good with different ethical systems.

Intended Outcome Ethics: It is often known as utilitarianism. It brings a practical solution to the problems and focuses on the results to be achieved. Its most important representative is John Stuart Mill. According to this ethical system, it is necessary to determine what the possible consequences and effects may be in order to decide whether an action is right or wrong. It is emphasized that on behalf of all parties that will be affected by the results of an action, the individual who makes the decision should be an independent observer. This ethics is built on seeking pleasure and avoiding pain. What makes an action "good" according to Mill is that it provides happiness with the benefit it brings. He argues that happiness, which is the criterion of "good", is the happiness of all humanity, not only of the actor but of everyone concerned. The practical approach to the problems of the intended outcome ethics and the utilitarian approach, which requires individuals to be taken into consideration in the decision made, can help to solve daily problems.

Table 1. Ethical System and Definitions

Ethical Systems	Definitions
Intended Outcome Ethics	The moral correctness of an action is determined by its intended results.
Rule Ethics	The moral correctness of an action is determined by standards and laws.
Social Contract Ethics	The moral correctness of an action is determined by the norms and traditions of a society.
Personal Ethics	The moral correctness of an action is determined by one's conscience.
Social Life Ethics	Social and ecological rules that must be followed in one's life.

Rule ethics: There are rules to be followed in daily life. Some of these are written rules. These are called laws. Criminal sanctions for non-compliance are in question. However, not every ethical behavior is legal and not every legal thing is ethical. True ethics is above what is legal. Rule ethics was developed by Immanuel Kant. This understanding led by Kant says to all people, "Behave in the same way as you want everyone to behave."

Social contract ethics: Adopting the principles and standards generally accepted by society as an individual can be beneficial in resolving the conflicts brought about by living in society. Its forerunner is Jean Jack Rousseau. The most important fact for him is the balance between authority and freedom. The rules that regulate the relations between individuals and society and that are obligatory to follow are called the rule of law. It is argued that the rules of law are put forward as a result of people's reconciliation with each other. People have to abide by the rules of law they set by a contract called "social contract".

According to the basic principles of Rousseau's ethical understanding,

1- When people live alone and primitive in nature, they disappear. For this, they need to become a society and be governed by agreed principles.
2- The transition from natural life to civilized life gives a moral character to actions and instincts that were not needed before.
3- It is important to act in accordance with the general procedures of the society.
4- The standards of the community help the individual decide what to do.

Personal ethics is determined by one's conscience. Martin Buber, the most important representative of the personal ethics system, argues that the source of personal ethics is the sound coming from within the individual. Personal ethics determines one's individual stance in society. It is the reactions and attitudes of the individual against the incidences that he/she goes through and that happens around him.

According to this;

1. Personal ethics defends the reality and truth not only come from intelligence but also from the individual himself.
2. Truth and ethical behavior exist in everyone and is the sound coming from within the individual.
3. Personal ethics enables the individual to return to himself.
4. The behavior that will lead the individual to the goal should be ethical.

5. Life has no specific formula, anything can happen.
6. Since the individual is in a group, he does not tolerate injustice.
7. As the individual becomes free, he develops his own ethical standards.
8. Since individuals develop throughout their lives, they also acquire new values and beliefs.

Social life ethics: It is a set of rules that one must follow in the relationships established in social life. According to this;

1. People maintain their relationships with each other within the framework of mutual respect and courtesy rules.
2. All parties act in an awareness of their rights and responsibilities required by their duties.
3. Pressure methods such as the use of influence and power cannot be used.
4. Colleagues, students, and employees are not directed in any interest (Kutlu,2015, pp.46).

The Effect of Business Ethics and Business Ethical Conditions on Employees

What is right and what is false or wrong has been discussed since the day the world was founded. It is natural that the conclusions reached will vary according to geopolitics, religion, morality, and mode of administration. Thus, by trying different legal rules, it tried to appreciate the right or wrong within the framework of the law and to base the verdict on state power.

What is good in working life? What are the ethical rules based on business life?

In working life; work ethic is intertwined in this direction. In other words, in cultures where work ethics are high, work ethics has also developed substantially. The figures showing that the sales volume and market values of the American industrial organizations in the top ranks are several trillion dollars shows not only the importance of business ethics but also whom it concerns.

There are of course moral problems that may arise in businesses, and Tierney lists what these could be as follows:

- Whether you will accept an expensive weekend holiday gift from one of the companies which you bought goods from.
- When communicating the research results to your customers, you will tell them the facts or what they want to hear.
- Will you sell a product that your customer really needs or that will bring you more commissions?
- Whether you will play over the year-end figures to show that the targets have been achieved.
- Whether you would want your staff to hide certain information.

In addition to the aforementioned problems, Tierney looks for an answer to the question of “What would you do if any of the following happened to you?”:

- You notice that one of your products is faulty. The product manager knows about the situation, but he hasn’t said anything.
- You notice that one of your superiors is using company materials for personal purposes.
- One of your superiors is constantly pressuring you to mislead the customer.

- Your general manager's brother-in-law sells goods to your company. He invites you to a fancy golf club, and you know he's sending defective supplies to the company.

The degree of implementation of ethical rules is different in every business. With Downey's definition, corporate culture is formed as "values, beliefs, and behaviors shared by a company". (Stephan M. Downey, "The Relationship Between Corporate Culture and Corporate Identity," ***Public Relations Quarterly***, Vol. 31, No.4 (Winter,1986/87, pp.7). This culture is the fundamental invariant of the corporate vision. The corporate culture can be ethics-oriented, or only profit can be pursued. For example; Bill Hewlett's friend and business partner David Packard said after his death: "The biggest thing he left behind for the company was the set of ethical rules known as the Hewlett-Packard Style." The core ideology that has been guiding Hewlett-Packard since its foundation more than 50 years ago includes its dedication to the individual, its commitment to providing enough quality and reliability, its social responsibility (Packard himself left Hewlett-Packard shares worth $ 4.3 billion to a charity) and the opinion that the company exists to make a technical contribution to humanity's progress and prosperity (Türel,1999, pp.32).

Business Ethics and Institutionalization of Ethics

It is observed that the concept of business ethics came into prominence in the periods when moral philosophy was replaced by law and philosophy of law (Hegel,1991, pp.7) along with social modernization and democratization. From the last quarter of the 20th century onwards, it is observed that it has been used more frequently, awareness of business ethics has increased, and business ethics has become a separate field of study (Freelance, 2002, pp. 2).

Business ethics is a concept closely related to the management of people, justice, honesty and equality (Maclagan, 1995, pp.85). It includes the values that form the basis of human relations and specific behaviors, rules and norms developed for special groups (Tansal, 2002, pp.10).

In organizations, ethical principles can be institutionalized in different ways. For example, according to Nan Demars, there are three ways to institutionalize ethical values in an organization:

1. Company awareness,
2. Not compromising on the principle that everyone is responsible for their own behavior,
3. Continuous communication on ethical issues.

The roles to be created depending on the size of the organization will be very useful for managing ethics in the workplace. These roles are respectively a top management that fully supports the program, an ethics committee at the level of the board, ethics board, ethics expert, complaint follower (Acar, 2000, pp.3). Senior management should believe in the necessity of ethics, show leadership, do preparatory work, pursue a long-term perspective, take a comprehensive approach (TUSİAD, 2009, pp.4). If the top management does not fully support the program, the employees will definitely realize this, and this will make them feel that they are faced with hypocrisy. The important thing is that the top management continues to consistently lead this program (Acar, 2000, pp.4).

We can also evaluate corporate business ethics with the concept of corporate social responsibility. We can think of corporate social responsibility as an organization's ethical and responsible behavior towards all stakeholders, both internal and external, taking decisions and implementing them in this direction.

One of the reasons for the commitment between business ethics and social responsibility concepts is the understanding of social responsibility that develops with public pressure for businesses to work ethically. The second reason is knowing that good morality is identical to good work. The third reason is that the discussions about ethics and business life for managers are not perceived as a weakness but as a natural extension of today's enterprises' efforts to reach high quality. The other reason is the need to balance the interests of some interest groups. Because business ethics is built on values such as freedom and responsibility towards interest groups.

Individual responsibility is also important here. Being aware of their own responsibilities brings along issues such as environmental awareness. Businesses are institutions that have responsibilities towards their internal and external environment.

When we analyzed historically, in the 1950s when the concept of social responsibility first appeared, this concept was perceived as the business's fulfillment of its responsibilities towards employees, shareholders, and customers, and today, the concept of responsibility has been expanded and some factors have influenced this.

Globalization has placed social responsibilities on businesses, and human resources practices here have increased the interest in social responsibility issues such as environment, health, and safety.

- National and international organizations have developed conventions, codes of conduct, and guidelines on social responsibility issues.
- The development of communication technologies has facilitated the monitoring of the activities in the enterprises.
- Support for responsible commercial activities has increased day by day.
- Corporate governance, transparency, accountability and the necessity of ethical standards have become important due to the unethical transactions performed by the enterprises.
- The awareness of societies has increased in terms of businesses fulfilling their social and environmental responsibilities.
- It is understood that the issues regarding social responsibility cannot be subject to legal regulations.
- It has been realized that corporate social responsibility practices will be a source of image and reputation for businesses.

Research Studies and Findings on Business Ethics

Human Resources Management policies and practices should be considered in conjunction with moral issues that require HRM's functioning and careful handling of HR managers while building the culture of an organization, as it affects the lives of every job candidate or employee. It is the universal moral responsibility of individuals and organizations to create an organizational culture, to achieve goals that benefit customers and society, or not to harm human relations, and to create value in the short and long term (Caldwell & Anderson, 2018). (Bennington, 2007) draws attention to the importance of acting correctly in recruitment for ethical human resources management, equal pay for equal work, fair policies, and avoiding invalid and discriminatory selection practices. Foremost among these, equal opportunity in employment is a basic human right.

The sample of this research consists of supervisors working as managers or specialists in the HR units of 120 companies participating in the Antalya Employment Fair held between 15-18 March 2018 and operating in Antalya or unit supervisors who carry out the personnel affairs and operations of the business. The questionnaires used as a data collection tool were answered by the company representatives who agreed to participate in the research. While creating the sample cluster, one participant to represent each company was tried to be included in the research study, and as a result, the final number of participants who agreed to participate in the research was reached. Therefore, in order to investigate ethical attitudes in the business life, it is aimed to take the opinions of people who have an important position in terms of realizing the interests of both the employee and the employer in accordance with the purposeful sampling design (Sencer, 1989, pp. 386), not from a universe consisting of all enterprises.

In addition to the factual and demographic questions, the questionnaire of the research study also included the Ethical Values Assessment (EVA) —Short Form developed by Padilla Walker and Jensen, (2016) and translated into Turkish by Ayfer Dost Gözkan, and Ethical Trends Scale developed by Koçyiğit and Karadağ (2017). In addition to these scales, the one-item job satisfaction scale developed by Scarpello and Campbell (1983) was also added to the survey questionnaire. Wanous et al. (1997) state that single-item job satisfaction scales present highly compatible results with multi-item scales (Wanous et al., 1997); (Aktaran & Orhan, 2011 pp.165).

RESULTS

Labor Relations and Ethics: On Mainstream Human Resources Management Perspective

Table 2. Findings Regarding Demographic Characteristics of Observation Units

		Community Ethics	Religious Ethics	Autonomy Ethics
Frequency	Current data	109	109	109
	Lost data	0	0	0
Average		4,5798	3,9702	4,6972
Median		4,6000	4,2500	5,0000
Standard deviation		,37707	,91460	,46157

It is seen that the human resources managers and unit supervisors participating in the research study have autonomy ethics (4.69), community ethics (4.57) and religious ethics (3.97) respectively. According to this, the autonomy ethic values that include taking responsibility for the individual's actions, being physically and psychologically good, pursuing their own goals, being fair, reciprocity, respecting the rights of individuals emerged as the most intense values in most of the participants. In the second place, the values of social ethics are observed in terms of ensuring justice and equality in society, living together, building social cohesion and revealing the role and status in society as a member of the society. Religious ethics, on the other hand, was in the third rank as reaching spiritual salvation, not sinning, living life with sacred values, and following the creator's orders.

When it was examined whether the gender of the participants in the study caused a difference in values ethics orientation, it was seen that the p-value was above the significance level of 0.05, therefore

Table 3. Correspondence of the Values Ethics Scale in the Sample

	Frequency	Percent
Gender distribution		
Woman	55	50,5
Male	54	49,5
Total	109	100
Age distribution		
Between 20-29 years old	39	35,8
30-39 years old	41	37,6
Between 40-49 years old	24	22,0
Between 50-59 years old	5	4,6
Total	109	100
Job		
HR Manager	34	31,2
HR Specialist	30	27,5
Unit manager / supervisor outside HR	45	41,3
Total	109	100
Working time in the current workplace		
Less than 1 year	21	19,3
1-5 years	60	55,0
6-9 years	10	9,2
10-15 years	11	10,1
16-20 years	3	2,8
21 and over	4	3,7
Total	109	100
Monthly fee charged		
Between 1000-2000 TL	30	27,5
Between 2001-3000 TL	35	32,1
Between 3001-4000 TL	23	21,1
Between 4001-5000 TL	9	8,3
Between 5001-6000 TL	4	3,7
6001 TL and above	8	7,3
Total	109	100
Satisfaction from work		
I am extremely dissatisfied	1	,9
I am not satisfied	1	,9
undecided	17	15,6
satisfied	27	24,8
I am completely satisfied	63	57,8
Total	109	100

Table 4. Gender Difference and Values Scale T-Test Results

Gender of the Interviewer		Number	Average	Standard Deviation	Standard Deviation Error	t	P
Community Ethics	Woman	55	4,5491	,40409	,05449	-,858	,393
	Male	54	4,6111	,34840	,04741	-,859	,392
Religious Ethics	Woman	55	3,8182	,94337	,12720	-1,768	,080
	Male	54	4,1250	,86568	,11780	-1,770	,080
Autonomy Ethics	Woman	55	4,6545	,43962	,05928	-,975	,332
	Male	54	4,7407	,48312	,06574	-,974	,332

there was no significant difference in the values ethics orientation between men and women. On the other hand, it is possible to say that men are grouped at higher levels in terms of community, religion and autonomy ethics.

Table 5. Positions and Values Ethics in the Workplace ANOVA Test Results

		Average	Standard Deviation	Standard Error	F	Sig.
Community Ethics	HR Manager	4,5471	,33866	,05808	0,9360	0,3950
	HR Specialist	4,6600	,35292	,06443		
	Officer of Unit Managers other than HR	4,5511	,41812	,06233		
	Total	4,5798	,37707	,03612		
Religious Ethics	HR Manager	3,9706	,91859	,15754	0,1810	0,8350
	HR Specialist	3,8917	1,08811	,19866		
	Officer of Unit Managers other than HR	4,0222	,79562	,11860		
	Total	3,9702	,91460	,08760		
Autonomy Ethics	HR Manager	4,6912	,52245	,08960	1,1820	0,3110
	HR Specialist	4,6000	,44334	,08094		
	Officer of Unit Managers other than HR	4,7667	,42104	,06276		
	Total	4,6972	,46157	04421		

In the analysis of whether monthly income and values ethics tendencies differ, it is seen that α value is greater than 0.05. Accordingly, there is no difference between the wage groupings received in terms of values ethics.

Table 6. Age Distribution and Values Ethics ANOVA Test Results

		Average	Standard Deviation	Standard Error	F	Sig.
Community Ethics	Between 20-29 years old	4,5846	,41393	,06628	0,13	0,94
	Between 30-39 years old	4,5902	,36318	,05672		
	Between 40-49 years old	4,5417	,36106	,07370		
	Between 50-59 years old	4,6400	,35777	,16000		
	Total	4,5798	,37707	,03612		
Religious Ethics	Between 20-29 years old	3,7949	1,05268	,16856	0,92	0,44
	Between 30-39 years old	4,1341	,84428	,13185		
	Between 40-49 years old	3,9792	,77641	,15848		
	Between 50-59 years old	3,9500	,94207	,42131		
	Total	3,9702	,91460	,08760		
Autonomy Ethics	Between 20-29 years old	4,6667	,36873	,05904	0,38	0,77
	Between 30-39 years old	4,6707	,56579	,08836		
	Between 40-49 years old	4,7708	,41649	08501		
	Between 50-59 years old	4,8000	,44721	,20000		
	Total	4,6972	,46157	04421		

In the ANOVA test carried out, it was observed that there was no statistically significant difference in the values ethics trends between the age groups. However, it is observed that community ethics values are predominantly preferred in the 50-59 age group, religious ethics tendency in the 30-39 age group and autonomy ethics in the 50- 59 age group. Accordingly, people at older ages may be making their ethical assessments based on autonomy and by giving importance to cooperation and harmony with others with the sense of autonomy provided by their experiences.

When we look at whether ethical perceptions differ by gender or not, it is seen that ethical perceptions do not present a statistically significant difference between gender distributions. However, unlike the values ethics approach given in Table 3, it can be seen that women take a more intense approach in all ethical trend areas compared to men (Aksoy & Koçancı, 2019, pp.359).

Table 7. Gender Differentiation and Ethical Trends Scale T-Test

Gender of the Interviewer		Frequency	Average	Standard Deviation	Standard Deviation Error	t	P
Deontological Ethics	Woman	55	3,1727	,38222	,05154	-1,721	,088
	Male	54	3,3130	,46502	,06328	-1,718	,089
Teleological Ethics	Woman	55	2,9636	,49223	,06637	-,114	,910
	Male	54	2,9745	,51014	,06942	-,113	,910
Virtue Ethics	Woman	55	3,0932	,36421	,04911	-,981	,329
	Male	54	3,1736	,48468	,06596	-,978	,330

BUSINESS ETHICS AND ETHICAL PROBLEMS

Development of Business Ethics and Distinguishing Features

Business ethics remains an important issue that concerns all businesses at all levels. Corporate governance scandals that have examples such as "Enron" and "WorldCom" have led to a rapid increase in ethical laws and regulations, especially in the United States and other world countries. Although the complex relationship between ethics and strategic management goes back to ancient times, today's conditions make this relationship even more imperative. For example, in the USA, although the Sarbanes-Oxley Act came into force in 2002, ethical scandals about businesses continue and social stakeholders' expectations for businesses to be more ethical are increasing. On the other hand, because ethics is a difficult issue to understand and put into practice, it has been a matter that business executives prefer to violate.

From the technological point of view, with the arrival of the Information Age, which has gained momentum with the innovations in the field of internet and telecommunications, the businesses and the way businesses are carried out in economies, industries, and companies are changing. Online technologies facilitate corporate change. While speed, scope, economies of scale and efficiency change through information technologies, there are different results for consumers and employees.

From a political perspective, the collapse of communist regimes and the rise of global terrorism are changing the structure of trade and business partners. National and organizational strategic associations blur political boundaries. Electronic democracy changes the political thoughts and behaviors of individuals and groups. On the other hand, the transparency of international issues has exposed introvert systems and ideologies to open. Liberalization of key industries (public services, airlines, telecommunications, insurance) has changed the structure of competition, accelerated mergers and acquisitions, triggered monopolization and oligopolization. Security issues have been threatened since 11 September and civil rights have further come to the fore.

From a demographic perspective, the workforce in businesses continues to be affected by various environments. Employers and employees faced the effectiveness and reliability of factors such as mobbing, sexual harassment, discrimination, and career changes.

Business ethics studies have developed in five different phases. This is a development that includes before the 1960s, 1960s, 1970s, 1980s, 1990s and after the 21st century.

- **Before the 1960s**

Looking towards the middle of the 20th century, business ethics consisted only of complying with the law (no deception and fraud) and making money for stakeholders for public companies (Bowie & Schneider, pp. 10). In this period, ethical issues that are business-related were mostly discussed theologically.

- **After 1970**

In the 1970s, the subject of ethics started to be handled under the title of "social responsibility" of enterprises. Business ethics started to develop in the 1970s as a research area. Philosophers were also included in this field using ethical theories and philosophical analyzes to establish the discipline of business ethics. In light of these developments, businesses have started to worry about their images in society and realized that they have to approach ethical problems directly.

In the late 1970s, important ethical issues such as bribery, deceptive advertising, product reliability, and the environment emerged.

- **The 1980s**

In the 1980s, business academics and experts accepted business ethics as a field of study. In particular, "ethical principles" and "ethical programs" have been established in many large enterprises in the USA and ethics has started to be taught in graduate programs in the academic field. Growing institutions supported business ethics studies. Business ethics organizations have managed to have thousands of members. In this period, business ethics has gained an important place in leading companies such as General Electric Co., The Chase Manhattan Corporation, General Motors, Atlantic Richfield's Co., Caterpillar Inc., and S.C. Johnson & Son, Inc. on a global scale.

- **2000 and later**

During this period, a number of corporate scandals and bankruptcies around the world have seriously undermined investors' trust. The global economic crisis, which started in 2008 and still has impacts, has deeply affected developing and third world countries. All these problems emphasized the necessity of effective corporate governance.

Causes of Unethical Behavior

Unethical behaviors arise as a result of individual perceptions and social degeneration in businesses. It refers to behavioral problems that occur for various reasons within the organization.

A wide range of ethical problems and unethical behaviors are encountered in organizations. No matter for what reasons and at what level, unethical behavior negatively affects the quality of organizational life, the motivation, performance, loyalty, and satisfaction of employees.

When the socio-cultural and economic types of unethical behaviors (corruption, favoritism, bribery, etc.) are examined, it is seen that these are generally derived from external reasons, and when psychological types (selfishness, torture, violence, sexual harassment ... etc.) are examined, it is seen that these are caused by personal reasons.

An international survey conducted among 300 large companies in 1987 revealed that over 80% of top executives saw the following issues as moral issues encountered in businesses.

- Employees' conflicts of interest 91%
- Inappropriate gifts 91%
- Sexual harassment 91%
- Disapproved payments 85%
- Mandatory approval procedures 84%

In another study conducted in the 1990s, the following situations were identified.

- Taking bribes or offering bribes
- Stealing, embezzlement from the firm
- Removing complainants or whistleblowers from the company
- Leaking out company information or secrets
- Using company tools and equipment for personal purposes.

In addition, it is for the individual to act only by thinking about his own interests, to think and manage all the developments based on himself, and to ignore the rights and interests of others. The selfishness perspective provides an interpretation of the extent to which incidents will affect his earnings and losses.

Sometimes it emerges as an ethical dilemma. An ethical dilemma is a conflict of two or more values. If one of these values is preserved, the other one cannot, or in order to protect one or more, it is imperative to ignore the others.

For example;

- Everyone's doing it. Just like an accountant who helps his client to evade taxes saying that his other colleagues also do it.
- Legal is moral. Like a person who finds a lost item does not want to deliver the item to anyone unless the owner of the item shows up and proves that he is the owner.
- Possibility of revealing and punishment. If the unethical behavior exhibited is likely to be revealed and punished in the end, the person may give up this behavior.

A moral dilemma can be defined as a moral problem that occurs when each of the conflicting parties raises an indisputable incident and that is not easily resolved.

- Recently, a formal approach is foreseen in the resolution of dilemmas.
- Identifying valid situations
- Determination of the moral problem based on a situation
- Identification of the person, people to be affected by a problem and how they will be affected
- Identification of alternative behaviors of affected people in problem-solving
- Determination of the possible outcome for each alternative
- Choosing the most appropriate behavior (Kutlu, 2015, pp.116)

TEDMER conducted a research study called "Approach of Turkish Workforce to Ethics" in 2002 and 2005.

Some summary results of the research are presented below:

- The rate of having heard the concept of "business ethics" in the Turkish workforce is over 60% (64.2% in 2002, 63% in 2005).
- The rate of those who stated that their organizations have business ethics standards, whether written or not, is 36.3% and 29.2% in 2002 and 28.6% and 25% in 2005, respectively. So, the total 66% (36.3 + 29.2) ratio in 2002 became 54% in 2005.
- The rate of those who stated that there is no separate training on work ethics behavior standards in the organization worked at is 62% in 2002 and 76% in 2005. In other words, separate and special training on business ethics does not seem to be a common practice.
- The proportion of respondents who stated that they observed behavior against work ethics in their organization was 35% in 2002 and 25% in 2005.
- The percentage of people who thought that the companies they work at have business ethics is close to each other in research studies conducted in 2002 (70%) and 2005 (67%).
- The three most important features for a company to be ethical are "Stealing, theft, or fraud", "Valuing employees", and "Paying obligations being realized on time". Although it is interesting that the feature of "valuing customers" can only be in the fourth place, this may be due to the possibility that the participants saw this issue within the scope of "the company not to commit fraud". The rankings in both 2002 and 2005 are similar.
- Automotive, consumer durables and tourism sectors took the first three places in the ethical evaluation of the sectors in both 2002 and 2005 studies.

According to the "Turkey Personnel Management Association" (PERYÖN);

Ethical principles and rules are defined under the following headings:

- Personal Integrity and Consistency
- Professional Responsibility
- Development Awareness
- Equitable and Fair Approach Awareness
- The Consciousness of Observing Balance of Interest
- Use of Information
- Social Responsibility

Examination of the violations of these principles is also carried out by the Ethics Committee.

At Turkey's Business Ethics Summit 2019 held under the theme "Business Ethics, Law and Justice Relationship" by IGIAD (Turkey Venture Economics and Business Ethics Society), it was aimed to create sensitivity about business ethics problems and to develop suggestions and solutions for what can be done to overcome these problems. The other aim was to share the best practices for business ethics and to demonstrate that business ethics is possible and feasible with concrete examples and to encourage the business world. It was underlined that the major deficiency of the business world in Turkey is the insufficient legal support in business ethics practices.

In the survey conducted in 2018, questions were asked on the relationship between business ethics and law, and 90 percent of the respondents stated that there are legal gaps in business ethics practices and this gap should be filled with laws. It is also stated that companies are obliged to protect and respect human rights.

Topics Covered in the Paper

Constitutional Court Member Prof. Dr. Engin Yildirim stated that companies have a responsibility to protect and respect human rights and that they have a responsibility to compensate if there has been any damage due to their activities.

Noting that with the increase in globalization since the 1990s, multinational companies have some responsibilities in the field of human rights, Yıldırım also said:

"Access to the Internet was recognized as a constitutional right in the Finnish Constitution. Water, electricity activities concern millions of people. Negativity stemming from activities of these can negatively affect the rights of these people, such as vitality, guaranteed under the Constitution. The United Nations (UN) High Commissioner for Human Rights conducted a research study to examine more than 300 alleged human rights violations. They found that companies were directly responsible for 60 percent of these 300 claims. Again, 159 case studies in 66 countries were examined and it was seen that multinational companies had negative activities and had negative effects on human rights. Companies have started to develop social responsibility projects based on volunteering."

Pointing out that the sensitivity of the companies in the field of human rights is not an ordinary social benefit, but the fulfillment of the debt to the society, Yıldırım said, "There are 3 basic trivets, and these are protection, respect, and compensation. Companies are obliged to protect and respect human rights. They are also responsible for compensating the damage that is due to their activities. This is also included in the UN Business and Human Rights Guidelines, which consist of 8 main and 24 complementary principles, adopted in 2013. This does not have a binding legal aspect, it is a voluntary practice, but currently, Business and Human Rights Contract studies are ongoing, and these are internationally binding within the framework of the UN Human Rights Council." (Turkey Business Ethics Summit, 2019).

Şekib Avdagiç, Chairman of the Board of the Istanbul Chamber of Commerce, said, "Since the agricultural revolution, which is the symbol of the transition to settled life, that is, approximately 10 thousand years, there are three indispensable elements of human life. These are morality, law, and justice, including business ethics." (Turkey Business Ethics Summit, 2019).

Turkey Business Ethics Survey 2018 results were announced by the Istanbul Chamber of Commerce (ITO). According to this, lying is one of the most common business ethics problems. Secondly, it has been observed that the evasion rates in the workplace have increased significantly compared to the previous study. In the evaluation made in terms of employees, it was seen that lying appeared again in the first place. While coming to work late and not working properly are also highly perceived as business ethics problems, evasion has increased slightly compared to previous research studies. Issues such as theft and bribery are among the business ethics problems that are said to be much less common.

When the situation between generations is evaluated;

It was stated that the members of the generation Y, consisting of people born between 1982 and 2000, perceived more problems than the members of Generation X, who were born between the years 1965 and 1980, and that the elderly may have been in a state of taking for granted (Turkey Business Ethics, 2018, Survey).

Professional Corruption and Weakening of Professional Ties

Corruption is a concept that shows the deterioration in social behavior patterns.

The occurrence of value confusion in the society and the benefit of utilitarianism getting ahead of everything over time also affect the implementation of ethical rules. In conflict at the personal social level, people with values different from social values find themselves in ethical conflict when they oppose the value system of society. Ethical conflicts may also arise due to differences in values between countries.

Professional corruption has affected business life as in every field of life. These reasons are:

- Overly ambitious, selfish and greedy behavior
- Not showing enough sensitivity
- Incomplete assessment
- Lack of planning
- The motive to protect friends in good faith
- Not knowing laws, rules, and methods
- Legal delay in business life parallel to technological and social developments
- Ideological and political discrimination (Başpınar & Çakıroğlu, 2011, pp. 153)

As a result, the concept of burnout emerges.

The concept of burnout was first introduced by Freudenberg in 1974 as a "professional danger" and entered the literature. Later, this concept was defined by Maslach and Jackson in 1981 as a syndrome that manifests itself in different ways, such as depersonalization of people they encounter due to their jobs, emotional exhaustion, decrease in personal accomplishment and competence feelings (Kutlu, 2015, pp.105).

Individual dissatisfaction and unhappiness increased. Life satisfaction was first introduced by Neugarten in 1961. Life satisfaction includes individuals' lives and all aspects of their lives. Life satisfaction is associated with age, gender, working and business conditions, education level, religion, race, income level, marriage and family life, social life, personality traits, biological factors.

The negligence of job quality. Quality is an issue that we are familiar with every day. Quality briefly means exceeding expectations. It is a dynamic concept. Because people tend to go up every time their expectations are met. A product that has been described as "high-quality" for a few years can be described as "ordinary" today and "unacceptable" tomorrow. First of all, we have to accept that ensuring continuity in service and product quality can be achieved with management quality. Improving management quality can be achieved by implementing some principles. One of the most important tools for improving management quality is managing with processes and data. Sharing performance-related information with employees and improving the competencies of the employees in line with the target and empowering the employees ensure continuous improvement of performance (Başpınar & Çakıroğlu, 2011, pp.163).

While the profession connects people, on one hand, it separates them and produces the social hierarchy on the other. The addition of the individual to the society and his integration, in other words, socialization takes place through his profession. It is a social consequence in the concept called self, and it is all of the feelings, ideas, intentions, and evaluations that make the man himself and distinguish him from the others. Socialization, which is available in the profession, also provides the gaining of personality and self-determination. However, the current labor market makes it risky for the individual to adapt and connect to his job as a profession. Thus, the individual's ties with the profession are weakened and it is difficult to identify himself with the profession. In the flexible labor market, the fact that employees have to change jobs in order to protect and maintain their earnings and having to be mobilized between new tasks and places cause the approach to the profession to come out of a long-term perception. Therefore, it damages the commitment to the profession and causes self-harm. As a result, the image of the traditional profession changes completely (Kutlu, 2015, pp. 109).

GENERATION GAPS AND NEW WORKING FORMS

Transformations Created by New Working Forms (The Concept of "Leisure Time")

The transformation of working is not only determined by the organization of the work. The reorganization of the work, that is, reshaping it according to the new conditions, and being able to make an intellectual transformation regarding its meaning in terms of the employee, is related to the reorganization of many areas in the life of the individual. This relationality necessitated some arrangements for the life of the employee other than business life, which is clearly seen in the differentiation of management styles. On one hand, while changes regarding the form of production are observed, on the other hand, it is observed that the shaping is made in which consumption has become a norm.

The responses regarding the discussions that the control of the employee over time, spatial autonomy and domination over labor differ with informational development are descriptive of these semantic transformations.

Norbert Elias (2000: 1-60) defines time as one of the tools that people have to learn in order to determine their place in the course of incidents from a certain stage of social development, enabling certain tasks to be met and achieving human-specific goals. He states that in the course of incidents, situations that will remain ambiguous on their own, moments and periodicals will be determined with the help of "time". Today, "time" is used to fulfill a wide variety of different tasks, to organize social relations and activities and coordinate with each other. In this context, just as in language and knowledge, we can talk about the individualization of a social phenomenon that adapts to the community by applying control to each user by the time.

Elias (2000: 1-60), pointed out that the societies of the modern era established their relations with time in different ways, and developed a unique sensitivity towards time, with the rapid development of social roles and functions as a result of increasingly differentiating and multiplying and increasingly integrated goals and activities. It creates social external coercion or pressure represented by clocks, calendars, and even plans showing vehicle take-off times, and by internalizing this pressure within a certain process, individuals in these societies control themselves and suppress themselves, adjusting the activities and behaviors according to a certain time matrix. It appears to have resulted in the consequences of their restrictions.

Time is a symbol of the synthesis that has been learned socially. As a characteristic feature seen in the people of developed industrial societies, it is the civilization process that lies behind the phenomenon of adjusting and organizing oneself according to social time. In simple societies, where these social habits and social time needs are much less than in industrial societies, it is understood that this phenomenon is a symptom of a civilization process by looking at the absence of a time sensitivity that is awake, alert and spread to the whole society every minute (Elias, 2000, pp.36).

Leisure was a time after productive activities and was necessary for the reproduction of the work. In a way, it had supportive and complementary use and it had a functional or administrative function (Juniu, cited in 2000, pp.69). (Ottoman & Kaya, 2014, pp. 3). Leisure belongs mostly to the modern period. Social and cultural life, which changed with industrialism, created autonomous living spaces in itself. Capitalism has organized not only work as a system of production, but also leisure time, and has trapped work in a strict coding, and stripped leisure time from its terminological meaning and has been in an effort to organize it as a profitable area. In this context, leisure time has moved away from the initiative of the individual and gone under the influence of institutional devices and instrumental hegemony.

Leisure is no longer the time of individual calmness, intellectual depth, and free choices. It has become the address of commodities, fictional life experiences, shopping events, package entertainment, media shows, and escapist tendencies that capitalism produces to increase profits (Aytaç, 2005, pp.1).

Puritan business ethics described leisure time as "stray" and "wasted" time.

Lafargue criticizes the "right to work" bourgeoisie principle where the employee has the right to live and defends the "right to leisure" under the name "Right to Laziness".

Russell likewise criticizes the notion of modern work in his work titled "Praise for Wandering" (1990, pp.111).

The capitalist system organizes leisure time as a way to eliminate the alienation and stress produced by the business process. The leisure time, which is accepted as the preparation for the upcoming working hours, is arranged according to the study. Deluze and Guattari (1990) explain that capitalism, through its contradictions and crises, has been discovering new points of opening and breathing and that it uses it for capitalist continuity.

Today's leisure time is an area where the individual has no belonging over time and its borders are drawn. As humanity continues to exist, production and consumption relations will continue with new forms.

Generational Differences and Business Life

It is a well-known fact that progress and distance can be achieved with the experience and knowledge that a generation will transfer to the next generation in terms of social life. It is stated that the changes in the generations are the forces acting in a certain process (Comte, 1974, pp. 635-641). In later years, researcher Karl Mannheim (Mannheim, 1998) conducted a comprehensive and systematic research study on generations. In his study, Mannheim described generations as a community of people with common habits and common cultures sharing these values. Jean-Claude Lagree (Lagree, 1991, pp. 7) defined the concept of generation as communities that lived in the same historical period, that were affected by the same incidents and had the same social identity.

It is not correct to classify and evaluate the generations only in terms of their period of birth. It is also necessary to define the emotions, thoughts, and experiences of the individuals who make up the group of generations. Each generation has its own distinctive characteristics, value judgments, and attitudes, strengths and weaknesses, apart from existing life and work habits, their way of perceiving life and different communication styles. These generations are defined as "Generation X" between 1965-1979, "Generation Y" between 1980-1999, and "Z Generation" between 2000-2021 according to the years they were born in.

Generation theory has defined Generation X as people who were born in 1960-70, care about making a career, educated, prefer watching movies instead of reading books, emotional, well-groomed, open to change, professionally focused on performance, who have similar characteristics in different geographies. These people are portrayed as an idealist, contentious and loyal generation (Senbir, 2004, pp.24). This generation, defined as a generation that follows the X generation rules, has a strong belonging, respects authority, and stands out with its loyalty and diligence (www.mostar.com.tr), has witnessed many inventions and discoveries. X generation residents who have opened their eyes to the world by the roller washing machine, transistor radio, cassette player and turntable have experienced many transformations.

Generation Y is a generation born between 1980-1999, knowing technological innovations, growing up with computer and arcade games, meeting mobile phones during adolescence, developing social aspects, adopting flexibility in business environment and working hours, taking care of expressing themselves more than career, and taking an active role in decision making (Howe and Strauss, 1992). Generation Y has the feature of a generation where the intergenerational difference is felt the most. Because they love to be independent, they are fond of their freedom and are different in their business life. They do not like to work between the specified working hours. So, they need to focus on work rather than on working hours. In this case, it is important to make them part of the job. Organizational commitment of generation Y is less compared to generation X.

Generation Z refers to people born in 2000 and later. Their capacity to access information is much higher thanks to technology. Generation Z, which started getting educated at an early age, shows faster mental development. Their individualism is stronger than the X and Y generations. Their most obvious way of socializing is social media. Generation Z, accepted as the children of the digital age, is known to have the ability to do many things at the same time. Generation Z has made wearable, portable, etc. technology products an indispensable part of daily life (Senbir, 2004, pp.27-28; Williams, 2010, pp.12). Generation Z likes to use the internet and mobile technologies. They play with iPads instead of toys and grow with technology. They are active in every field with smartphones, iPads or tablet computers that are common today. They especially prefer to socialize via the internet.

Table 8. Technology Uses and Likes of X, Y, and Z Generations

	Generation X 1960-1979	**Generation Y 1980-1999**	**Generation Z 2000 – 2000+**
Iconic Technology	Video, Walkman, IBM, PC	DVD, Internet, Email, SMS, Gameboy, Xbox, iPod	Google, Facebook, Twitter, Instagram, iPad, iPhone, Android
Popular Culture	Ripped jeans Extreme colors Piercings	Hat Men's Cosmetics Overhead Style	Tight pants V-neck Wearable technology
Purchasing Motivations	Inter-brand Preference Specializations	No brand loyalty Friend Reference	Brand Obsession Trends
Ideal Leader Feature	Directive Authoritarian	Empowering Collaborative	Inspiring Co-Creator
Marketing Method	Direct Marketing Balance Sheet	Viral References Electronic Marketing	Interactive Campaigns Brand Interest
Training Methods	Spontaneous Interactive Comfortable environment Table-order	Multiple Sensory Image Cafe Style Music-Multi Model	Student Focused Kinesthetic Salon Style Multiple stimuli

Table 9. Characteristics and Values of X, Y, and Z Generation

Generations	Generation X	Generation Y	Generation Z
Properties and Values	Loyalty feelings are variable Respectful to authority Sensitive to society Business motivations are high Abstinent Disconcerting Low relationship with technology	Feeling of loyalty Hard to accept authority Fond of independence Very often changing jobs Individualist Growing with technology	Collaborator Creative In technology

Source: Deneçli, C. & Deneçli S. Sherbet by Pulse, Activity by Generation: Marketing of Entertainment.

Almost all of the Z generation is an active social media user and the majority say that the e-commerce sites' advertisements on social media are effective in their shopping decisions. Although they seem wasteful, they prefer to save instead of spending money. They follow the campaigns closely and they are the most active users of opportunity sites. One of the most distinctive features of the Z generation is that they are researchers. 80 percent of these people make research before shopping and make price comparisons (Epnext.com). According to the research, Generation Z, born inside the digital world and growing in the environments, is most active on YouTube with 97 percent. One of the most remarkable results of the research is the impact of the digital world on the social life of children belonging to Generation Z. Digitalization changes the way they behave.

Figure 1. Computer usage rates of people at home in Turkey

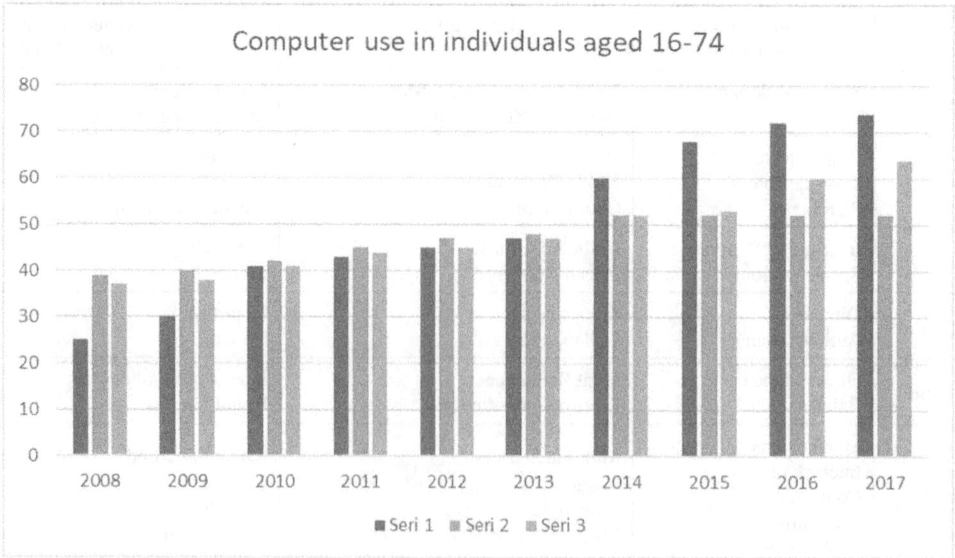

ETHICAL CLIMATE

An Ethical Climate at Work and Unethical Behavior Analysis

The ethical climate is the organization's common perceptions of how to deal with ethical issues and what is ethically correct behavior. Victor and Cullen (1988) emphasized that factors such as social norms, organizational behaviors and patterns, and business-specific factors influence and enhance the ethical climate of the organization. An important source of ethical climate was prepared by Reidenbach and Robin (Venom et al, 2003, pp. 148). The authors categorized the climates encountered in organizations in five steps starting from the lowest level according to their morale maturity degrees: immorality, legality, responsiveness, rising ethics, and improved ethics.

The ethical climate of a business is a shared set of thoughts about which behavior is correct and how ethical problems can be overcome (Savran, 2007, pp.16). This set affects organizational decisions at all levels and conditions. Factors such as personal interest, corporate profit, activity effectiveness, individual friendships, interests and interest areas of teams, social responsibility, personal ethics, rules, standards, procedures, laws, professional codes and so on are effective in the formation of different ethical climates in businesses (Savran, 2007, pp.16-17). Creating an ethical climate in a business means that business ethics is formally, precisely and clearly a normal part and order of the business in daily business life, as well as spreading ethics in the business's foreign policies, senior management, all business activities and daily decision making (Savran, 2007, pp. 18).

Ethical Leadership and Ethical Leadership Typology

Enderle (1987), one of the first authors who used the concept of ethical leadership in the literature, used ethical leadership under the name of "Managerial ethical leadership". Enderle has defined managerial ethical leadership as making decisions with responsibility in complex situations and divided it into two: personalized leadership and corporate leadership. Employees follow the instructions, remain loyal to their employers, and get paid for the honest work they perform maintaining their ethical standards. However, when it comes to ethical behavior, questions such as "What happens if employers produce harmful products due to insufficient control?", "Who should employees go to if employers ignore concerns?" come to mind. If there are no ethical standards, the dilemma forces employees to make a decision and unconsciously harm others may be in question.

Internet and information technologies, globalization, liberalization, and mergers have accelerated the rate of change and uncertainty in organizations. An ethical decision is defined as a decision to be accepted by a broad society, both legally and morally.

Three approaches, namely "Utilitarian", "Justice" and "Fundamental Rights", can be mentioned in making ethical decisions.

- Utilitarian approach: It deals with the decisions taken by looking at the results of the action, rather than focusing on whether the action itself is good or bad.
- Justice approach: This approach questions whether the decisions taken are fair. The justice here is based on the balanced distribution of not only benefits but also penalties and burdens within the organization.

- Fundamental rights approach: Based on the assumption that individuals have inherited rights, it mentions that businesses, employees and managers should be respectful of these rights.

The absence of ethical leadership in businesses can lead managers not to notice unethical incidents within the organization. From the perspective of abuse of duty, if there are no complaints within an organization, managers should be suspicious of this situation. Failure in ethical leadership generally requires the CEO, team leader or other rank managers to be scrutinized in the aspects outlined below.

- Ethical blindness: They do not address ethical issues due to carelessness and inadequacy.
- Ethical dumbness: They do not use ethical language or firsts. It is a non-straightforward perspective on values.
- Ethical inconsistency: They cannot see the discrepancy between the values they say they follow.
- Ethical paralysis: They do not act according to their values because they are afraid of the consequences of their actions or because of ignorance.
- Ethical hypocrisy: They do not act adhering to the values they advocate. They transfer the work they do not want or cannot do to others.
- Ethical schizophrenia: They have no consistent values.
- Ethical indifference: They think they cannot do anything wrong because they know who they are.

Different suggestions and precautions have been expressed for businesses regarding the above-mentioned issues.

- **Adopting a proactive and reactive approach:** In the management of unethical behavior, managers take two types of approaches, proactive and reactive. The proactive approach involves preventive measures that have been introduced before unethical behavior has emerged. These include policies such as increasing organizational commitment, reorganizing administrative processes, reviewing hiring methods, strengthening staff, organizing training programs, and improving the quality of working life.
- **Creating a strong ethical culture:** In organizations with a strong culture, it will be easier and shorter to create ethical codes and turn them into behaviors. When creating organizational culture; ethical codes, organizational policies or organizational plans can be used.
- **Legal regulations:** In order for legal regulations to be effective, laws must be actively used.
- **Creating an effective reward/penalty system:** The reward system is based on rewarding employees who display ethical behavior. Conflicts can occur between rewarded ethical behavior and employees' ethical values. Different rewarding systems can be used to prevent this and provide motivation for employees. This will positively affect employees.

Figure 2. Ethical Leadership Typology

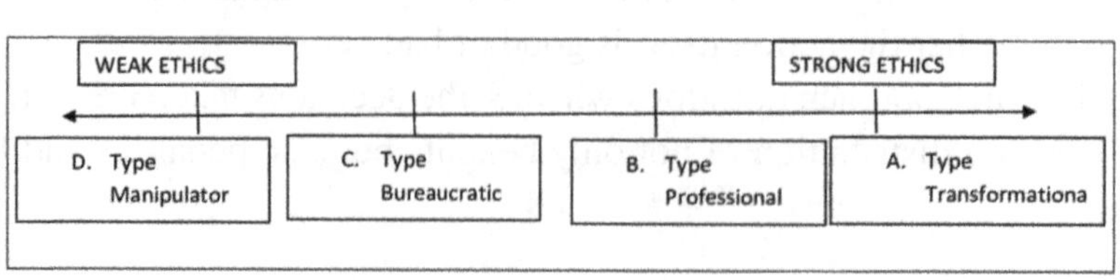

- **Ethical Leadership Typology:** Weiss examined the ethical leader with a weak and strong ethical aspect and addressed the ethical leader in 4 categories.

- **Manipulator Leader:** Manipulator leadership style is based on Machiavellian ethics, which deems leadership immoral. So, the result obtained confirms the power behind the objectives of the tools and manipulator used to achieve this result. This is an egoistic and essentially economic form of leadership.
- **Bureaucratic Leader:** It is rule-based moral leadership. It is based on the theories of the famous German sociologist Max Weber. The driving force behind this style is effectiveness rather than efficiency.
- **Professional Leader:** The professional leader aims to do something right. The basis of this theory is the principles of Peter Drucker and these allow managers to see experts as professionals with the tools to ensure that the task is completed effectively.
- **Transformational Leader:** The basis of this style is based on James Burns' theory and personal ethics. The transformational leader bases his effectiveness on his relationships with his followers. Also, this style focuses on the charisma, energy, and excitement that the leader adds to their relationship (Karabayır, 2015, pp. 86).

Ethical Code

Having a written code of ethics is important for companies. Ethical code expressing universal truth in business life is a set of written rules prepared by the institution to guide the current and future behavior of the institution's managers and employees.

In the process of creating ethical codes, it is necessary to set out from corporate management principles, mission, vision, and values. These basic principles, which will establish only the basic infrastructure, help companies to get started. With the addition of responsibilities, standards, and rules, the parts of the whole are completed. If the content of ethical codes is considered as the layers of a pyramid, "mission" is the top layer of this pyramid, while "standards and rules" are located at the bottom layer. Details and technical information given as we go from the bottom to the top layer of the pyramid are reduced. In some cases, all layers of an ethical code may not be included in a single document. Companies wishing to explain one or more of these layers in more detail may choose to prepare separate documents. In any case, however, these layers form the company's ethical backbone together.

The principles of business ethics include various codes of ethics for inter-business relations and non-business relations, apart from general principles. Ethical codes for the enterprise: these codes are seen in the form of responsibilities against the employee and responsibilities of the employee. It is seen that businesses consider suitable working conditions for human health and for efficient work as their primary responsibilities against their employees. The expected responsibilities of the employees are to refrain from behavior that would put the workplace and himself in a difficult situation. In addition, other responsibilities expected from employees are not using the facilities for their own benefit, showing diligence in the use of tangible or abstract assets of the workplace, and in particular, complying with the law.

Ethical codes for non-business relations: These codes include the responsibilities of the business to its own customers and stakeholders as well as to the community in which it is involved. Businesses have various responsibilities against their customers, stakeholders and the community. The protection of confidential or proprietary information produced or existing in the institution is one of the responsibilities of the enterprises.

Because if such information is shared, the enterprise or its stakeholders will be damaged. (Çetintaş,2016, pp.17).

Ethical Rules About Professions

Within the scope of determining the ethical rules related to the professions in our country, the Vocational Qualifications Authority was established with the Law No. 5544 dated 21.09.2006. Based on national and international professional standards, the purpose of the institution is to establish and operate the national qualification system necessary to determine the principles of national qualifications in technical and professional fields and to carry out activities related to inspection, measurement and evaluation, documentation and certification. Accordingly, the Regulation on the Preparation of National Occupational Standards for implementation has been prepared and put into effect. According to article 5/2 of this regulation, proficiency levels related to the profession's standards of which are to be determined must comply with the qualification levels adopted by the European Union and the European Competence Framework for Lifelong Learning (AYÇ), adopted by the European Parliament and Council on 23 April 2008. In addition, each professional chamber has adopted the principles of professional ethics that its members must comply with and have made regulations (Kolçak, 2013, pp.185).

National Professional Standard

(UMS) is the minimum norm that indicates what the required knowledge, skills, behaviors, and attitudes accepted by the Vocational Qualifications Authority are for the successful execution of a profession.

Determination of Professions

The occupations standards of which are to be prepared are determined by the Board of Directors, taking into account the priority needs of the labor market and educational institutions and the recommendations of the sector committees.

Some examples of ethical rules for professions are:

Ethical Rules Related to the Profession of Teaching:

The Principles of Teaching Profession Ethics have been tried to be stated below in a general and basic framework.

- **Professionalism:** Professionalism as an ethical rule is used to mean that a teacher's job has the required teaching profession standards.
  - To have gained knowledge, skills, and attitudes about the task completely,
  - To provide the service on time and flawlessly,
  - Continuous self-criticism and evaluation in order to execute the profession better,
  - To provide quality in service,
  - To use his personal talents to the end when performing his duty,

  - To be able to use the discretionary power in a reasonable measure regarding his duty,
  - To be able to continuously improve the methods and techniques of his profession.
- **Responsibility in Service:** Responsibility is a concept that requires the creation of these measures as well as the compliance with professional and ethical criteria. The basis of responsibility is the obligation to use authority.
- **Justice:** The teacher has to be fair in all kinds of actions and to provide justice in the relations between students.
- **Equality:** Equality includes determining the limits to be applied in the distribution of benefits, issues, and services.
- **Providing a Healthy and Safe Environment:** One of the basic rights of a student is to be in a healthy and safe school-classroom environment.

Honesty- Integrity, Trust, Impartiality, Professional Commitment and Continuous Development, Respect, Effective Use of Resources also indicate ethical rules about the teaching profession.

To Give You Information About the Ethical Rules in the Tourism Sector

They have started working on some internationally binding principles to minimize the negative impact of the industry on cultural heritage and the environment, and to maximize the benefit from tourism. The first work on this subject was performed in 1997 at the General Meeting of the World Tourism Organization in Istanbul. The 10-point Global Tourism Ethics principles were adopted at the General Assembly meeting held in Santiago in October 1999.

According to this;

- Contributions of Tourism to the Principles of Mutual Respect and Understanding Among People and Communities
- Tourism as an Individual and Collective Satisfaction Tool
- Tourism as a Continuous Development Factor
- Tourism as a Benefit Generating Benefit for Host Countries and Communities
- Investors' Obligations in the Development of Tourism
- Right to Tourism
- Freedom of Tourist Movement
- Rights of Tourism Employees and Entrepreneurs
- Applicability of Global Ethical Articles to Tourism

In addition to the issues listed above, it should be noted that there are ethical rules that professional staff guiding in the tourism sector must follow (Kolçak, 2013, pp. 193).

Evaluation of the Relationship Between Employees' Age and Business Ethics

Considering the age of employees and partners, 7.6% of the partners are under the age of 25, 26.9% are between the ages of 25-34, 24.1% are between the ages of 35-44, 20.3% are between 45 -54 years old, 21% are 55 years or older. When we look at the average age in the table, it is seen that the employees have an average age of 33 years old and the partners have an average age of 40 years old. Partners need

Table 10. Distribution of the sample by age variable

Variable		Partner		Working	
		N	%	N	%
Age	Under 25	**22**	**7,6**	**23**	**12,0**
	25-43 years old	78	26,9	78	40,8
	35-44 years old	70	24,1	47	24,6
	45-54 years old	59	20,3	32	16,8
	Age 55 and over	61	21,0	11	5,8
	TOTAL	290	100,0	191	100,0

Table 11. Averages of Employees' Perceptions of Business Ethics Sub-Dimensions by Age

		N	X	Ss
Employee Ethics	Under 25	23	4,3739	,27833
	25-43 years old	78	4,3256	,41916
	35-44 years old	47	4,2383	,50500
	45-54 years old	32	4,3875	,47908
	Age 55 and over	11	4,1273	,48392
	TOTAL	191	4,3089	,44328
Customer Ethics	Under 25	23	3,9022	,72197
	25-43 years old	78	3,9231	,66974
	35-44 years old	47	3,9096	,54530
	45-54 years old	32	4,0312	,61155
	Age 55 and over	11	3,6818	,44848
	TOTAL	191	3,9215	,62543
Partner/ Colleague Ethics	Under 25	23	4,2609	,58620
	25-43 years old	78	4,3397	,38005
	35-44 years old	47	4,2340	,35891
	45-54 years old	32	4,3516	,45729
	Age 55 and over	11	4,0682	,56003
	TOTAL	191	4,2906	,43072
Environmental Ethics	Under 25	23	4,1304	,38585
	25-43 years old	78	4,1752	,45813
	35-44 years old	47	4,0709	,40490
	45-54 years old	32	4,0625	,37328
	Age 55 and over	11	4,0909	,33635
	TOTAL	191	4,1204	,41610

(İştar,2015, pp.188).

to gain a certain experience to grow their companies and start recruiting staff. It was determined from the results that the employees are younger than the partners as expected.

In addition, it was determined whether the relationship between the age of the employees and the sub-variables of business ethics was significant. In this case, frequency value percentages of the employees' perception pertaining to work ethics and levels with their scores are given according to the age variable.

According to the Research Data

The global youth unemployment rate recorded the largest annual increase in 2009, far exceeding the rates in other age groups. 75.8 million young people are unemployed.

In 2010, the global youth unemployment rate of 12.6% surpassed the adult unemployment rate of 4.8%.

Today, about 152 million young workers live in households that are below the poverty line ($ 1.25 per day) and they make up 24% of the total number of working poor people.

Young people who can find jobs have to accept extremely low wages. Some employers use this as an opportunity to exploit young people. PATH, 24, INDIA

In 2010, the total youth unemployment rate was 25.5% in the Middle East and 23.8% in North Africa. Young female unemployment is particularly remarkable in these regions: 39.4 in the Middle East and 34.1 in North Africa.

In 2010, 56.3% of young men joined the workforce worldwide, compared to 40.8% of young women. The global unemployment rate for young women is 12.9%, while for young men it is 12.5%.

In developed economies, the European Union and East Asia, men have lower unemployment rates compared to women.

In the past, young women were excluded from some jobs. But in recent years, young women have become even more preferable. Therefore, in some job offers, female candidates' applications will be often supported in recruitment. YOUSSOUPHA, 23, SENEGAL

In developing countries, unemployment is more common among educated youth, which leads to educated unemployment problems.

Employees' job satisfaction may differ according to their age, expectation, and position in the organization.

Job satisfaction is all of the thoughts, beliefs, and emotions that individuals have about the organizations they work with (Özdemir, 2003, pp.160).

The attitudes of the employees in the human resources management process are considered as an extremely important factor because the organization is directly affected. The physiological, psychological and social needs of individuals are the basis of job satisfaction. Individual needs constitute the most important point in employee satisfaction. In addition, factors such as the job selection, the job itself, its location, type, the level of knowledge it requires, its purpose, physical conditions, wages, employee relations, security, etc. are also considered important variables affecting satisfaction. It is understood that business management should not only meet the economic needs but the social needs in order to ensure the satisfaction of its employees (Sabuncuoğlu, 2005, pp.19-20).

In addition, of course, it is seen that today's businesses face business ethics problems regardless of their field of activity. The solution to these problems requires that both business managers and employees act in accordance with business ethics.

In the survey conducted, 3 of the 90 people in question work as managers, 17 of them are directors and 70 of them are authorized personnel. When the educational status of the respondents is examined, 7.8% (7 people) have an associate degree, 77.8% (70 people) are undergraduates, 13.3% (12 people) are graduates, 1.1% (1 person) received a doctorate education. The distribution of employees by gender is as follows: 83.3% (75 persons) male, 16.7% (15 persons) female. Considering the distribution of the surveyed employees by age groups, 18.9% (17 people) are 25 or younger, 67.8% (61 people) are between 26-34 years old, 8.9% (8 people) are between the ages of 35-44 and 4.4% (4 people) are 45 years or older.

Percentage distribution of statements of company practices which are sub-factor of business ethics is given. There is 57.8% of employees in the institution who stated that their "business ethics is stated in writing". The rate of those who agree with the statement that "business ethics in their institution takes place as a coercive rule" is 40%.

52.2% of employees disagreed with the statement "I encounter situations within the company that I think is not in line with business ethics. "75.6% of employees say that "it is stated by clear rules that behaviors that aren't in line with business ethics will not be tolerated."

As regards the working environment, 58.9% of employees feel that "there is no environment within the organization that incites or sets the ground for unwanted behavior." The proportion of those who think that "there is a safe and healthy business environment in the institution" is 62.2%.

80% of the employees agreed with the statement "business ethics values increase the success of the business' strategy". 43.3% of employees think that "the ideas they propose to develop the business ethics model in the institution are rewarded." Business management's valuing of employees' ideas about business ethics facilitates business ethics practices and helps ensure compliance with business strategy.

40% of employees feel that "it is not in any way appropriate for employees in the institution to receive and give gifts." The gifts given should not be intended to affect the other party related to any business or agreement in which the institution is involved. 76.6% of employees said: "the frequent use of mechanisms that reward honesty motivates us in our work". According to another study conducted in this field, the most important moral values of the employees are honesty and integrity, and secondly, the sense of personal responsibility. Participants in the study also stated that someone with confidence and self-esteem could easily direct himself to other values (Yilmaz, 2009, pp. 316).

71.1% of the employees agree with the statement "Customer satisfaction is at the forefront for our company." This statement shows that businesses and employees are aware of their "responsibility against customer satisfaction" (Steel, 2010, pp.7).

The Role and Effects of Transformational Leadership in Placing Business Ethics

It is evident that transformational leadership must also include ethical behavior. Similarly, Burns, too, indicates that a leader must be morally high and create a climate of trust within the organization in order to qualify as a "transformational" leader (Williams et al., 1999, pp. 898).

The fact that transformational leaders have a vision and can act as representatives of change will make this situation easier. It is clear that these leaders should also be open to changes and developments

Table 12. Data on Business Ethics Sub-Factors and Employee Satisfaction

Company Applications	Percentage Distribution of Answers to Questions About Firm Practices							
	Attending		Unstable		Disagrees		Total	
	n	%	n	%	n	%	n	%
Work ethic is stated in writing at our institution.	52	57,8	23	25,6	15	16,7	90	100
I encounter situations in my institution that I think are not suitable for business ethics.	22	24,5	21	23,3	47	52	90	100
Business ethics takes place in our institution as a compelling rule	36	40,0	28	31,1	26	28,9	90	100
There is no environment in the institution where I work that sometimes pushes us towards unwanted behavior or prepares the ground.	53	58,9	15	16,7	22	24,5	90	100
Frequent use of mechanisms that reward honesty motivates us in our work.	69	76,6	8	8,9	13	14,5	90	100
Customer satisfaction is at the forefront for our institution.	64	71,1	20	22,2	6	6,6	90	100
It is stated with clear rules that behaviors that are not in compliance with work ethics will not be despised.	68	75,6	10	11,1	12	13,3	90	100
It is not deemed appropriate for the employees in our institution to exchange gifts.	36	40,0	24	26,7	30	33,4	90	100
The business ethics values of the business increase the success of the business strategy.	72	80,0	10	11,1	8	8,9	90	100
The ideas proposed by the employees to improve the business ethics model in the institution are rewarded.	39	43,3	32	35,6	15	21,1	90	100

in the internal and external environment and be strong persons with the ability to investigate, analyze and interpret.

In order to bring about a formal structure of change in business ethics in organizations, senior management must carry out some of the following activities (Connock &Johns, 1995, pp. 200-221).

- The "ethical principles" that will guide ethical practices in the organization should be determined.
- An "ethics committee" should be established to ensure the implementation and maintenance of the determined ethical principles.
- An "ethical training program" should be prepared and executed to ensure the adoption of these ethical principles among the stakeholders of the organization.
- In order to maintain all practices in the organization in accordance with the established ethical principles, effective coordination and communication must be ensured both inside and outside the organization.

It will not be sufficient alone to make a change in business ethics in the organizational structure. At the same time, this structure should be mobilized, and the stakeholders of the organization should internalize ethical practices, that is, reflect them on their behavior (Waldersee, 1997, pp.262). To this end, the organization's top management should perform some of the activities mentioned below in the leadership dimension:

1. First of all, organizational strategies for the implementation of business ethics should be developed, these should be put into practice and converted into shared values.
2. The organization's top management should be a model for demonstrating ethical behavior in organizational practices and this should be visible in practices.
3. A systematic reward/punishment system should be established to encourage and spread ethical practices and all communication tools should be used in this regard.
4. The performance evaluation system should be restructured to evaluate the ethical behavior of the employees.
5. Objectives and targets related to ethical practices should be determined and measurements and evaluations should be made to determine whether these are achieved or not.
6. On ethical issues, the participation of organization stakeholders in management should be encouraged.

Considering the leadership practices mentioned above all together, it can be seen that these overlap with the characteristics of the transformational leadership style.

Transformational leaders can create a model for their audience to demonstrate ethical behavior and mobilize them with the help of some symbols. Similarly, with this leadership style, it may be easier to encourage viewers in the creativity dimension, to enable them to conceptualize, understand, analyze and solve the ethical problems they face. Finally, transformational leadership can have an impact on the individual's emphasis on increasing the desire of viewers to use their full potential in an ethical sense.

SOLUTIONS AND RECOMMENDATIONS

In order to prevent discomfort within the organization, solutions can be brought into consideration by considering the employer's rules and the wishes of the employee.

For this;

Measures to reduce conflict should be taken in the organization.

Early warning system:

- Is responsibility put on one person?
- Is there a distinction between employees?
- Are there a lot of wage differences?
- Are there more people who take leave from employees?
- Does the manager have a bias regarding the employee?
- Is there a general unhappiness in the employees?
- How does the manager make his observations?

The manager must manage the conflict. Problems should be examined, not bias, and talking to neutral supervisors to prevent conflicts can also be a solution. Conflict groups can be warned. The manager should have a strong leader qualification and employees should be warned to respect each other when necessary. A professional management plan or procedures should be determined. Employees will be successful in the job they are happy with. It should increase motivation by making displacements when necessary.

Role ambiguities should be removed. Instead of giving more than one task to an employee, it should be ensured that it progresses in the specified area.

The source of the conflict should be destroyed by examining the conflicts.

Frequent face-to-face communication should be provided.

FUTURE RESEARCH DIRECTIONS

Business life is a part of social life and it has to develop principles compatible with value judgments in society. The concept of business ethics has recently become a concept that all organizations are closely interested in. The effect of globalization forces organizations to operate within the framework of ethical principles. In future studies, emphasis should be placed on what financial benefits business ethics can bring to the firm. Its relationship to social performance should be evaluated.

Business ethics audit; It will improve relations between employees, customers, suppliers or investors and prevent potential conflicts. In this regard, the commitment of the board of directors and top managers to the process should be ensured. The interpersonal relations of an enterprise in a corporate structure will be within the framework of ethical rules and will strengthen its loyalty. Having ethical standards is a necessity in the corporate process. Expressing the standards clearly also maintains people's commitment to the process.

Voluntary practices, on the other hand, cover the beliefs, values of a business and the obligations in contracts that it voluntarily signs.

In general, there are differences between the age group, experience process, salary expectations and people in the business. When evaluating a person's performance in a business, age, experience, education should be carried out within the framework of ethical rules without ignoring them. There will be no equal pay for everyone in the code of ethics. But when there is a separation of people by the administration, interpersonal conflicts also increase. To prevent this, it is necessary to make assessments objectively. In performance measurements, it is more appropriate to evaluate people by giving them separate definition codes. In these definitions, people should be evaluated by looking at age, education, experience process.

In the same way, it is important for the employee to respect the job and to carry out the work in the job descriptions.

In this context, it is important for businesses to carry out their job descriptions as managers and employees within the framework of ethical code and systematic rules and to keep up with changing conditions and conditions.

CONCLUSION

In general, to know what is right and wrong at work and Business ethics, which means doing the right thing, is a determining factor in the long-term success of businesses together in the globalizing world economy.

The impact of globalization has led companies to ethical principles.

Employees should also listen to what they expect from the company before being hired. In which working environment does he want to work? How are your teammates? How are social activities carried out within the company?

Employers should also express their expectations from the employee. The expectations should be examined and the most appropriate decision should be taken.

Working times and expectations of groups should be separated according to different age groups.

Employee performance should be evaluated, and their expectations answered.

Wage and performance evaluations should be made and put into practice.

Personal development training should also be provided to employees at the workplace.

In order to provide the best performance in work environments, the subordinate parent relationship must be carried out correctly.

The continuity of the existing worker should be ensured and the expectations should be measured every year rather than constantly changing workers.

REFERENCES

Akdoğan, A., & Demirtaş, Ö. (2014). The Impact of Ethical Leadership Behaviors on Ethical Climate: The Mediating Role of Organizational Political Perceptions. *Afyon Kocatepe University Journal of Economics and Administrative Sciences*, *16*(1), 107–124. doi:10.5578/jeas.7819

Aksoy, B., & Koçancı, M. (2019). Labor Relations and Ethics: A Study on the Mainstream Human Resources Management Perspective. *Suleyman Demirel University Journal of Faculty of Economics & Administrative Sciences*, *24*(2).

Arslan, M., & Berkman, Ü. (2009). *Turkey and the World Business Ethics and Ethical Management*. TUSIAD.

Başpınar, N., & Çakıroğlu, D. (2011). *Professional Ethics*. Nobel Publications.

Bolat, T., & Seymen, O. A. (2003). An Evaluation on the Effects of "Transformational Leadership Style" on the Establishment of Business Ethics in Organizations. Balıkesir University Journal of Social Sciences Institute, 6(9), 59-85.

Çelik, S. (2010). The relationship of business ethics practices with employee satisfaction. *Business Ethics Journal, 3*(5), 21-40.

Çetintaş, H. B. (2016). Examination of Corporate Ethical Principles in the Context of Code of Ethics: Istanbul Chamber of Industry, Turkey's Top 500 Industrial Enterprises Case. *Electronic Turkish Studies, 11*(8).

Çiftçioğlu, B. A. (2011). Business and Ethics. Beta Publications.

Karabay, M. (2015). *Ethical and Ethical Leadership in Business*. Beta Publications.

Karabay, M. E. (2010). *A Phenomenon of Strategic Competition in Financial Markets in the Transition to Knowledge Economy: An Application on the Banking Sector* (Unpublished PhD thesis). Marmara University Banking and Insurance Institute Banking Department.

Kuçükoğlu, M. T. (2012). Ethical Values and Institutionalization of Ethics. *Journal of Law and Economics Studies, 4*(1), 177–185.

Kurtuluş, B. (2005). Business Ethics: Past and Present. *Journal of Social Politics Conferences*, (50), 737-759.

Kutlu, H. A. (2015). *Professional Ethics*. Nobel Publications.

Ma, Z. (2010). *The Status of Contemporary Business Ethics Research: Present and Future*. Academic Press.

Mustafa, B. (2011). Relationship between ethical climate, organizational trust and individual performance. *Atatürk University Journal of Economic and Administrative Sciences, 25*(1), 171–192.

Samsum, N. (2017). Discussions on the changing meaning and current state of the work. *Journal of Open Education Applications and Research, 3*(3), 160–210.

Taş, H. Y., Demirdöğmez, M., & Küçükoğlu, M. (2017). Possible effects of Generation Z, our future, on working life. *OPUS International Journal of Society Research, 7*(13), 1031–1048.

Weiss, J. W. (2003). Business Ethics, A Stakeholder and Issues Management Approach. Thomson-South Western.

Yıldırım, G., & Kadıoğlu, S. (2007). Basic Concepts of Ethics and Medical Ethics, C.U. *Journal of the Faculty of Medicine, 29*(2), 7–12.

ADDITIONAL READING

İştar, E. (2015). *Business Ethics in the Sustainability of Business Partners*. Academician Publisher.

Johnson, C. E. (2018). *Organizational Ethics: A Practical Approach*. Sage Publisher.

KEY TERMS AND DEFINITIONS

Business Ethics (Also Company Ethics): It is a form of applied ethics or professional ethics that examines ethical principles and ethical or ethical issues that arise in a business environment. It applies to all aspects of business behavior and concerns the behavior of individuals and all organizations.

Ethical Climate: It is the culture of an institution related to right and wrong problems. The governance that exists in an organization stems from values, norms and habits. The ethical climate stems from a firm's past and leadership.

Ethical Dilemma or Ethical Paradox: It is a question of making a decision between the two possible moral imperatives, both of which are unacceptable or undesirable. Complexity arises from situational conflict where obedience will lead to someone else's passing.

Ethical Leadership: It is leadership that is guided by respect for ethical beliefs and values and the dignity and rights of others. So, it's about concepts like trust, honesty, thought, charisma and justice.

Ethical Perception: It means the individual's perception of the action.

Ethical Principle: It is a fundamental philosophical attitude that leads directly or indirectly to an ethical standard. The term "ethics" simply means "moral principles that govern the behavior of a person or group."

Ethical System: In its simplest form, it is an ethical system of moral principles. It affects the way people make decisions and live their lives. Ethics deals with what is good for individuals and society and is also defined as moral philosophy.

Generational Differences: It is a generation gap, a phrase that is often used when referring to generational differences: A generation gap, or a generational gap, is the difference of opinion about belief, policy or values between one generation and another.

Unethical Behavior: A person is an act other than what is considered morally correct or appropriate for the profession or industry. They can act unethical like individuals, businesses, professionals and politicians.

Ethical Code: A code of ethics also referred to as an "ethical code," may encompass areas such as business ethics, a code of professional practice and an employee code of conduct.

APPENDIX

List of Abbreviations

HRM: Human Resources Management
HR: Human Resources
TEDMER: Ethical Values of Turkey Center Foundation
PERYON: Personnel Management Association of Turkey
IGIAD: Turkey Venture Economics and Business Ethical Society
BM: United Nations
ITO: Istanbul Chamber of Commerce
CEO: Chief Executive Officer
AYC: European Competence Framework in Lifelong Learning
UMS: National Occupational Standard

Chapter 13
The Historical Importance of the Concept of Business Ethics and Evaluation in Terms of Employees in Human Resources

Zafer Adiguzel
https://orcid.org/0000-0001-8743-356X
Istanbul Medipol University, Turkey

ABSTRACT

To provide a better and happier society, the framework of certain principles and the history of the concept of ethics aims to direct the people's behaviours constituting that society. The first known philosophers discussed the concept of ethics at that time. The continuation of discussions about ethics to the present day shows that the importance of the image in human life has never changed and has not lost its actuality. In fact, it isn't easy to show a specific and bordered area of business ethics. In this case, it is possible to narrow the definition of business ethics by excluding some important elements or extending it to a wide area that is completely encompassing. In this study, the historical importance of business ethics and human resources are evaluated.

INTRODUCTION

The history of the concept of business ethics is very new according to the concept of ethics. However, especially with the beginning of business life globalization, the concept has encountered intense interest both in academic terms and in terms of practice. In parallel with this interest, business ethics became an academic field after the 1980s with trainings starting in many departments, especially human resources management (HRM). The reason for the interest in business ethics since it was first added to the agenda of organizations can be explained by certain factors. First of all, the "ethical erosion" that has begun to occur with the increase of unethical behaviors emerging in the business world is one of the factors that explains the social aspect of this interest. The consequences of ethical erosion, which can be as high as

DOI: 10.4018/978-1-7998-4117-3.ch013

endangering public life, increases the community's awareness of organizational activities. Awareness in this direction strengthens the likelihood that organizations that do not operate in accordance with business ethics will face sanctioning by the community. However, the idea that organizations that continue their activities within the framework of business ethics will be successful in the long term and that this idea is supported by academic studies keeps the concept on the agenda of the organizations.

Business ethics is a collection of principles and rules based on beliefs about what is "right", what is "wrong", what is "good", and what is "bad" in relation to organizational behavior. When the concepts of business ethics and social responsibility are compared to each other, it can be said that business ethics briefly expresses intra-organizational discipline. When it comes to business ethics, the good-bad, right-wrong, and justified-unfair practices related to business life are examined. The tools used in the study of ethics are a set of principles, rules, values, and ideas about what to do. Developments in the business world are increasingly bringing organizations' activities in alignment with the principles of business ethics, ensuring that these activities become more necessity than choice. For this purpose, these principles, a guide in the execution of organizational activities, should be applied for all activities. In this context, the principles of business ethics provide an important framework for organizations that want to carry out HRM effectively, as in other organizational functions. Organizations wishing to conduct HRM in line with ethical principles should benefit from the guidance of ethical principles in determining all policies and strategies related to human resources and in making decisions and reflecting these decisions within activities. This is the only way that organizations will be able to gain long-term advantages especially as a result of motivating human resources with ethical practices. In the end, positive employee behavior within the organization will be strengthened and this motivation will improve employee performances.

Business ethics represents the corporate or organizational ethics of a business organization. Since the organization's culture has made the business organization a cultural unit, the business organization also becomes an ethical whole. In other words, as well as individual ethical tendencies and philosophies, the working ethics, and professional ethics of the individuals working in the business are influenced; as well, the business has a unique ethical tendency and philosophy. However, professional and operational ethics are not completely independent areas. Organizations that institutionalize the principles of business ethics and reflect these ethics within their activities will be able to maintain positive organizational imagery in front of the public. Therefore, organizational life continuity and success will be supported. Thus, as with other functions, the performance of organizations that can effectively manage the relationship between HRM and business ethics can be expected to improve in the long term. Additionally, as a result of efforts to develop a framework for business ethics practices in HRM, business ethics practices in HRM will make a meaningful contribution to the literature. Therefore, the situation is presented in a realistic way from the point of reflection of the subject theoretically discussed in the study, and thus, by addressing the historical importance of the concept of ethics, it is aimed to make a meaningful contribution to the practices of business ethics in HRM.

DEFINITION OF ETHICS

There is no clarity on the definition of ethics in dictionaries: The science of honor is defined as the whole of ethics, morality, behaviors that must be followed or avoided among various professional branches. Ethics from Latin ethics and Greek ethikos are related to character and can be referred to as behavioral science, and different perspectives of different scientists lead to different definitions of ethics. When looking at some definitions, ethics is understood to be a set of moral principles or values that are considered right or wrong, fair or unjust, appropriate and inappropriate, and where what is right is considered ethical, and what is wrong is considered unethical (Pratley, 1995). Ethics is a discipline or living space that usually deals with moral duty or obligation. Both ethics and individual and social life is flowing and changing at a very rapid pace. And, as a result, the value systems that will measure and evaluate their lifestyles can be passed through the filter of criticism when necessary. Furthermore, what they are based on, and whether they are built on an incorrect foundation or whether or not these systems can easily keep up with the speed of change is one of the most difficult questions to answer today, leading to a severe crisis of value (Vogel, 1991). Recognition is given that a person or group is a collection of rules containing moral values and principles that guide their behavior about what is right and what is wrong (Harvey, 1994). Specifically, this is a discipline that investigates the conduct and actions of human beings as a special problem area and examines the existence qualities of this field and the principles (values) governing this field, examining whether a human being's qualifications are dependent or independent (Weiss, 1994). The reason why ethical definitions are different is due to the difference in ethical values, religious beliefs, the emotions of the person who defines them, and the role of the individual in the branch and society in which he/she is interested. Considering all these effects and the definitions made up to now, ethics can be defined as a set of values that morally directs an individuals' behaviors and relationships in terms of good and bad and right and wrong in all kinds of groups and societies in which he or she takes part (Tutan 2006).

Ethics and Morality; As it is possible to see that the words ethics and morality are used interchangeably when the literature is examined, it is obvious that during the definition of both concepts, the definition of the other concept is mentioned. Clearly, it is not possible to distinguish between these two concepts with strict boundaries. One of the reasons why these two concepts are used interchangeably is that there are similar aspects of both concepts in addition to the intertwining of the concepts. Morality is people's actions and beliefs about the good and bad of the characters and their reasoning for right and wrong (Hartman 1998). Morality is the consciousness of the individual about the values that form and shape the individual's behavior (Özgener 2004). It refers to the sum of human relations with other beings according to certain norms, principles, rules, and values that regulate and make sense of the actions of man towards these beings and through these relations (Sternberg 2000). But, there is a difference between morality and ethics in terms of breadth and narrowness, theory, and practice. Morality, as a discipline, is a set of rules reflected in the practice of the everyday life of ethics. Morality is in everyday life, while thinking about how individuals should live in fine details; whereas, ethics requires a more abstract and theoretical perspective. We can explain the differences between ethics and morality more clearly with examples. For example, the subject of literary science is "beautiful literature", which is researched and classified from different angles (e.g. linguistic, formal content). Literary science analyzes the literary texts in terms of certain regular structure elements and forms in order to reach general expressions on novels, dramas, and poems, and tries to critically evaluate novels, dramas, and poems through these rules. The novelist does not do science, although he or she will mainly benefit from the knowledge of the science of literature (Başarır, 2006). Morality determines what people do what they should do ethically. There is a clear difference between them: between being a rule and creating rules.

Business Ethics Concept

Business ethics deals with right and wrong behaviors that prevail in the business world. In other words, business ethics honesty, keeping promises, respecting nature, righteous behavior, opposing injustices are interested in such values. However, ethical values such as honesty, trust, respect, and justice are general concepts. Therefore, it is very difficult to determine with precise lines (Thiroux and Krasemann 1980). Business ethics examines three different topics (Tutan 2006):

- Systematic Topics: Ethical issues related to economic, political, legal, and other social systems related to business operations.
- Business Related Topics: Business policies, procedures, organizational structure, business activities, etc. Ethical issues related to all issues within the business.
- Individual Topics: includes individual activities in the business. These are ethical issues related to decisions, behaviours or individual character.

Definition of Business Ethics: It is very difficult to show a specific and bordered area of business ethics. In this case, it is possible to narrow the definition of the field of business ethics by excluding some important elements or to extend it to a wider area that is completely encompassing and encompassing. For most people, business ethics, which are difficult to define and understand, can be defined as "honesty, trust, respect, fairness when conducting economic activities, and acting in a way that supports society." In another definition, business ethics is expressed as ethical principles and standards that guide the business world (Bowie and Beauchamp 2001). According to Taka, business ethics is a discipline that examines the relations between economic activities and the realization of honesty and justice in societies (to create a harmonious and balanced society) and activities that contribute to the realization of justice in these societies as well as economic practices (White 2001). Business ethics is a collection of principles and rules based on beliefs about what is "right", what is "wrong", what is "good" and what is "bad" in relation to organizational behavior. When the concepts of business ethics and social responsibility are compared to each other, it can be said that business ethics briefly expresses intra-organizational discipline. When it comes to business ethics, it is mentioned that good, bad, right, wrong, justified, and unfair practices related to business life are examined. The tools used in the study of ethics are a set of principles, rules, values, and ideas about what to do. Judgments based on principles and standards related to ethical behavior generally arise from religion and philosophical thought, cultural values, laws, and human conscience (Ardichvili et al. 2009). Business ethics, social environment, personal ethics, and organizational characteristics can be seen as a whole (Gini and Marcoux 2005). Ethical common sense is a concept that depends on existing ethical values. Descriptive ethics, on the other hand, focuses on uncovering ethical problems and attitudes and behaviors in the business world and uses methods of behavioral sciences. The most important feature that distinguishes business ethics from the philosophy of theoretical ethics is its interest in the current values in society (DesJardins and DesJardins 2009). Another commonly assumed issue of business ethics is that a business is lawful and respectful, and that compliance with business ethics is the same thing. However, obeying the law is not a sufficient condition for business ethics. It's ethically something that needs to be done, a very different concept from the law and other regulations. The violation of the law is a threat to the existence of businesses itself. In the short term, breaking the rule of law may provide some advantages, but breaking the law in a long-term and competitive free environment will have the same effect as cutting off the branch it rides. The question is: "what if the rule of law is not ethically correct?" Sometimes the requirements of the law can be absurd, inefficient, or represent nepotism. In such a case, efforts should be made to change the legal regulation that is ethically incomplete or wrong to the extent permitted by the system.

The Importance of Business Ethics: Today, the concept of business ethics has become so important in developing internal business ethics rules, in publishing international journals related to business ethics, and in establishing research units in universities (De George 2005). Whether an organization operates in the economic, cultural, or political spheres, it has to be completely transparent in its relations with all segments of society. It is not possible to prevent corruption as well as injustices in organizations without transparency. For this, the ethical principles that distinguish right from wrong, good from bad, and beautiful ugly must be spoken, discussed, and applied flawlessly in all organizations (Ferrell and Fraedrich 2015). There are many answers to a question of why business ethics is necessary, taking into account the globalizing world economy. It is possible to sort these as follows (Arslan 2005):

- The world population is growing and with this increase, the need for new employment is growing. If new jobs are not created, the unemployed and the poor will resort to social and political violence to gain a share of the wealth, thereby threatening civilization.
- Newly developed biological and military technologies are under the control of the business world. If they are not dealt with responsibly, there will be a threat that could destroy the world.
- As a result of globalization, the work of people from different cultures in multinational enterprises has brought new ethical problems. In such environments, people should understand each other and respect each other.
- There is a demand for more democracy and human rights all over the world, and these principles are becoming important in business life. Ethnicity, language, religion, gender, etc. should not be discriminated against in business.
- Increased environmental pollution necessitates an environmentally conscious business world.
- How different religions affect culture and values in the business world is also investigated by the discipline of business ethics.
- Developing countries must fight corruption and strive for a clean economic life.
- In recent years, it is observed that business ethics has gained importance due to various reasons. It is possible to summarize these reasons as follows:
- increased public pressure and improved social responsibility in the ethical business of businesses,
- Recognizing that "A good job is Identical to Good Ethics" and understanding that legal limits are not sufficient for a good job description,
- Starting to see business ethics and ethical discussions as a natural extension of excellence and quality and as a basic condition of longevity, rather than being seen as a threat and weakness for managers,
- Increasing the intensity of competition and changing the quality,
- The understanding of business ethics produced according to a narrower economic living space and more limited variables is not sufficient to comprehend and regulate the business relations today.

Therefore, unlike previous generations, this generation should gain sensitivity not only in material elements but also in ethical elements. Because the transition from the agricultural society to the industrial society involved experiencing a rapid technological renewal, generations could not produce the ethical values required by that time; on the contrary, they consumed a period of time by carving out the nonconforming aspects of the anthropological ethics produced by the previous generation. However, rapid change has filled this gap with undesirable and ethical crisis-causing behaviors in societies and brought production as a problem of today's generations.

Scope of Business Ethics: There are three main areas that constitute work ethic: work ethic, professional ethic, and business organizations' ethics (Schwartz 2007). When it comes to work ethic, the first thing that comes to mind is personal attitudes and behaviors developed toeards work. However, these attitudes and behaviors will of course be influenced by the culture and values of the society in which they live. As well as the work ethic of individuals, groups, and societies may have their own working ethics (Hartman et al. 2008). Just as one society's attitude towards business can differ in another society, there may be different approaches between different layers of the same society. While some societies create positive attitudes towards work, some societies may give more importance to rest and entertainment than work (Singer 2011).

Business ethics represents the corporate or organizational ethics of a business organization. Since the organization's culture has made the business organization a cultural unit, the business organization also becomes an ethical whole. In other words, as well as individual ethical tendencies and philosophies, working ethics, and the professional ethics of the individuals working in the business, has a unique ethical tendency and philosophy. Work, professional, and business organizations' ethics are not completely independent of each other. For example, the fact that a physician arrives at work on time shows his work ethics. It is a requirement of professional ethics that the same physician does not prescribe medicines to his patients unnecessarily in accordance with the wishes of the pharmaceutical enterprises. However, these two may also, simultaneously, be the business ethics or policy of that particular health care business. In this case, the ethics of work, professional ethics, and management ethics are intertwined (Audi 2009).

Historical Development of Business Ethics

Business ethics is not a specific problem. Throughout history, people have not ignored the code of ethics in their business relationships. In ancient Greek and Roman civilizations, philosophers did not view commercial activities positively, thinking that it could lead to corruption and social disintegration in society. In these kinds of slavery civilizations dominated by aristocracy, activities based on manual labor and physical effort were underestimated. Job and work are known as a features of slaves and lower classes. This attitude was continued in commercial activities (Yilmaz 2006). Despite the negative attitude of the Greek and Roman worlds, this negative attitude towards business and commercial activities in antiquity was not the same in every society. Although Jewish scriptures draw attention to the dangers and evils of income and wealth inequalities, it is accepted that accumulating wealth is not contrary to general ethics. This, as Weber pointed out, paved the way for capitalism (Arslan 2005).

Christianity has also played an important role in the development of business ethics. Christianity is a celestial religion with universal ethics that instills tolerance and plays a unifying role in the socially depressed society of the age. According to Jewish ethics, Christian ethics has a softness of style. Christianity, ethics from the pretension of ethics, ethics within the framework of reason, and conscience tried to interpret these ethics. There are many ethical principles in the Bible, such as the first search for one's own flaws and avoiding vanity. It is undoubtedly true that Islam, as well as the influence of Christianity, has had important effects on business ethics.

It is also acknowledged by many thinkers that Islamic ethics and the rich heritage of Islamic civilization play an important role in the formation of business ethics principles and give new insights. The third principle in Islam after ethics, faith and worship, is founded on the principle of righteousness, restraint, and virtue. The rules of Islamic ethics are life-enriching. Ideal for Islamic ethics is to create a healthy society free of extremism and following a middle path (Özgener 2004). Business and trade activities were encouraged in the sources of Islam and it was stated that these activities should be based on truth and honesty not only against Muslims but also towards all humanity: To give zakat, to spend the goods in the way of God, not to take an interest, to avoid cheating, profiteering, speculation, and monopoly, to make money from halal jobs, to contribute to society instead of accumulating excess wealth, to protect the poor, to maintain equality against classifications and stratification, to live away from the show, to live in trade to have professional knowledge and competence, to pay attention to the limits of halal-haram, to observe the rights of workers and employers, not to waste, not to lie, not to swear during trade, to start early in the morning, not to hinder worship, not to borrow unless obliged, not to squeeze the debtor, to avoid black market and unfair competition, be brave and enterprising, not to be arrogant, not to try to take advantage of legal gaps, not to abuse family and partnership relations, all are the principles of preventing competition that can be given as an example of Islamic business ethics (Torlak et al. 2008).

Today, there are many businesses in the world that apply not only three major religions but also other religions such as Buddhism, Hinduism, and Confucianism (not selling certain products, not working on certain days, etc.), which are fully in line with religious principles. From the point of view of business ethics, Japanese traditions and beliefs are the basis of business practices, especially in Japan, achieving worldwide success for many businesses (Torlak et al. 2008).

(1900-1920) Between Ethics and Business: Research on business ethics began in the early 20th century with the weakening of traditional liberal understanding in Europe and America and the strengthening of the socialist trend. Increasing debates between the socialist and liberal movement have also led to the discussion of the objectives and the contributions or harms of businesses and management to society, and thus ethical values in business have begun to be questioned. Between the years 1904-1918, it was emphasized that business ethics courses should be a good ethical basis. Between the years 1908-1915 the Scheffield Scientific School in Britain, in teaching trade ethics courses in business have needed to determine the basic business issues (Dogan 2007). Between 1900 and 1920, those interested in ethics in the United States discussed ethical issues such as improving working conditions for women and children, workers' right to compensation, and providing fair information in advertising. For this purpose, the Office of Better Business was established in the USA and business ethics courses started to take place in both courses and course programs in universities. In the 1920s, with the so-called progress movement in the United States, living wages were defined as income that could be sufficient for education, recreation, health, and retirement. Businesses have been asked to be careful about unfair price increases and other practices that could jeopardize a family's living wage (Kirel 2000).

(1920-1950) Between Professionalism and Business Ethics: During the period between 1920 and 1950, issues such as determination of business and professional ethics principles, development of standards in trade activities, the realization of advertisements, problems with bankers, insurers, and lawyers were examined. During this period, two new developments involving business ethics occurred in the Western world. The first of these is the number of occupations, which were limited to medicine, law, pastor, and academician. The second is that managers are separated from business owners and have become a completely separate professional group (Dogan 2007). With the complete separation of management from business ownership, courses, conferences and academic publications for managers helped to establish the professional identity of management, while articles on business and professional ethics began to fall within the scope of these publications and courses. The Institute of Industrial Management, established

in Britain in 1920, represents a milestone in this development process. With the launch of this institution, the discipline of business ethics began to be seen entirely within the scope of management and became independent from theology, philosophy, and economics. Again, during this period, two fundamental distinctions emerged in business ethics. These are the distinction between ethical problems arising from the internal activities of the business and the ethical problems that arise as a result of the external activities of the business. Borden and Hoper's books on Banking and Business Ethics were published in Chicago in 1921, creating the first theory and systematic work in this field. This study is also the first book on business ethics taught in universities (Arslan 2005). In the 1930s, the New Deal program came, blaming businesses in particular for the problems facing the United States. Businesses have been asked to work more closely with the state to increase family income. In the 1950s, The New Deal was reshaped as a fair deal by U.S. president Harry S. Truman. This program was aimed at civil rights and ethical issues, such as the environmental responsibilities of enterprises (Bayram 2005).

(1950-1970) Between: Business Ethics and Growing Complexity: The 1960s were the years when there was a lot of work in business ethics. Business ethics is defined in English as "Ethics in the Workplace" or "Business Ethics" and in the United States under the name "Social Responsibility of businesses." During these years, large enterprises and international enterprises producing mass production in Europe and the USA started to occupy an important place in the business world. But, the growth of business volumes has brought many problems. One of the most important studies conducted in the 1960s, a study of work ethics of classical value, was conducted by Raymond Baumhort S. J. in 1967. Baumhort in 1967, sought answers to people's questions about what ethics means to them in order to learn about their ethical attitudes at work (Kinran 2006). The answers are briefly as follows: 50% said that the feelings of ethics in the workplace were to do what they said was right, 25% said that what was religiously correct was also ethically correct, and 18% defined ethics as the golden rule at work (Kinran 2006). At the beginning of this period, ethical problems were still treated as a personal issue. Consumer rights, advertisements, and sales development techniques were examined. However, in the second half of this period, the findings of the organization theory were used and the business was considered as an ethical unit or as a whole. The 1960s were a decade in which work in the field of business ethics dramatically increased. During this period, many field studies were conducted and doctoral dissertations and books were written. In 1953, 5.6% of US faculties and colleges included business ethics in their programs, while in 1967 this ratio increased to 17%. But what is more important than the number of business ethics courses is that the content of the courses has expanded. Unlike secular business ethics in the United States, the connection between church and theology occupied an important place in business ethics in Europe between 1950 and 1970. Writers interested in business ethics in Britain have often been linked to the Anglican Church. Organizations involving employees and administrators were established within the Catholic Church (Arslan 2005). Significant changes in the business world also took place in the 1960s. These changes created many ecological problems such as pollution and nuclear waste due to the development of industrialization. However, it was observed that consumers also gave greater importance to ethics due to the change in marketing logic during these years. In 1962, a four-point declaration was issued by President John F. Kennedy regarding the right of protection, right of information, right to choose, and the right to be informed by President John F. Kennedy in the United States (Bektas and Köseoğlu 2008).

(1970-1990) Between: Initiatives to Bring Order in Business Ethics: By 1977, the ethical theories of businesses began to emerge. In the following periods, business ethics courses were shown in management schools, and business ethics centers were opened in universities. In these business ethics centers, the responsibilities of enterprises towards society and the requirements of this social responsibility were

examined. During this period, concepts related to business and labour sociology were also discussed within the scope of business ethics. For example; concepts such as values, strategy, structure, systems, skills, and culture were examined within the business ethics, and the ethical dimension of business policies emerged. After the 1980s, businesses began to play important roles in a society like never before, and therefore the importance of business ethics increased accordingly. During this period, international organizations on business ethics began to be established. The most important of these is the European Organisation for Business Ethics, founded in the Netherlands in 1987. After the second half of the 1980s, the number of researchers in the field of business ethics rapidly increased. In 1980, the U.S. Public Service Code of Ethics was designed, published, and distributed by "The Ethics Resource Center - ERC". The American Institute of Certified Public Accountants (AICPA), established in 1887, also established a Code of Professional Conduct in 1988 for accountants who were members. This code explains the responsibilities of accountants to the public, customers, and other colleagues (Özgener 2004). A 1980 study conducted in the United States stated that only 8% of the major businesses on the Fortune 500 list had an ethical code. This ratio increased to 77% in 1985 and then to 90% in 1990 (Yildiz 2007).

(From 1990 to the Present) Global Business Ethics: In the last decade of the twentieth century, technology has developed with the Internet becoming an indispensable part of our lives. With the collapse of the Soviet Union in 1993, the socialist system collapsed, and the market economy entered China, where the socialist state still exists. Then the volume of business increased, and the world's population began to threaten the environment in a serious way. These developments have added to the scope of business ethics, specifically the environmental problems caused by businesses. Topics such as overcoming cultural differences and discrimination in multinational and global businesses as a result of globalization have also become among the topics of business ethics research. By entering the twenty-first century, business ethics became a growing issue at the global level (Wray-Bliss, 2009). In some countries, such as the Netherlands, France, and the UK, consultancy services in the field of business ethics have shown significant improvement. These services are focused on developing and reviewing behavioral codes for businesses and industries. In the early 1900s, the number of businesses with codes of ethics in Germany, France, and the UK was estimated at approximately 40%. In 1999, the Q-RES Project (Quality of Business Social and Ethical Responsibility Project) was designed by a group of businesses, professional organizations, consulting firms, and non-profit organizations led by the Center for Ethics, Law & Economics - CELE, Castellanza University in Italy. The aim of this project was to introduce a management model that focuses on the social and ethical responsibility of businesses, including the concept of social connection between businesses and interest groups. Projects equivalent to the Q-RES project can also be found in other countries of Europe. For example, there is the SIGMA project in the UK and Values Management Systems in Germany. In particular, the importance of social and ethical responsibilities for businesses increased with the publication of the European Commission's "Green Paper on Business Social Responsibility on Social Responsibility" in July 2001. This book focuses on the eligibility criteria of the codes of business codes and the issues aimed at establishing whether businesses comply with voluntary standards in meeting the minimum requirements set by law in planning and implementation (Özgener 2004).

Since the 2000s, business ethics has taken part in all the discourses associated with management and has even become the main actor and paradigm, topping the popular literature. It was determined that business ethics increased the brand value of enterprises and therefore the market value in the medium and long term. The rhetoric of "good ethics is good business" a slogan. In the 2000s, important organizations related to business ethics started to operate. Branches of these centers were established in many countries and application-oriented studies carried out on a global scale with forums and meetings being organized: As an example of this expansion, the Japanese Association of Business Ethics Studies, Institute of Korean Business Ethics, Latin American Business Ethics Network / Sector, African Business Ethics Network, the South African Institute of Ethics, Turkey Ethics Center, St. St. Petersburg Business Ethics Center and Chinese Business Ethics Center (Berkman and Arslan 2009). By the 2000s, business ethics in the became a workspace that dealt with the ethics of business activities. The field of business ethics simply establishes systematic relationships between concepts such as ethics, responsibility, and decision-making in businesses, rather than focusing on what to do or what not to do in certain situations. The process of ethical decision-making in business activities is dealt with from philosophical, economic, sociological, psychological, and religious perspectives. Managers and researchers working on this issue have sought to develop systematic guidelines that can help individuals and organizations make ethical decisions (Beschorner, 2006).

In the future, business ethics will not revolve around an international ethics agreement, as accepted business ethics vary from one culture to another. Therefore, universal business ethics standards are out of the question. Instead, multinational businesses will be caught between ethical management and ethical conflicts. It will create more diversity in future ethical issues and bring more effort to managers who will need to better understand the ethical decision-making process (Jose and Thibodeaux 1999).

HUMAN RESOURCE MANAGEMENT AND BUSINESS ETHICS

It is a fact that in order for organizations to succeed in a global competitive environment, they must value the human resources within them. Thus, the strategic feature of human resources, which is emphasized more as a result of the developments in today's business world, is accepted as the cornerstone of making a difference against competitors. Correct positioning is directly related to the importance that organizations attribute to the HRM function (Kolodinsky 2006). In this sense, it can be said that there is no such thing as an organization that creates and implements policies in line with the highest level of human resources potential.

Historical Development of Human Resources Management

In order to better understand the concept of HRM, it is necessary to examine the factors that have brought this understanding to life and enabled the concept to mature. In the 18th century, with the industrial revolution, a deep power asymmetry was established between employees and employers who were part of mass production. Employers in the face of this worsening situation began to organize and collectively seek their rights. Collective bargaining, which emerged as an outcome of the organization of employees, brought the concept of personnel management (PM) to the agenda of organizations for the first time (Winstanley et al. 1996a). II. After World War II, humanity learned the concept of Social Relations, which contrasted with the strict automation of the concept of Scientific Management, which Peter Taylor had proposed years ago. Hawthorne Studies, which started at Western Electrics in 1923, proved that employees are social assets and that communication and motivation on their productivity are of great importance (Price 2007).

By introducing new dimensions in explaining human behavior, Hawthorne research has played a very important role in the development of the concept of PM and the subsequent development of HRM. As a result of this research, the importance of the person ignored by the Taylorist movement was understood and efforts to "humanize the workplace" began to spread. PM, which gained popularity during this period, aimed to increase the productivity levels of organizations (Üsdiken and Wasti 2002). Especially after the developments in World War II, organizations increased their awareness of the concept of PM (Akdemir 2005) and for the first time, PM departments started to take part in the organizational structure. The main purpose of PM understanding, which carries the traces of this period, was to create a labor force capacity to support the achievement of the objectives of the organizations (Deckop 2006).

By the 1980s, the PM concept had begun to change. So much so that these years are considered a turning point for HRM. Multiple factors then brought HRM understanding to the agenda of the organization (Keser 2002; Smilansky 2002): Changes in technology, economic structure, demographic characteristics of employees, cultural elements, tendency of organizations to shrink, maturation of flexible working order, expansion of Total Quality Management (TQM) understanding, strengthening of productivity and employee relationship, need of holistic perspective towards employees, globalization, establishment of multinational organizations and intense competitive pressure. It can be said human resources, which was seen as a production factor until this period, became the value of production in the new understanding. In the 1980s, although organizations began to change the names of the personnel departments to the human resources department, they could not isolate themselves from the personnel processes. In addition, new concepts such as rewarding, performance evaluation and career management began to emerge among the functions (Schumann 2001).

This transformation in working life was also felt in academic studies. In this period, it is noteworthy that in the studies related to PM discipline, the concepts of Personnel and HRM or only HRM were used instead of PM (Hartog and Winstanley 2002). The question of whether this change of name is actually a change of understanding and how this new concept is filled is discussed by the researchers. In fact, researchers have tried to put forward different items about the characteristics that differentiate HRM and PM (Rose 2007).

In the 1990s, scientists added the word strategy to the term HRM and the concept of strategic integration began to stand out in the HRM literature (Chukwunonso 2009). In this sense, a relationship between the strategic human resource management (SHRM) concept and the organizational strategies, structure, task and HRM were established (Keçecioğlu 2006). As HRM functions and practices were handled in a more holistic perspective in accordance with the general organizational strategy, HRM moved to a different dimension (Winstanley et al. 1996b).

The most important factor that initiated the transformation of HRM understanding to SHRM was the fact that organizations searched for a competitive advantage (Simmons 2008). The concept of SHRM has been developed within the framework of the idea that human resources can meet the strategic needs of organizations, can be managed more effectively, and organizational performance can be affected (Benligiray et al. 2010). In most of the studies carried out in the field, models have been developed to demonstrate this relationship (Boxall and Purcell 2007).

Today, the HRM department is considered as a strategic function that adds value to organizations (Aghazadeh 2003). Parallel to the emphasis on the strategic feature of HRM, the positions of HR managers within the organization are strengthened (Budd and Scoville 2005), and their exit towards senior management is accelerating. With HR managers starting to take part in senior management, the importance of the decisions made has increased. It can be said that the necessity of HR managers to act within the framework of business ethics principles is due to this importance. Accordingly, HR managers can be expected to be the party that initiated the change in the organisation's adoption of the principles of business ethics. Studies conducted in foreign literature show that the understanding of SHRM has been adopted in management and that the strategic aspect of human resources is strengthened. It is observed that many organizations do not adopt the SHRM approach due to various factors (such as an inability to complete institutionalization or economic instability). The fact that most organisations do not even understand HRM yet is seen as the most important reason for this (Winstanley and Hartog 2002).

Scope of Human Resources Management

When the historical development process is analyzed, it is seen that Personnel Management (PM), which focuses on recording employee personal affairs, shifted its axis due to various micro and macro factors. In the new understanding expressed as HRM, where the employees are placed in the center, people find value as an entity that needs to be developed (Gospel 2009). The factors that reveal the maturation of HRM also require that employees be managed effectively. Other micro and macro variables, especially in intense competition conditions, emphasize the meaning that are placed on the human being within the organization and put organizations in new pursuits. The strongest motivator of these searches is the desire to benefit from one's strategic characteristics at the highest level. The belief that the return on investment in human beings will be much greater than the cost of the organization makes HRM the focus function of the organizations. For the HRM function, which is so important to organizations, researchers make different definitions. Some of these definitions include, in its broadest sense, that HRM can be defined as a strategic and coherent approach developed for the effective management of the most valuable asset of an organization that is, the people working there. In other words, HRM is the management of all activities carried out in an organization to gain, develop, motivate and retain a high-performance workforce (Pinnington et al. 2007). HRM is the determination of the policies and strategies related to the management of human resources that provide a competitive advantage to organizations and are reflected in their activities (Schuler 2000; Danley et al. 1996). According to another definition, HRM is the efficient use of employees to reach the strategic goals of the organization and to satisfy the individual needs of the employees (Greenwood 2002). What the definitions have in common is that HRM is conducted with a holistic understanding of responding to both employee needs and organizational objectives. Now, the starting point of the definition hinges on the assumption that organizations can achieve high productivity targets by providing motivation to employees and giving them a competitive advantage over competitors. To sum up, HRM is a strategic function that can attract a potentially qualified workforce to the organization, maximize the efficiency of the existing workforce, ensure the highest level of human resources in accordance with legal regulations and ethical principles, and conduct practices that encourage the continuity of the workforce.

Objectives of Human Resources Management: The developments in the micro and macro environments of the organizations have made the role of HRM on organizational success more prominent. For this reason, it is very important to determine the objectives of the HRM, which has become increasingly strategic units, in terms of organizational success. The objectives and characteristics of HRM are influenced by factors such as economic, political and social conditions, industrial relations, community culture, habits, organizational structure and technology of the country in which it operates (Cascio 2015). In light of this, HRM's objectives can be summarized as follows;

- To provide support for the formation and implementation of the organization's strategy,
- To create a competitive advantage that cannot be easily imitated by competitors,

- To adapt to the changing environment and to improve organizational efficiency,
- To invest in human resources and try to make the most of this resource,
- To provide the necessary support to the realization of organizational objectives as a result of more effective management of employees,
- To support the provision of justice within the organization,
- To develop the qualifications and abilities of the employees in line with the objectives of the organization,
- To provide labor flexibility,
- To improve the level of service satisfaction and commitment by improving the quality of working life,
- To solve the legal problems arising in employee-employer relations,
- To create an effective organizational culture and to support the creativity of the employees,
- In addition to the activities carried out in relation to the employees, it is also trying to find solutions to problems related to ecology and the environment (Guest 2007; Burke and Ng 2006). As we have seen, the objectives of the HRM are shaped by a holistic perspective. In fulfilling these objectives, the HRM has responsibilities towards different parties and the most important one is the management of the organization. In addition, the responsibilities imposed on employees, laws, and society make the decisions and activities of HRM more complex.

Human Resources Management Functions: HRM carries out multiple activities: finding and selecting human resources within the framework of human resource planning, training and development, career management, performance evaluation, remuneration, strengthening the organizational culture, increasing the motivation of employees, ensuring health and safety and resolving conflicts regarding industrial relations (Lussier and Hendon 2017). In carrying out these activities, HRM acts with the logic of function (Ahammad 2017). Like the pieces of a puzzle, HRM functions carry the traces of a whole. For this reason, each interconnected part is equally important in completing the whole. In its simplest form, HRM functions can be defined as the practices performed in relation to the management and development of human resources in organizations (Lawler and Boudreau 2009). According to the HRM approach, employees are the most valuable assets of organizations and humans are the most strategic resource with a feature that can be improved compared to other organizational resources (Mehrabad and Brojeny 2007). Different researchers also include functions such as motivation, health and safety, working relationships and communication. In this sense, it is not exactly a consensus in expressing HRM applications on a function basis. More comprehensive information about HRM functions will be included later in the study.

Human Resources Management and Business Ethics: The trends that arise in working life make the organisations operating in line with the ethical framework increasingly transformed from a choice to a necessity (Heim 1991). In this direction, it can be said that business ethics principles provide an important framework for HRM function in determining the policies that will enable the most efficient use of human resources and in implementing these policies (Hartman 1998). Although its history dates back to the past, the issue of business ethics was seen as a management discipline in the 1960s (Derr and McNamara 2003). It is generally accepted that HRM emerged as an extension of the PM in the early 1980s. Therefore, the introduction of the concepts of business ethics and HRM as a discipline is relatively long. The relationship between HRM and business ethics and the subject of studies is more recent (Winstanley et al. 1996b). The relationship between HRM and business ethics is also becoming an issue of interest to academics. However, it is seen that the relationship between HRM and business ethics is not studied as much as business ethics. Since 1990, more than 4500 books and articles have been published in the field of business ethics. However, a small number of these publications have discussed the ethical relationship with HRM practices (Wells and Schminke 2001). Although some studies have been taken PM into account, the issue has been neglected and under-discussed (Greenwood and De Cieri 2007). Research conducted at a limited level and trying to shed light on this relationship generally focuses on the following topics:

Building the philosophical ethical dimension in the HRM debates (Kamoche 2007),

Managing the relationship between HRM and business ethics at optimum level (Berenbeim 2010),

- Ethical dilemmas faced by HR managers and resolution of these dilemmas by considering organizational benefit (Kamoche 2007),
- Employment relations, discrimination, employee rights and duties (Pinnington et al. 2007).

At the micro level, individual practices or HRAs as a whole are discussed ethically, and individual employee rights and responsibilities are questioned in terms of business ethics (Winstanley and Woodall 2000).

Work done at the macro level focuses on the relationship between HRM, performance, and classification of HRC practices to open up ethical debate (Greenwood and De Cieri 2007).

HRM policies and practices all have an ethical dimension (Woodall & Winstanley, 2000). In other words, there is a close relationship between HRM and business ethics (Deckop 2006). Starting from the recruitment and orientation process of HRM, it can be said that it has a strategic role in teaching and adopting ethical codes in all training and development studies and transforming them into behaviors (Pinnington et al. 2007). In terms of business ethics, it also covers organizational behavior issues. The ignorance of the ethical elements of the enterprise causes the emergence of organizational behavior issues in the enterprises. Unethical behavior also covers issues of organizational behavior and becomes the common denominator of these two disciplines.

Mobbing: Mobbing comes from the root of the word mob. The word Mob is an irregular crowd that uses illegal violence. It comes from the word mobile vulgus, which means unstable crowd in Latin. Many concepts (psychological harassment, psychological violence, spiritual abuse, etc.) are used instead of mobbing (Leymann 1996). Mobbing was first noted in the 1960s by Austrian researcher Konrad Lorenz in his book, on aggression, where he found some propositions suggesting that animal conflicts were similar in human beings (Lorenz 1966). One of these propositions that caused criticism and discussions at that time was that some animals in the flock came together and used the concept of mobbing to surround one of them and expel it (Sheehan 2004).

Whistleblowing: Whistleblowing is reported as internal abuse and corruption, dealing with ethical issues, employee disclosure of workplace abuse or illegal activities, and civil virtue, moral reaction and conscientious objection to unethical misconduct in organizations (Near and Miceli 1985). The concept of Whistleblowing had a negative connotation in its first years (Larmer 1992). In fact, whistleblowing, which means to inform the colleagues and the public with the whistle the British police stole when they see a criminal on the street, was used for the first time in a computer article in a computer magazine in 1971 and was criticized by the author. The author says "How can the employee make his/her own institution ridiculous and tramp his/her own profession and company by giving trumps to others?" (Miceli et al. 2008).

Discrimination: Recruitment, the initial function of human resources, requires making decisions that will cause many moral problems (Stallworth et al. 2001). The most common problem in Western countries is the practice of discrimination against ethnic minorities while recruiting. For example, in the United States, a federal law known as the "equal opportunities law" for whites who generally discriminated against negroes, thereby imposing a certain proportion of black, female or ethnic minorities in business (Coombs and King 2005).

The Role of Human Resources Management in Institutionalizing Business Ethics: The HRM assumes responsibility for conducting its functions and practices within the framework of ethics, while at the same time the department takes a central position in the institutionalization of business ethics within the organization. This responsibility, undertaken by the HRM, emphasizes the strategic character of the department and shows the importance of its role in establishing business ethics awareness (Velasquez and Velazquez 2002). The most important factor that drives HRM to assume a key responsibility in institutionalizing business ethics is that ethics is related to human behavior. Thus, the management of human behavior is the main task of HRM (Vuuren and Eiselen 2006). The roles that HRM undertakes in this direction can be expressed as follows:

- HRM should implement ethical employee behavior policies that will increase the productivity and profitability of organizations.
- The HRM function should support the establishment of business ethics and become an organizational culture in an organization.
- HRM should assume responsibility for the development, implementation, and management of ethical codes.
- HRM should play an active role in raising an organizations' awareness of ethics, disseminating ethical behavior, and supporting ethical leadership.
- HRM should have responsibility for the preparation and execution of ethical programs (Hussein 2009; Bennington 2007; Weaver and Trevino 2001).

The basic requirement for the effective fulfilment of these responsibilities undertaken by HRM is to keep the ethical consciousness high. In this respect, HRM needs to take some measures both in terms of its own practices and other practices that will ensure the institutionalization of ethical values in the organization. These measures can be summarized as follows:

- First of all, the commitment of the candidates to ethical values should be evaluated during the selection and placement process. During the application, interview and reference evaluation stages, the ethical values of the candidates and their compliance with the corporate culture should be examined.
- Adherence to ethical values and proven ethical behavior should be seen as a prerequisite for reaching certain levels in career planning. This criterion should be clearly known to all employees.
- Ethical values should be continuously emphasized continuously in training and development activities. In addition to knowledge and skills training, attitude development training should also be included.
- Within the organization, all kinds of communication tools and employees should be given messages emphasizing the benefits of compliance with ethical rules, openness, and honesty.

- Ethics should be respected in HRM decisions and practices, such as recruitment, promotion, reward, remuneration, performance measurement, training.
- The private lives, personal rights, freedoms and personalities of all managers and employees should be treated with respect. This understanding should be placed throughout the entire organization.
- Responsibility should be taken to identify unethical behaviors in the organization, to find out the reasons for these behaviors, to find the responsible parties, and to deal with the problems caused by them.
- In the performance management process, objectives, performance standards, competencies and valuation criteria should be ensured to contain ethical values.
- Senior management should be encouraged to adopt the principles of business ethics in their activities.
- In determining HRM policies, ethical codes should be chosen as a road map, ethical codes should be taught to employees, and guidance should be undertaken in implementing these codes (Ferrel and Fraedrich 2015; Sayli and Kızıldağ 2007).

Roles of Human Resources Managers in Institutionalizing Business Ethics: There is no consensus among researchers on how to establish an ethical framework in HRM policies and practices. The main consensus is that HR managers play a key role in conducting ethical HRM practices (Woodall and Winstanley 2000). HR managers have the responsibility to consider ethical issues in their decisions regarding all HRM functions and practices and to apply ethical principles in these decisions (Bennington 2007; Walsh 2007). Along with this responsibility, HR managers also play a central role in implementing ethical codes, creating a strong ethical culture, and strengthening the values of ethical employees. At the same time, HR managers guide other managers regarding the need to achieve economic efficiency within the framework of ethical principles (Barrett 1999). HR managers are aware of the responsibilities of HR managers in two different studies conducted to determine the awareness of the responsibility of institutionalizing business ethics (Vuuren and Eiselen 2006). As a matter of fact, other studies reveal that most HR managers play an active role in the management of business ethics (Vickers 2005; Heim 1991). In the light of this awareness, HR managers reflect their responsibilities to the organizational life by playing different roles. Wiley (1998) tried to determine the roles that HR managers will play in this direction. The roles expressed in the research are as follows:

- Consultant: Consulting on ethical policies and standards,
- Stimulant: Monitoring the compliance of conduct and actions with laws and policies,
- Educator: Providing and educating regarding ethical policies,
- Defenders: Protecting employees from management retaliation,
- Searching: Investigate unethical situations and complaints,
- Interrogator: Examining managers' decisions in terms of ethical dimension,
- Organizational: Protecting the organization against the dangers that may arise by other institutions.

Model: Ethics is a role model. As you can see, there are many different roles that HR managers can play in ethics management. However, the ability of HR managers to play these roles will depend largely on the importance that senior management attaches to the HRM function and the level of authority they give them. Therefore, HR managers who are not authorized should not be expected to assume this responsibility (Kamoche 2007).

Ethical Responsibilities of Human Resources Managers to Employees: Employees constitute the most important group of stakeholders that directly affect the organisation's success and are affected by organizational activities (Pinnington et al. 2007). Evaluating the organization as a whole is the most important feature that distinguishes employees from other stakeholders (Greenwood and De Cieri 2007). The idea that the expectations and evaluations of employees regarding ethical practices will be reflected in their performance puts employees in a focus position in ethical management (Jose and Thibodeaux 1999). Thus, ethical sensitivity is expected to be strong in employee-related practices and the responsibility to demonstrate this sensitivity falls largely to HR managers. Recently, organizations have been under more pressure to develop HRM strategies and policies that protect the rights of employees. Today's employee profile is more educated, conscious, and open to learning compared to the past (Burke and Ng 2006), thereby increasing the willingness of employees to participate in the decisions, and the demand to be more respectful and fair to them (Schwoerer et al. 1995). At the same time, the pressure directed at organizations by society creates image anxiety and makes behaving more ethically in the management of employees an inevitability (Weaver and Trevino 2001). There are many practices that show that HR managers are fulfilling their ethical responsibilities towards employees. First of all, HR managers should be aware that they have the responsibility to apply ethical principles in all their functions, decisions and practices: adopting the principle of justice in the determination of HRM policies, respecting the employees, protecting their fundamental rights and privacy, not using the employee information beyond their purposes, not discriminating, making overtime more than paying wages, being honest, paying wages, equal pay, equal opportunity, Ensuring equality will be an expression of the awareness of HR managers to use ethical responsibility (Aswathappa 2005; Bennington 2007).

Ethical Practices in Human Resources Management: In the literature, it is generally accepted that the HRM function conducted within the ethical framework will positively affect the organizational performance (Svensson et al. 2010). However, this acceptance cannot be adequately tested by empirical studies. This general acceptance regarding the relationship between HRM and business ethics is questioned by some researchers. According to Lafer (2005), ethical behavior in essence should not be exhibited in anticipation of any obligation or interest. Therefore, the motivation of an administrator's ethical behavior should not be for performance purposes. According to him, the display of ethical conduct should not be based on any cost-income calculation but that ethical behavior must be demonstrated without an expectation because it really should be. While this idea is accepted, it can be given a different perspective. In our opinion, organizations should not resort to business ethics as a means of improving profitability and performance. On the contrary, business ethics should be regarded as a purpose in ensuring the necessity of living in society and the continuity of organizational life. For this reason, instead of approaching business ethics with an image concern and instrumentalizing it, the perspective that accepts business ethics as a way of doing business should be adopted. After this discussion, the benefits of HRM practices, which are generally carried out in an ethical framework, can be summarized as follows:

- According to the CEOs of 63% of the Fortune 500 organizations, strong ethical practices bring strategic advantages (Buckley et al. 2001).
- The positive image of the organization is supported by society and the possibility of harm to the reputation of the organization that may occur as a result of unethical behavior decreases (Berenbeim 2010).
- The level of trust in relations within and outside the organization increases.

- The society's expectations for ethical organizational practices are answered. It also contributes to raising the ethical consciousness of the society.
- It supports organizational life by giving legitimacy to organizations.
- It inhibits the negative organizational behavior of the employees and strengthens the positive employee behavior.
- The organization's attractiveness for potential employee candidates increases, also increasing the likelihood of more qualified candidates applying to the organization.
- The labor turnover rate decreases with the provision of employee continuity.
- Ethical violations related to employee management are prevented. Therefore, lawsuits that may be filed in this direction and the costs that can be brought with them are prevented.

Unethical Behaviors in Human Resources Management: Although the research shows that organizations have increased their awareness of business ethics, unethical behaviors continue to be exhibited in today's modern organizations. This reveals the fact that there is a problem in reflecting ethical principles in the activities of the organization (Zey-Ferrell and Ferrell 1982). This ethical erosion in organizations feeds on whichever organizational activity it feeds on, and as a result, the quality of working life deteriorates and organizations face high cost risks (Kaptein 2008). A study of unethical behavior by Robertson and Schlegelmilch (1993), Wooten (2001) involving more than 700 organisations found that the most ethically problematic issues are related to HRM. Another study, which was conducted with the participation of more than 900 organisations, found that 6 of the 10 most important ethical problems presented in organizations were related to HRM (Fusilier et al. 1996; Wooten 2001). As research findings show, most of the problems faced by organizations are related to human behavior (Kapu 2009; O'Higgins and Kelleher 2005). This argument obliges the conduct of HRM in an ethical framework for the success of the organization. Unethical behaviors in HRM are evaluated as a multifaceted situation that includes individual, professional, and organizational factors (Reiss and Mitra 1998). A study conducted with the participation of 1078 HR managers found that the attitude of senior management and auditors, individual values and performance pressures, respectively, led to unethical behavior (Danley et al. 1996). In addition, the demands of employers from HR managers, insufficient knowledge or lack of commitment of HR managers, incomplete evaluation of the situation causing ethical violations and low ethical sensitivity may bring unethical behaviors. In addition, some studies have determined that there is a significant correlation between the ethical problems that arise in HRM and the scale of the organization. Accordingly, there are more unethical problems in large organizations than in small organizations (Bartels et al. 1998; Mithchell et al. 1996).

SOLUTIONS AND RECOMMENDATIONS

It is the common acceptance of researchers and managers that there is a strong relationship between HRM and business ethics and that this relationship is increasingly strengthened. However, the question of why this relationship is important, in other words, is not yet sufficiently clarified to ask why business ethics principles should be applied in the execution of HRM applications. The nature of HRM is that employees are used in accordance with the objectives of the organization. This usage is the main source on which HRM and ethical relationship is based. In other words, the use of human resources in this direction brings with it ethical issues and concerns (Greenwood 2002). Furthermore, the direct influence of human behavior and activities by the decisions taken by the organization necessitates such a relationship (Szalkowski and Jankowicz 1999). HR managers also have an obligation to be a role model in strengthening the ethical behavior of employees and preventing unethical behavior (Mauro et al. 1999). The fact that employees follow the example of their managers in their behavior and attitude stipulates the necessity of this obligation. Thus, studies in the literature support the assumption that there is a correlation between the ethical values and behavior of managers and the behavior of employees (Mcdonald 2009; O'Higgins and Kelleher 2005). The way organizations treat their employees is an important indicator of the extent to which they embrace their work ethics. Therefore, an organization that respects ethical principles in its activities should treat its employees as fairly as other interested persons (Kirel 2000). In addition, the image of the organization in front of the society, which can be defined as an external factor, is also important in establishing the relationship between business ethics and HRM. Business ethics requires employees to be respectful and fair to their employees and to be valued as human beings outside the qualifications of employees within the framework of ethical principles (Arslan 2001). With this awareness, HRM should bring ethical values about individuals to the workplace and make them inseparable elements of the workplace (Akgeyik 2009).

FUTURE RESEARCH DIRECTIONS

It can be said that HR management is a profession that has to take account of responsibilities to different parties due to the roles it covers. Because HR managers are performing a profession that requires a high level of role diversity. This diversity of roles often leads to ethical dilemmas (Becker 2010). In this respect, it can be said that HR managers are in conflict between their expert judgments about how to implement the best practices and their responsibilities to fulfill their employers' expectations (Scott 2016). The development of professional codes is one of the most important steps taken in the realization of professionalization (Wooten 2001). The ethical codes developed in this direction aim to establish ethical standards among professional members (Budd and Scoville 2005). With the increase in the importance of HRM, it is observed that in recent times HR management has been rapidly turning to professionalization and the HRY organizations have begun to explain ethical codes. Organizations that disclose ethical codes in this direction ask HR managers to consider the principles of ethical codes in their decisions (Greenwood 2007). The Society for Human Resource Management, which has more than 100,000 members in the United States, is one of these organizations that discloses ethical code. With the ethical code created, the HR Association aims to develop ethical practices and support the ethical achievements of organizations, which are the basic professional responsibilities of HR managers (Bennington 2007). When the content of the code of ethics is examined, it is seen that the responsibilities of HR managers to their professionalism and then to the organization, employees, and shareholders they work with are emphasized. HR managers are expected to base their principles of respect, honesty, justice and trust in fulfilling these responsibilities. These and similar ethical principles should be encouraged to guide relations with all parties responsible. The concept of business ethics is an issue on the agenda of both practitioners and academicians. Especially after the 1980s, the concept of business ethics became the subject of intensive academic studies. As the subject is popular, the literature needs to be enriched to this extent. However, it is observed that there is a significant gap in the field of study. Thus, there are a very limited number of studies investigating the relationship between HRM and business ethics. In the domestic and foreign literature survey on the relationship between business ethics practices and organizational performance in HRM, it was determined that there was a similar gap. Therefore, it should be aimed to contribute to the literature with business ethics practices in HRM.

CONCLUSION

Although there is a consensus among researchers that there is a close and strong relationship between HRM and business ethics, the main problem in this relationship is whether or not the ethical framework in HRM can be established and how ethical principles are expressed in HRM. In other words, it is discussed whether ethical principles that all organizations can apply on cultural differences and other diversity are possible. According to Legge (2000), the number of people in the world is the first of its worlds. The relativity of postmodernism and its emphasis on differences create a paradox in the creation of universal ethical principles. Therefore, postmodernism provides different openings in the face of the problem of common ethical principles, behaviors, and the possibility of applications. Furthermore, researchers who argue that universal ethical principles are not possible also base their ideas with relativistic theory. In this direction, the question of ethics -according to whom or by what means makes the issue ambiguous and ignites the debate. Researchers discuss the possibility of the best ethical standards in HRM subjects and functions in a post-modern business world based on different ethical theories (Budd and Scoville 2005; Cole 2002). Schumann (2001) tried to shed light on this issue. In this direction, Schumann argued that relativist theory is inconsistent in itself. The researcher examined issues such as discrimination and sexual harassment in employment in the light of different ethical theories and demonstrated that these behaviors are unethical in all of the underlying ethical theories. Based on this finding obtained in the study, Schumann demonstrated that an ethical framework and principles that can be accepted at the universal level can be established in HRM. All of the principles in HRM policies and functions are of great importance. Employees' assessments of these principles will directly affect their behavior. For example, the principle of justice forms a face to business ethics (Ferrell and Fraedrich, 2015). According to the principle of justice, their contribution to decisions concerning employees should be taken into account (Arslan 2001). Thus, employees who believe there is injustice in practice and decisions will have less confidence in the organisation (Armstrong 2006). At the same time, injustice, discrimination, disrespect, etc. will adversely affect the satisfaction, commitment and participation of the employees. Each type of human activity (scientific, artistic, educational, etc.) corresponds to a certain type of professional ethics (Ferris et al. 2002). Before defining the ethics of the profession, the concept of the profession must be put forward because in order for any job to be counted as a profession, the field of interest must also include ethical values and principles. In other words, there are three basic elements called "competence" that a professional element is expected to have: knowledge, skills and correct business habits and attitudes. This means that in a profession, as well as knowledge and skills, professional values and ethical principles are defined (Lussier and Hendon 2019). Professional ethics regulates the relations of the individuals who make up the professional group with each other and with the community. Professional ethics is a specific occupational group that establishes and maintains (in relation to the profession), orders the members of the profession, compels them to behave in a certain way, and limits their personal tendencies. It is a set of professional principles that exclude inadequate members from the profession, regulates in-service competition, and aims to protect service ideals. The question of which principles may be used to carry out the HRM function within the ethical framework is reviewed by the researchers. In discussions, it is generally accepted that the ethical framework in HRM can be established with the principles of equality of opportunity, honesty, fairness and honesty (Beardwell and Claydon 2007; Buhler 2002). Different principles that can be established in an ethical framework can also be mentioned in HRM, and it is possible to make different assessments in this direction. However, these principles encourage employees to be at an equal distance to eliminate discrimination, to provide them with output in terms of their efforts, and to treat their individual rights and freedoms with respect.

REFERENCES

Aghazadeh, S.-M. (2003). The Future of Human resource Management. *Work Study*, *52*(4), 201–207. doi:10.1108/00438020310479045

Ahammad, T. (2017). Personnel Management to Human Resource Management (HRM): How HRM Functions? *Journal of Modern Accounting and Auditing*, *13*(9), 412–420.

Akdemir, B. (2005). Küresel rekabet ortamında insan kaynakları yönetiminin değişen rolü: Stratejik insan kaynakları yönetimi. *Sosyal Siyaset Konferansları Dergisi*, (49).

Akgeyik, T. (2009). *Türkiye'de iş etiği: İnsan kaynakları yönetimi boyutuyla*. ITO Yayınları.

Ardichvili, A., Mitchell, J. A., & Jondle, D. (2009). Characteristics of ethical business cultures. *Journal of Business Ethics*, *85*(4), 445–451. doi:10.100710551-008-9782-4

Armstrong, M. (2006). *A handbook of human resource management practice*. Kogan Page Publishers.

Arslan, M. (2001). *İş ve meslek ahlakı*. Nobel yayın dağıtım.

Arslan, M. (2005). *Is ve Meslek Ahlaki* [Business and Professional Ethics]. Ankara: Siyasal yayinevi.

Aswathappa, K. E. M. A. L. (2005). *Human resource and personnel management*. Tata McGraw-Hill Education.

Audi, R. (2009). *Business ethics and ethical business*. Oxford University Press.

Barrett, E. (1999). Justice in the workplace? Normative ethics and the critique of human resource management. *Personnel Review*, *28*(4), 307–318. doi:10.1108/00483489910273633

Bartels, K. K., Harrick, E., Martell, K., & Strickland, D. (1998). The relationship between ethical climate and ethical problems within human resource management. *Journal of Business Ethics*, *17*(7), 799–804. doi:10.1023/A:1005817401688

Başarır, Ç. (2006). *İnsan kaynakları yönetiminde etiksel ilkelerin yerleştirilmesi ve bir uygulama* (Master's thesis). Balıkesir Üniversitesi Sosyal Bilimler Enstitüsü.

Bayram, S. (2005). *İşletme Etiği ve Çalışanların Şirketlerinde Mevcut Olan Etik Uygulamalara İlişkin Algılarının Ölçümüne Yönelik Bir Araştırma*. Yayınlanmış Yüksek Lisans Tezi, Uludağ Üniversitesi SBE.

Beardwell, J., & Claydon, T. (Eds.). (2007). *Human resource management: A contemporary approach*. Pearson Education.

Becker, W. (2010). Ethics in human resources: An exercise involving new employees. *Journal of Human Resources Education*, *4*(4).

Bektaş, Ç., & Köseoğlu, M. A. (2008). Iş etiği ve iş etiğinin yayilim süreci. *Süleyman Demirel Üniversitesi İktisadi ve İdari Bilimler Fakültesi Dergisi*, *13*(1), 145–158.

Benligiray, S., Geylan, A., & Duman, E. (2010). İnsan Kaynakları Yönetiminin Stratejik Olarak Yönlendirilmesinin Finansal Performansı Etkileyip Etkilemediğinin Analizi. *Anadolu Üniversitesi Sosyal Bilimler Dergisi*, *10*(1), 61–84.

Bennington, L. (2007). HR managers as ethics agents of the state. *Human resource management: Ethics and employment*, 137.

Berenbeim, R. E. (2010). Utilizing HR and ethics and compliance collaboration to promote an ethical culture. *Employment Relations Today*, *37*(1), 17–26. doi:10.1002/ert.20281

Berkman, Ü., & Arslan, M. (2009). Dünyada ve Türkiye'de iş etiği ve etik yönetimi. İstanbul: Tüsiad Yayınları.

Beschorner, T. (2006). Ethical theory and business practices: The case of discourse ethics. *Journal of Business Ethics*, *66*(1), 127–139. doi:10.100710551-006-9049-x

Bowie, N. E., & Beauchamp, T. L. (Eds.). (2001). *Ethical theory and business*. Prentice Hall.

Boxall, P., & Purcell, J. (2007). Strategic management and human resources: the pursuit of productivity, flexibility, and legitimacy. *Human resource management: Ethics and employment*, 66.

Buckley, M. R., Beu, D. S., Frink, D. D., Howard, J. L., Berkson, H., Mobbs, T. A., & Ferris, G. R. (2001). Ethical issues in human resources systems. *Human Resource Management Review*, *11*(1-2), 11–29. doi:10.1016/S1053-4822(00)00038-3

Budd, J. W., & Scoville, J. G. (Eds.). (2005). *The ethics of human resources and industrial relations*. Cornell University Press.

Buhler, P. (2002). *Human resources management: All the information you need to manage your staff and meet your business objectives*. Simon and Schuster.

Burke, R. J., & Ng, E. (2006). The changing nature of work and organizations: Implications for human resource management. *Human Resource Management Review*, *16*(2), 86–94. doi:10.1016/j.hrmr.2006.03.006

Cascio, W. F. (2015). *Managing human resources*. McGraw-Hill.

Chukwunonso, F. (2009). *The development of human resource management from a historical perspective and its implications for the human resource manager*. Academic Press.

Cole, G. A. (2002). *Personnel and human resource management*. Cengage Learning EMEA.

Coombs, A. A. T., & King, R. K. (2005). Workplace discrimination: Experiences of practicing physicians. *Journal of the National Medical Association*, *97*(4), 467. PMID:15868767

Danley, J., Harrick, E., Schaefer, D., Strickland, D., & Sullivan, G. (1996). HR's view of ethics in the work place: Are the barbarians at the gate? *Journal of Business Ethics, 15*(3), 273–285. doi:10.1007/BF00382953

De George, R. T. (2005). *A history of business ethics.* Paper delivered at the third biennial global business ethics conference, Santa Clara, CA.

Deckop, J. R. (Ed.). (2006). *Human resource management ethics.* IAP.

Derr, P., & McNamara, E. (2003). *Case studies in environmental ethics.* Rowman & Littlefield Publishers.

DesJardins, J. R., & DesJardins, J. R. (2009). *An introduction to business ethics.* McGraw-Hill Higher Education.

Doğan, N. (2007). *İş Etiğinin Kurumsal Sosyal Sorumluluktaki Rolü ve Bir Araştırma.* Yayınlanmış Yüksek Lisans Tezi, Marmara Üniversitesi SBE.

Ferrell, O. C., & Fraedrich, J. (2015). *Business ethics: Ethical decision making & cases.* Nelson Education.

Ferris, G. R., Buckley, M. R., & Fedor, D. B. (Eds.). (2002). *Human resources management: Perspectives, context, functions, and outcomes.* Prentice Hall.

Fusilier, M. R., Aby, C. D. Jr, Worley, J. K., & Elliott, S. (1996). Perceived seriousness of business ethics issues. *Business & Professional Ethics Journal, 15*(1), 67–78. doi:10.5840/bpej19961519

Gini, A., & Marcoux, A. M. (Eds.). (2005). *Case studies in business ethics.* Pearson Prentice Hall.

Gospel, H. (2009). Human Resources Management: A Historical. The SAGE handbook of human resource management, 12.

Greenwood, M. (2007). Ethics and human resource management. *Human Resource Management in Australia, 3e*, 562–587.

Greenwood, M., & De Cieri, H. (2007). Stakeholder theory and the ethics of HRM. *Human resource management: Ethics and employment*, 119.

Greenwood, M. R. (2002). Ethics and HRM: A review and conceptual analysis. *Journal of Business Ethics, 36*(3), 261–278. doi:10.1023/A:1014090411946

Guest, D. E. (2007). HRM and performance: can partnership address the ethical dilemmas? *Human Resource Management: Ethics and Employment*, 52.

Hartman, L. P. (1998). *Perspectives in business ethics.* Irwin/McGraw-Hill.

Hartman, L. P., DesJardins, J. R., & MacDonald, C. (2008). *Business ethics: Decision making for personal integrity and social responsibility.* McGraw-Hill.

Hartog, M., & Winstanley, D. (2002). Ethics and human resource management: Professional development and practice. *Business & Professional Ethics Journal, 21*(2), 3–9. doi:10.5840/bpej200221210

Harvey, B. (Ed.). (1994). *Business ethics: A European approach.* Prentice Hall.

Heim, K. M. (1991). *Human resources management: Ethics in personnel.* Allerton Park Institute.

Hussein, M. (2009). Hiring and firing with ethics. *Human Resource Management International Digest, 17*(4), 37–40. doi:10.1108/09670730910963334

Jose, A., & Thibodeaux, M. S. (1999). Institutionalization of ethics: The perspective of managers. *Journal of Business Ethics, 22*(2), 133–143. doi:10.1023/A:1006027423495

Kamoche, K. (2007). 15 Strategy, knowledge. *Human Resource Management: Ethics and Employment: Ethics and Employment*, 252.

Kaptein, M. (2008). Developing a measure of unethical behavior in the workplace: A stakeholder perspective. *Journal of Management, 34*(5), 978–1008. doi:10.1177/0149206308318614

Kapu, H. (2009). *Akademik bir disiplin olarak iş ahlakı. İşletmelerde iş etiği.* İstanbul Ticaret Odası Yayınları.

Keçecioğlu, T. (2006). *Stratejik insan kaynakları yönetimi.* Sistem yayıncılık.

Keser, A. (2002). Değişen Yönleriyle Personel Yönetimi: İnsan Kaynakları Yönetimi. *ISGUC The Journal of Industrial Relations and Human Resources, 4*(1).

Kinran, S. B. (2006). *İş Ahlakı ve Etik Değerlerin Türk İşletmeciliğindeki Yeri.* Yayınlanmış Yüksek Lisans Tezi, İstanbul Üniversitesi SBE.

Kirel, Ç. (2000). *Örgütlerde Etik Davranışlar.* Yönetimi ve Bir Uygulama Çalışması, Anadolu Üniversitesi Yayınları.

Kolodinsky, R. W. (2006). Wisdom, Ethics and Human Resources Management. *Human resource management ethics*, 47-69.

Lafer, G. (2005). The critical failure of workplace ethics. *The ethics of human resources and labor relations*, 273-298.

Larmer, R. A. (1992). Whistleblowing and employee loyalty. *Journal of Business Ethics, 11*(2), 125–128. doi:10.1007/BF00872319

Lawler, E., & Boudreau, J. W. (2009). *Achieving excellence in human resources management: An assessment of human resource functions*. Stanford University Press. doi:10.1515/9780804771238

Legge, K. (2000). The ethical context of HRM: the ethical organisation in the boundaryless world. *Ethical issues in contemporary human resource management*, 23-40.

Leymann, H. (1996). The content and development of mobbing at work. *European Journal of Work and Organizational Psychology, 5*(2), 165–184. doi:10.1080/13594329608414853

Lorenz, K. (1966). *Evolution and modification of behavior*. Methuen.

Lussier, R. N., & Hendon, J. R. (2017). *Human resource management: Functions, applications, and skill development*. Sage publications.

Lussier, R. N., & Hendon, J. R. (2019). *Fundamentals of Human Resource Management: Functions, Applications, and Skill Development*. SAGE Publications, Incorporated.

Mauro, N., Natale, S. M., & Libertella, A. F. (1999). Personal values, business ethics and strategic development. *Cross Cultural Management*, *6*(2), 22–28. doi:10.1108/13527609910796942

Mcdonald, G. M. (2009). An anthology of codes of ethics. *European Business Review*, *21*(4), 344–372. doi:10.1108/09555340910970445

Mehrabad, M. S., & Brojeny, M. F. (2007). The development of an expert system for effective selection and appointment of the jobs applicants in human resource management. *Computers & Industrial Engineering*, *53*(2), 306–312. doi:10.1016/j.cie.2007.06.023

Miceli, M. P., Near, J. P., & Dworkin, T. M. (2008). *Whistle-blowing in organizations*. Psychology Press. doi:10.4324/9780203809495

Mitchell, T. R., Daniels, D., Hopper, H., George-Falvy, J., & Ferris, G. R. (1996). Perceived correlates of illegal behavior in organizations. *Journal of Business Ethics*, *15*(4), 439–455. doi:10.1007/BF00380364

Near, J. P., & Miceli, M. P. (1985). Organizational dissidence: The case of whistle-blowing. *Journal of Business Ethics*, *4*(1), 1–16. doi:10.1007/BF00382668

O'Higgins, E., & Kelleher, B. (2005). Comparative perspectives on the ethical orientations of human resources, marketing and finance functional managers. *Journal of Business Ethics*, *56*(3), 275–288. doi:10.100710551-004-3898-y

Özgener, Ş. (2004). *İş Ahlakının Temelleri*. Yönetsel Bir Yaklaşım, Nobel Yayın Dağıtım.

Pinnington, A., Macklin, R., & Campbell, T. (Eds.). (2007). *Human resource management: Ethics and employment*. Oxford University Press.

Pratley, P. (1995). *The essence of business ethics* (Vol. 111). Prentice Hall.

Price, A. (2007). *Human resource management in a business context*. Cengage Learning EMEA.

Reiss, M. C., & Mitra, K. (1998). The effects of individual difference factors on the acceptability of ethical and unethical workplace behaviors. *Journal of Business Ethics*, *17*(14), 1581–1593. doi:10.1023/A:1005742408725

Robertson, D. C., & Schlegelmilch, B. B. (1993). Corporate institutionalization of ethics in the United States and Great Britain. *Journal of Business Ethics*, *12*(4), 301–312. doi:10.1007/BF01666534

Rose, A. (2007). Ethics and human resource management. *Newyork. McGrow-Hill*, *37*, 38.

Sayli, H., & Kızıldağ, D. (2007). Yönetsel Etik ve Yönetsel Etiğin Oluşmasında İnsan Kaynakları Yönetiminin Rolünü Belirlemeye Yönelik Bir Analiz. *Afyon Kocatepe Üniversitesi Sosyal Bilimler Dergisi*, *9*(1), 231–251.

Schuler, R. S. (2000). The internationalization of human resource management. *Journal of International Management*, *6*(3), 239–260. doi:10.1016/S1075-4253(00)00025-9

Schumann, P. L. (2001). A moral principles framework for human resource management ethics. *Human Resource Management Review*, *11*(1-2), 93–111. doi:10.1016/S1053-4822(00)00042-5

Schwartz, M. (2007). The "business ethics" of management theory. *Journal of Management History*, *13*(1), 43–54. doi:10.1108/17511340710715160

Schwoerer, C. E., May, D. R., & Rosen, B. (1995). Organizational characteristics and HRM policies on rights: Exploring the patterns of connections. *Journal of Business Ethics*, *14*(7), 531–549. doi:10.1007/BF00871982

Scott, E. (2016). Ethics and human resource management. In *Practicing professional ethics in economics and public policy* (pp. 215–221). Springer. doi:10.1007/978-94-017-7306-5_11

Sheehan, M. (2004). Workplace mobbing: A proactive response. In *Workplace Mobbing Conference* (pp. 1-11). Academic Press.

Simmons, J. (2008). Ethics and morality in human resource management. *Social Responsibility Journal*, *4*(1/2), 8–23. doi:10.1108/17471110810856794

Singer, P. (2011). *Practical ethics*. Cambridge University Press. doi:10.1017/CBO9780511975950

Smilansky, J. (2002). *Yeni İnsan Kaynakları, Çev*. Derya Atakan, Epsilon Yayıncılık.

Stallworth, L. E., McPherson, T., & Rute, L. (2001). Discrimination in the workplace: How mediation can help.... *Dispute Resolution Journal*, *56*(1), 35.

Sternberg, E. (2000). *Just business: Business ethics in action*. Academic Press.

Svensson, G., Wood, G., & Callaghan, M. (2010). A corporate model of sustainable business practices: An ethical perspective. *Journal of World Business*, *45*(4), 336–345. doi:10.1016/j.jwb.2009.08.005

Szalkowski, A., & Jankowicz, D. (1999). The ethical problems of personnel management in a transition economy. *International Journal of Social Economics*, *26*(12), 1418–1427. doi:10.1108/03068299910248513

Thiroux, J. P., & Krasemann, K. W. (1980). *Ethics: Theory and practice*. Glencoe Publishing Company.

Torlak, Ö., Özdemir, Ş., & Erdemir, E. (2008). *İş Ahlakı Raporu*. İGİAD.

Tutan, A. A. (2006). İşletmelerde iş etiğinin değerlendirilmesi ve bir kamu kurumunda uygulanması, yayınlanmış yüksek lisans tezi. Selçuk Üniversitesi Fen Bilimleri Enstitüsü, Konya.

Üsdiken, B., & Wasti, S. A. (2002). Türkiye'de akademik bir inceleme alanı olarak personel veya" insan kaynakları" yönetimi, 1972-1999. *Amme Idaresi Dergisi*, *35*(3), 1–37.

Van Vuuren, L. J., & Eiselen, R. J. (2006). A role for HR in corporate ethics? South African practitioners' perspectives. *SA Journal of Human Resource Management*, *4*(3), 22–28. doi:10.4102ajhrm.v4i3.95

Velasquez, M. G., & Velazquez, M. (2002). *Business ethics: Concepts and cases* (Vol. 111). Prentice Hall.

Vickers, M. R. (2005). Business ethics and the HR role: Past, present, and future. *Human Resource Planning*, *28*(1), 26–33.

Vogel, D. (1991). Business ethics: New perspectives on old problems. *California Management Review*, *33*(4), 101–117. doi:10.2307/41166675

Walsh, A. J. (2007). HRM and the ethics of commodified work in a market economy. *Human resource management: Ethics and employment*, 102.

Weaver, G. R., & Trevino, L. K. (2001). The role of human resources in ethics/compliance management: A fairness perspective. *Human Resource Management Review*, *11*(1-2), 113–134. doi:10.1016/S1053-4822(00)00043-7

Weiss, J. W. (1994). Business ethics: A managerial, stakeholder approach (No. s 287). Belmont, CA: Wadsworth Publishing Company.

Wells, D., & Schminke, M. (2001). Ethical development and human resources training: An integrative framework. *Human Resource Management Review*, *11*(1-2), 135–158. doi:10.1016/S1053-4822(00)00044-9

White, G. W. (2001). Business ethics. *Journal of Business & Finance Librarianship*, *6*(4), 49–49. doi:10.1300/J109v06n04_05

Wiley, C. (1998). Reexaminating perceived ethics issues and ethics roles among employment managers. *Journal of Business Ethics*, *17*(2), 147–161. doi:10.1023/A:1005795731002

Winstanley, D., & Hartog, M. (2002). Ethics and human resource management: Introduction. *Business Ethics (Oxford, England)*, *11*(3), 200–201. doi:10.1111/1467-8608.00277

Winstanley, D., & Woodall, J. (2000). The ethical dimension of human resource management. *Human Resource Management Journal*, *10*(2), 5–20. doi:10.1111/j.1748-8583.2000.tb00017.x

Winstanley, D., Woodall, J., & Heery, E. (1996a). The agenda for ethics in human resource management. *Business Ethics: A European Review*, *5*(4), 187-194.

Winstanley, D., Woodall, J., & Heery, E. (1996b). Business ethics and human resource management: Themes and issues. *Personnel Review*, *25*(6), 5–12. doi:10.1108/00483489610148491

Woodall, J., & Winstanley, D. (Eds.). (2000). *Ethical issues in contemporary human resource management*. Macmillan Business.

Wooten, K. C. (2001). Ethical dilemmas in human resource management: An application of a multidimensional framework, a unifying taxonomy, and applicable codes. *Human Resource Management Review*, *11*(1-2), 159–175. doi:10.1016/S1053-4822(00)00045-0

Wray-Bliss, E. (2009). Ethics. In The Oxford handbook of critical management studies. OUP.

Yildiz, A. (2007). *KOBİ'lerde İş Ahlakı: Adapazarı Örneği. In Yayınlanmış Yüksek Lisans Tezi, Sakarya Üniversitesi SBE*. Sakarya.

Yilmaz, G. (2006). *Örgüt Kültürü ve İnsan Kaynakları Yönetimi*. Yayınlanmış Yüksek Lisans Tezi, Marmara Üniversitesi SBE.

Zey-Ferrell, M., & Ferrell, O. C. (1982). Role-set configuration and opportunity as predictors of unethical behavior in organizations. *Human Relations*, *35*(7), 587–604. doi:10.1177/001872678203500707

ADDITIONAL READING

Condrey, S. E. (Ed.). (2005). *Handbook of human resources management in government*. John Wiley & Sons.

Fine, S. A., & Cronshaw, S. F. (1999). *Functional job analysis: A foundation for human resources management*. Psychology Press. doi:10.4324/9781410602497

Gomez-Mejia, L. R., Balkin, D. B., & Cardy, R. L. (2007). *Managing human resources*. Pearson/Prentice Hall.

Langford, D., Fellows, R. F., Hancock, M. R., & Gale, A. W. (2014). *Human resources management in construction*. Routledge. doi:10.4324/9781315844695

Lupton, B., & Mathieson, H. (2006). *Human Resources Management in an International Context*. CIPD.

Martocchio, J. J. (Ed.). (2008). *Research in personnel and human resources management*. Emerald Group Publishing Limited. doi:10.1016/S0742-7301(2008)27

Nadler, L., & Nadler, Z. (1989). *Developing human resources*. Jossey-Bass.

Sims, R. R. (2002). *Organizational success through effective human resources management*. Greenwood Publishing Group.

KEY TERMS AND DEFINITIONS

Business Ethics: Interested in values such as honesty, keeping his word, respecting nature, behaving fairly, and opposing injustices.

Culture: Culture is the whole way of life of a society. They are patterns of emotions, thoughts, and movements of society.

Ethical: Honor science is the whole of moral and moral behaviors that must be followed or avoided among various professional branches.

Morality: It is the consciousness of the individual about the provisions of value, which forms the basis of the individual's behaviors and shapes the individual's behaviors.

Religion: Religion is the order that unites and provides beliefs in the form of rules, institutions, customs, and symbols.

Social Norms: It regulates the relationships of individuals within a certain group and guides their actions.

Value: Value is a concept that we gain from what we have been influenced by in the past and that we specify within our life choices.

Whistleblowing: Internal fraud and corruption, dealing with ethical issues, employee misuse or unlawful activities in the workplace, are all behaviors exhibited against unethical misconduct in organizations.

Chapter 14
Evaluation and Importance of Business Ethics in Terms of Organizational Culture

Zafer Adiguzel
https://orcid.org/0000-0001-8743-356X
Istanbul Medipol University, Turkey

ABSTRACT

Business ethics is a scientific discipline that analyzes and explains norms and values that guide the business world. When the books and academic researches about business ethics are examined, it is understood that interest in business ethics has started to increase. Among the most important reasons for business interest, ethics is the impact of economic development and globalization. The story of the economy and the adaptation to the global market with each passing day has brought unethical practices to the agenda. It is considered that models that improve the decision-making mechanisms of managers against corrupt practices are deemed necessary. In the study, many studies have been cited as references to explain business ethics' individual and organizational factors. It aims to contribute to the literature with the results and results of the academic research conducted on these models.

INTRODUCTION

Nowadays, in order for organizations to exist and survive in a competitive environment, behavioral and management styles based on organizational values have become one of the prerequisites. In recent years, it is becoming more important for organizations to assume social responsibility towards society, to adopt ethical values, and to be competitive and innovative in their adaptation to international business standards. There is an unwritten "standard of behavior" that indicates whether existing behaviors are in line with the team spirit. Every new member of the organization will understand the way the organization works, the way of dressing, what kind of behavior will put themselves in a difficult situation, how important honesty is, how the hierarchical system works and how they have to adapt to such behavior. There is a dominant understanding of organizational culture that distinguishes each organization from

DOI: 10.4018/978-1-7998-4117-3.ch014

other organizations and this must be adopted by all employees. Organizational culture, which means the common beliefs, values, norms, traditions, practices and the way of doing things that keep the members of the organization together; Achieving certain common goals, developing a sense of unity and integrity among the members of the organization, overcomes the uncertainties created by internal and external changes, and plays an important role in adapting to the changing environment. Ethical values and business ethics form the basis of organizational culture, define who the organization is, what it believes and how it tries to achieve it; determines the principles of strategy, policy, and behavior in this process; it shapes the beliefs, traditions, and ways of working of the organization.

The modern world has not come to the point where it came from the easy way. Throughout history, human beings have been influenced by ethical theories and created the modern world. Ethics has tried to provide the moral structure that human beings need by dealing with what and how to decide the right and wrong actions that have been going on for thousands of years. In this historical adventure, ethics has undergone a transformation from the philosophers' understanding of virtue in ancient times to the Kant (2001) approach to enlightenment. Ethics has undergone its first major transformation throughout history during the Bentham era (Bentham 1989). Bentham's theory of "utilitarianism" emerged from the ethical approach laid the foundation for the liberal economic views of Adam Smith (1723-1790), described as the founder of today's economic system (Smith 1987). The second major transformation in ethics has emerged with enlightenment. Kant, one of the greatest thinkers of the Enlightenment, led to the secularization of morality. Kant, who says that he made the Copernican revolution in ethics with the ethical approach he put forward, is undoubtedly still one of the greatest and most influential thinkers of our century. Kant's homework ethics not only shaped the individual morality of today's modern world, but also influenced the legal systems of states.

Business ethics has also been transformed according to the ages dominated by ethical theories. Business ethics has not been able to struggle for existence within the utilitarian approach and has been called an oxymoron in the Western world. Major economic crises and increased human rights violations have led to a paradigm shift in business ethics within the Western world. The business ethics that came to Kant's ethics underwent a major transformation and suffered road accidents in economic crises. Academicians who examined the causes of unethical practices with economic crises have increased their publications on business ethics. In order to prevent ethical violations, governments have begun to regulate their business life through legal legislation. Today's modern states have not been freed from these changes. With the establishment of ethics committees, ethical violations in the public are tried to be prevented. Ethics associations and foundations operating in the private sector strive to increase ethical awareness and serve to complete the moral and legal gaps in business life. Universities have not been exempted from the changes that have taken place. Vocational ethics or business ethics courses have been converted into compulsory or elective courses. The lack of academic literature on business ethics and the limited research in certain structures cause this field to be underdeveloped. This study was carried out in order to contribute to the literature of business ethics, to ensure that ethical issues remain on the agenda and to demonstrate the importance of organizational culture.

ETHICS CONCEPT

Each society has its own unique understanding of ethics, which depends on many variables such as culture, ethnicity and time, and varies from one society to another, even between regions and in proportion to time processes. Ethics emerges as a group and network of beliefs, values, norms, orders, prohibitions and designs that enter the life of a person, a group, a people, a social class, a nation, a cultural environment in a certain historical period and direct their actions. In this respect, ethics is in our lives on every side; it is something historically experienced at the personal and group / social level; it is found in every historical period, in every human community (Evans 1991). In the literature, it is observed that various definitions of ethics have been made. A group of researchers discuss ethics in the form of ethical standards and rules that an individual should follow; it is defined as a system of principles, values and standards that explain and define how individuals behave correctly (Gupta 2010). According to a definition, ethics is the most to investigate what is good and evil, what is right and wrong (Clegg et al. 2007). Similarly, Runes (1964) defines ethics as the expression of behaviors between the two parties with rights and justification standards. According to De George (2011), ethics are efforts to ensure morality. Ethics is, above all, the research and understanding of a desirable life. To put all activities and objectives in place with a wider perspective; what to do or not to do; what will or will not be desired; know what to have or not to have (Weiss 1994). A slightly more easily understood definition of the concept of ethics is as follows: Ethics is a discipline of philosophy that investigates the values, rules, right-wrong or good-bad that form the basis of the individual and social relations established by people. As can be seen in this definition, the concepts of ethics and morality are close concepts (Marcoux 2006).

Ethical and Moral Discrimination: The concepts of ethics and morality are often used interchangeably. These two terms are very closely related, and this distinction is not always fully valid in discussions or practice of ethics and morality (Buchholz and Rosenthal 1998). Morale to the beliefs and designs that dominate the conscious life of an individual, a people, a social class, morality to live them as a social phenomenon, the human attitudes directed according to these beliefs and designs and the actions directed according to these attitudes, true or wrong, good or bad judgments are called moralisch (Shaw and Barry 2015). Immanuel Kant (1724-1804), in his Critique of Practical Reason, defines morality as a doctrine that shows how we can make our happiness valuable, not how we make ourselves happy (Özgener 2004).

Although morality is a factual and historical concept, ethics is the name of the discipline of philosophy that is directed towards this phenomenon. Morality is used to characterize a person's actions. In general, morality understands both a certain form of behavior and judgments about the actions of ourselves and others. The other difference between morality and ethics is this: There's morality without a moral philosophy. In other words, morality exists before ethics. Because even the most primitive society has its own morals. In this sense, morality is self-formed and generally accepted prohibitions and assessments that regulate what people do in all areas of social life. Moral philosophy, on the other hand, was founded when it separated its basic concepts from the concepts of close branches of knowledge and brought them into a state of its own unique concepts. Indeed, the emergence of the concept of ethics BC. 4th century. The greatest philosopher of the period, Aristotle, good, virtue, freedom, happiness is considered to be the founder of ethics for the concept structure (Ural 2003). Many thinkers have no problem using the words morality and ethics interchangeably. However, the general trend is that these two concepts are separate (Kirel 2000). In business life, ethics is used instead of the concept of ethics, professional ethics principles or rules (Beauchamp et al. 2004). It is worth noting that the term "ethics" is used more in the business management literature (DesJardins and DesJardins 2009).

Importance and Characteristics of Ethics: The reasons for the need for ethics can be examined and explained from different angles. Among them, the moment is more emphasized, the peace and happiness of the individual and the possibility, construction, peace and continuation of public life. Because without ethics, public life is not possible, people can't live together. Without a common understanding of what is good and what is bad, there will be a complete commotion among people, instead of order and peace. There are two obvious reasons why human behavior is only subject to ethical consideration among living things: The first is the fact that people have a mind structure that can distinguish between good and evil, and the second is the fact that people live together side by side in society. In short, human beings are subject to ethical evaluation because they are intelligent and live in society (Petty and Hill 2005). When one considers the four characteristics of human being as having a mind, being free, being selfish and having to live together, it is better understood why ethics is only for people. One can use selfishness to satisfy others and to use others to harm others. Since violating the right of others harms living together, people themselves are adversely affected by such a situation. If so, there is a need for a rule and order to exercise the rights of freedom of individuals. Disciplines such as ethics, religion, and law determine the limits of human freedom through their arrangements. The concepts of shame, sin, and prohibition are, in a sense, the cornerstones of human freedom. Thanks to the regulations of ethics, religion, and law, it is possible for people to live together in peace. The existence of ethics can be thought of as a kind of natural law: Just as there is life where water is found, there is also ethics where people are located (Ardichvili et al. 2009). People need each other to live in. But in social life, there is a need for an ever-expanding system of values that everyone agrees on. These ethical values, which form the basis of social life, have to keep partners even when there is nothing in common with the parties who are competing and conflicting in society. No one can openly oppose them. For example, no one can recommend deception or fraud, not to be honesty. No one can value betrayal instead of loyalty, injustice instead of justice. Therefore, the concept of ethics is important for individuals and societies.

Factors Affecting Ethics

Culture: Culture is the whole way of life of a society. Patterns of the emotions, thoughts and movements of society form the culture (Hugman 2012). The subject of culture and ethics is thought to be intertwined and inseparable. Culture has a great impact on what is ethical and unethical, as well as affecting the ethical attitudes and behaviors of individuals. As a result, it is possible to find different ethical thoughts, attitudes and behaviors in different cultures (Ricoeur 1973).

Value: Value is an appraisal that we have earned from those we have influenced in the past and stated with our life choices (Anderson 1995). It is often mentioned not the only value that individuals develop, but values. This is because a person puts a value on every object or event. To value one thing is to choose some among others. The values we choose guide us to find the right approach when analyzing and evaluating the ethical problems we face (Spash 2002). Basic ethics-based values that are generally the same for everyone; honesty, integrity, help and respect for others, reliability, loyalty, justice and responsibility.

Law: Human; creates very different social-public structures while exploring its problems as a social-historical-cultural entity, determining their boundaries through analyzing them, and then looking for solutions to their problems; so that these structures carry society and/or the public in one way or another. One of these structures is called ethics and the other is called law. Both are born of necessities; both are based on values, or both are also a whole of values. Like ethics in law, the rules governing the relations between facts, actions and behaviors, including human beings, do not correspond to a certain existential situation in the outside world as a whole. The ethics of the law is an ideal, desirable situation. However, when you look at the human world, it is noticed that this is not the case for most of the situation. The ethical basis that lawyers have in their hands to protect public health or to perform the functions of treating deteriorating health is embodied in human rights. If the law is based on human rights based on the sum of ethical principles, it will not be just law/rule, etc. It seems extremely important that the subjects of the field of law pay attention to this point (Kaler and Chryssides, 1996).

Religion: Religion is the order that unites and provides beliefs in the form of rules, institutions, customs and symbols (Baum 2001). There is a strong relationship between religion and ethics. First of all, every religion has a certain moral dimension. Each religion brings a certain value system, informing people of what ideals, how to act and what kind of character they should create (Conroy and Emerson 2004). However, religious values are evident because they are in scripture, and do not change over time. Although ethical values include religious values, they vary according to the characteristics of society and renew themselves over time. Because there are other factors that affect society as much as religion (technology, economy... etc). For example, while religious values are the same for all Muslims, ethical values vary according to the society in which they are members.

Social Norms: In order for the society to survive, it is a necessity for members to cooperate with each other within a defined order and there are social norms in the element that will ensure this necessity (Dunfee 1991). Social norms regulate the relationships of individuals within a certain group and guide their actions. They are usually a reflection of values and are collectives because they are shared by all members of a group (Donaldson and Dunfee 1994). In order for the society to survive, it is a necessity for members to cooperate with each other within a defined order and there are social norms in the element that will ensure this necessity (Cragg 2000).

Definition of Business Ethics: The introduction of the concept of ethics into businesses and organizations has gained importance with the increasing disclosure of unethical behaviors. Initially, this importance has taken its place under the social responsibilities title of enterprises, and then it has been examined under the concept of business ethics (De George 1987). Business ethics, business life laws, it is stated that include the principles of social responsibility and relations with stakeholders; concept, to be fair to people and businesses to create good feelings about being expressed as to act (Randall and Gibson 1990). Business ethics, which enters into a applied ethical approach, not only examines moral norms and values, but also applies the results of the examination to institutions, technologies, way of doing business and its activities (DesJardins and DesJardins 2009). Examining business ethics, applied ethics, ethical rules and principles in a commercial context, examining various moral and ethical problems that may arise in the business environment, and any special duties or obligations of persons in business (Fisher and Lovell 2009). Another definition describes business ethics as a concept that expresses right and wrongs in the process of production and consumption of goods and services (Audi 2009). Based on these definitions, business ethics can be defined as ethical rules that individuals who influence the same business environment from customers to employees, partners to suppliers, combine ethical values in society with a common denominator (Paliwal 2006). All definitions that try to explain business ethics, which are the sum of ethical principles and standards that guide and guide the behaviors in the business world, include the existing rules, standards and ethical principles about what is wrong and right in a certain situation. Therefore, the suitability of a behavior is determined not only by the personal ethics and values of individuals, but also by the society including mass media, interest groups and organizations (Ferrell and Fraedrich 2015).

- Systematic issues are ethical, economic, political, legal and other social systems that relate to business operations.
- Issues related to the Company; company policies, procedures, organizational structure, business activities, etc. are ethical issues related to all issues within the company.

- Individual issues; individual activities in the enterprise. These are ethical issues related to decisions, behaviours or individual character (Bradburn 2001).

If we look at the definitions of business ethics and business ethics; that business ethics encompasses a set of rules that are relative, varying from society to society, and must be respected according to tradition, ethnicity, and religion; on the other hand, it is possible to say that business ethics includes the principles of thinking that is universal, valid for all and based on thinking and philosophy (Jones et al. 2005). Although these three topics examined by business ethics are seen as different from each other, there are points where they intersect. Systematic issues involving legal, economic, political and cultural impacts and issues related to the company are combined in the individual and enable individuals to exhibit ethical behavior with the participation of individual issues (Velasquez and Velazquez 2002). Individual ethics plays an important role in the formation of business ethics. Because in enterprises, decisions are often made by individuals at the management level. An administrator's erroneous decision or unethical behavior may damage the image of businesses (De George 2011).

Business ethics as an interdisciplinary academic field is often the subject of the work of philosophers, academics and sociologists (De George 2005). In this sense, the researchers approaching the subject within the framework of their own disciplines reveal different aspects of the field. However, the fact that business ethics is an interdisciplinary feature also makes the field controversial. Therefore, there is no consensus among researchers about how the concept of ethics can be adapted to organizational practices and how ethical dilemmas in organizations should be resolved (Ferrell and Fraedrich 2015). As an interdisciplinary field of study, the definitions made about business ethics differ within themselves. According to Budd and Scoville (2005), the adaptation of the philosophy of ethics in business ethics to the functions and practices of the organization is, in other words, the application of ethical principles in decision-making. This definition, which reflects a broad perspective, focuses on regulating the relationship between business players, organizations, employees, customers, suppliers, society and other stakeholders in the light of common principles. Business ethics, in line with the rules and standards that guide the working life, organizational practices and activities, right and wrong means to discuss and do the right thing (Megone and Robinson 2002). In general, business ethics, economics and business opportunities in the face of the opportunities provided by making prudent choices, honesty, trust, respect and fairness in all economic activities to act as a principle and to share the same environment while in contact with the environment is to support the community (Payne et al. 1997). According to another definition, business ethics is a managerial tool that affects the sensitivity of relationships in work life, designs strategic behaviors in these relationships, also plays a key role in strategic movements and minimizes the risks that threaten the life of organizations in a variable environmental structure (Kopperi 1999). Within the framework of the relationship between business ethics, ethical principles and the organization (Hartman 1996), it tries to find answers to the questions which behaviors are acceptable or which are unacceptable for a good working life (Sintonen and Takala 2002). According to the definition of Svensson and Wood (2003), business ethics tries to reveal what is acceptable and unacceptable in relation to the activities of organizations operating in a certain time and culture. This definition emphasizes cultural relativity and argues that a business ethics approach that ignores cultural differences in the business world would be meaningless. Most definitions of business ethics cover rules, standards and moral principles regarding right and wrong in certain situations (Parker 2003). In other words, they state that an activity where ethical values are not taken into account is not possible. In summary, business ethics can be defined as a management discipline that questions acceptable behaviors in the business life within the framework of the cultural elements shared by the society in which organizations operate, based on the relations between the actors and trying to regulate these relations in line with common principles.

Historical Development of Business Ethics: The historical development of business ethics dates back to ancient times to the present day. The writers and philosophers of the ancient Greek and Roman worlds did not take kindly to business and commercial activities. Because such activities assumed that the ambition to make money, mixed with lies and fraud, and thought that their civilization would eventually be degenerated from these societies after a relationship with barbarian merchants (Small 1993). In this type of slavery civilizations dominated by aristocracy, activities based on physical labor were underestimated. In addition to commercial activities, work and work were known to be an occupation of slaves and lower classes (Garrett 2004). The attitude developed against work and all kinds of commercial activities has not been the same in every society, and even the role of religions in this regard has been very important. The last six of the second group of the two Ten Commandments in the Torah are purely moral orders, the accumulation of wealth is not contrary to the general morality, the prohibition of interest is a regulation on business ethics and affects commercial activities (De George 2005). The role of religions in work and work ethics has been greater and more important than thought. Christianity saw trade as a service for people to overcome each other's shortcomings, and preached that Christian merchants should be honest with all Christian and non-Christian people and apply the same standards of business ethics. This feature contributes to the creation of a universal work ethic of Christianity (Audi 2010). The practices of Islamic religion on business ethics are as old as the date of the emergence of religion. In Islam, business and trade activities are encouraged and it is stated that only honesty and honesty should be taken as basis for all humanity.

Business Ethics in the World and Historical Development of Business Ethics: Economic, political and cultural processes are shaped according to the trends of business ethics. Taking these processes into consideration, the historical process of business ethics is handled in six stages as pre-1960 business ethics, 1970's business ethics, 1980's business ethics, 1990's business ethics and 2000's business ethics (Drover et al. 2012).

In the pre-1960 period, the role of religious ideologies was emphasized, assuming that religious and social ideas shaped business actions in the United States. In this period, the moral problems related to work were mostly dealt with in terms of theology. Religious leaders voiced the problems of fair wages, business practices and the morality of capitalism aloud (Ciulla 2011). Significant changes in the business world occurred in the 1960s. These changes cause many ecological problems such as pollution and nuclear waste due to the development of industrialization. In the United States, in 1962, President John F. Kennedy issued a four-point declaration on the Protection Of Consumer Rights, namely the right to protection, the right to information, the right to choice and the right to be informed. After Kennedy, as president, Johnson addressed the business community and stated that the responsibility of the American Government was to provide some degree of economic balance to its citizens. Activities that could destabilize the economy have been described as unethical and illegal. In these periods, the results of developments in the industry have emerged. Thus, it is stated that there should not be industrialization that negatively affects living conditions. On the other hand, consumers have been given certain rights by preventing the understanding of selling what I produce. Thus, it is ensured that the quality of life is indexed to business ethics (Jones et al. 2005).

The 70s were the years when business ethics emerged as a workspace. The principles put forward by philosophers and religious thinkers have found a number of applications in the business world during these years. Business ethics centers have been opened in universities and started to institutionalize as business ethics, and businesses are still more interested in their image in the public eye. Towards the end of the 1970s, basic ethical issues such as bribery, misleading advertisements, price deals, product and environmental safety emerged (Chryssides and Kaler 1993).

By the 1980s, corporate moral theories of enterprises began to emerge. The main argument of business ethics defenders is that businesses have responsibilities towards society and this social responsibility should be examined. It was during this period that business ethics courses were taken into university programs. The increasing number of institutions with different interests has increased the interest in business ethics studies. Consequently, business ethics has become a major concern of many businesses, particularly leading companies such as General Electric Co., General Motor, S. T. John & Son, etc. In these years, many studies were carried out in the name of institutionalization of ethics. Preparation of ethical codes, setting ethical standards of conduct. Moreover, in 1987, the first chair of European business ethics was established in the Netherlands.

In the 1990s, free trade and open competition now replaced limited trade and controlled competition. More businesses enter into partnerships with their competitors, and host governments continue their activities with outside assistance. Governments are now becoming an integrated part of the global strategies of enterprises. In 1991, the United States was billed federally as Sentencing Guidelines, a guide to businesses with illegal practices. In spite of the fact that all practices are focused on compliance with the law, enterprises have made efforts to create an ethical environment that will enable their employees to behave more carefully. Thus, business ethics experts responsible for the will of ethical programs were employed in the enterprises. Business ethics has become a systematic field of study rather than the question of what to do.

Business ethics in the 2000s; is becoming a workspace that deals with the ethics of business activities. The field of business ethics simply establishes systematic relationships between concepts such as ethics, responsibility and decision-making in businesses, rather than focusing on what to do or what not to do in certain situations. The process of ethical decision-making in business activities is dealt with from philosophical, economic, sociological, psychological and religious perspectives. Managers and researchers working on this issue have sought to develop systematic guidelines that can help individuals and organizations make ethical decisions. In other words, while business ethics issues are becoming more and more diversified in these years, business ethics principles are tried to be implemented and become the basic standard in all enterprises (Ghillyer 2008). There are no signs of unification of business ethics around an international ethics treaty in the future. Therefore, universal business codes are not in question. Instead, multinational businesses can be involved in ethical management and ethical conflicts. The future can be seen more diversity in ethical issues.

Examination of the Concept of Business Ethics in Theoretical Framework: The history of ethics in organizations dates back to the first organizations (Schwartz 2007). Accordingly, it can be said that business ethics has just begun to be discussed in relation to the concept of ethics (Gini and Marcoux 2005). Especially with the process of globalization, the concept has begun to attract the attention of the business world and recently the interest of managers and academics has increased (Bennington 2007). Since the 1960s, business ethics (Spence 2000), which has been the subject of academic studies extensively, has become an academic field of study after 1985 (Schwartz 2007). The literature on business ethics, which became an academic field of study, became richer rapidly as the subject became popular (Velasquez and Velazquez 2002). In parallel with the developments, business ethics has become a concept that has been taught in universities, especially in departments such as HRM, marketing, advertising and accounting (Hurn 2008). Organization and management scandals (e.g. Enron, Worldcom, Arthur Anderson), corruption, bribery, destruction of documents, misleading advertisements, increase in environmental pollution caused by organizations, differences (especially ethnic, gender and race) factors such as conflict between organization and personal values, deregulation on the economy, globalization, unification of organizations, technological developments, the desire of organizations to create positive image, and the increasing importance of human rights explain the reasons for the interest in business ethics (Valentine et al. 2009; Holme 2008). At the same time, the beliefs of the managers that the organizations having ethical policies will be successful in the long term brings the ethical issue into focus (Ferrell and Fraedrich 2015). Dangerous situations caused by organizational activities and especially caused by large organizations have increased the awareness of the society about organizational practices (Appelbaum et al. 2005). In this sense, the pressures of investors, customers, society and other interested parties force the organizations to operate within the framework of ethics (Koh and Boo 2004), and the ethical erosion that brings about the necessity to reorganize the working life within the framework of ethical principles (Pratley 1995).

The Purpose and Importance of Business Ethics

The starting point of the discussion of business ethics is organizations. As parties affected by organizational activities, customers, suppliers, employees, other stakeholders and the community form different sides of business ethics (Booth et al. 2005). The mutual relations between these parties are effective in the emergence of issues related to business ethics. Researchers who try to identify right and wrong, good and bad, beneficial and harmful behaviors and elements in the conduct of organizational activities aim to look at business ethics from the perspective of all parties (Weiss 2008). In the information economy of the 21st century, it sheds light on the visions of business ethics organizations and takes on the role of being their common sense in the way of realizing their missions. Prudent action requires considering the economic activities as well as the interests of other interests and society in the activities of organizations. Therefore, for organizational success and a better working life, organizations should be supported by stakeholders and have activities and practices in accordance with business ethics (Mayer 2001). Understanding and managing business ethics in today's business world, bringing together individuals of different values, is a critical issue for organizational success (Audi 2010). The issue of business ethics, which is thought to have a direct impact on the organization's effectiveness, continues to attract the attention of all organizations. The main purpose of business ethics, which has become an indispensable element for organizations, is to establish a set of facilitating rules in the products and services of organizations that give the public confidence (Keller et al. 2007). In this sense, in today's business world, where intense competition is experienced, business ethics and organizations have a protective feature (Wells 1994). By adopting business ethics, organizations are protected from the dangers of unethical behavior. In addition, the reputation of organizations that adopt business ethics in public is not damaged in this respect and positive organizational images are protected. Although there is no consensus among researchers about whether ethics and business relations can be established, whether business ethics is possible, or whether universal principles can be established in business life, researches and practices generally show that organizations organizing their activities within the framework of business ethics have benefits.

Different Perspectives on Business Ethics: What are the roles and functions of business ethics organizations in society? tries to find the answer to the question. The researchers found that organizations have no ethical responsibility; some argue that there is ethics in organizations, but this ethics differs from the ethics in other human activities. Some researchers argue that organizations and societies share the same ethical understanding and responsibility, so that organizations are a part of the moral community

(Kingsley Trezise 1996). It is better understood in light of the developments in the business world that business ethics is important for the success of the organization in recent times (Warren and Tweedale 2002). Organizations continue to organize their activities within the framework of business ethics day by day. These evaluations of business ethics are misleading and unrealistic. Opinions that reflect a narrow perspective are fundamentally united at the point where business and ethics cannot be considered together. For organizations that aim for long-term success and seek the support of their stakeholders, the principles of business ethics are of undeniable importance. Another misconception about business ethics is that some organizations think that operating in accordance with the law and operating within the framework of business ethics is the same thing. However, obeying the law and operating in accordance with business ethics are not the same thing. Because ethical values demand a morality above the minimum level of behavior determined by law (Holme 2008; Bennington 2007). In other words, laws are imperative for the regular continuation of society and business life. However, the laws regulate the basic aspects of business life by providing a specific framework. For example, there is a legal sanction, a penalty, for stealing from a workplace. But stealing from work time, in other words, there is no legal sanction for not using working time effectively. Therefore, it can be said that the principles of business ethics provide a framework that can be applied in all subjects and practices of working life.

Historical Development Process of Business Ethics

In the literature, the historical development process of business ethics is divided into different stages. Business ethics: Since the 1690s, a new era has begun in the world thanks to the steam power that has been used in production. In the new period called the industrial revolution, production methods shifted from arm strength to machine. This development has enabled production to be more abundant and cheaper than before. Production by making it faster and cheaper has led to large-scale production. The number of factories and companies with large capital has increased in the rapidly developing industrial sector. Parallel to the developments, labor has become cheaper in this period, and a workforce profile has emerged with heavy conditions and no social security. After the industrial revolution, the balance of labor supply and demand deteriorated to the detriment of workers and a huge gap between the effort spent and the income generated. Thus, during this period, a great power asymmetry occurred between employers and employees, and the dominance of employers over employees made them feel intensely.

Between 1900 and 1920, issues such as improving the working conditions of women and children employees, compensation rights arising from employees' work, and accurate information of consumers in advertisements related to products have been on the agenda of his/her life. In 1920, the Institute of Industrial Management was opened in Britain. The institute argued that management should be considered separate from organizational ownership and argued that business ethics should be seen in the field of business administration regardless of religion, philosophy and economics (Braybrooke 1983). This development has been an important step in establishing the relationship between business and ethics. With the expansion of trade union activities, efforts to improve the working conditions of the workers began to emerge during this period.

In the 1930s, pressures by trade unions to reduce working hours yielded results, and working conditions were relatively improved (Porter 2005). Until the 1960s, the concept of ethics in organizations was addressed from a religious and philosophical point of view (Ferrell and Fraedrich 2015) and developments took place on a religious basis (De George 1987). During this period, each religion tried to apply its own ethical elements in other aspects as well as working life. The religious and ethical questioning of capitalism, the payment of the workers' wages and other business activities are supported by religious elements (Enderle 1999).

When the developments are taken into consideration, it can be said that employees' rights against employers and minimum wage, erosion of ethical values against the materialist elements brought by capitalism and the efforts of society to correct the deteriorating economic situation are among the important developments of this period (De George 1987). In this respect, the ethical and business relations were established in the light of developments and the necessity of organizing the working life within the framework of ethical principles came up for the first time during this period.

Business Ethics in the 1960s: Factors such as the increase in the level of welfare and the increase in the level of education of individuals during the period of 1960s increased the awareness of consumers about the activities of the organization. In light of these developments, consumer movement took place in the United States in 1962 and as a result the Consumer Rights Declaration, which grants consumers a number of rights, came into force (Ferrell and Fraedrich 2015). The fact that consumers began to acquire certain rights has led to a shift in the attention of organizations focusing on mass production after the industrial revolution from production to consumer satisfaction. During this period, environmental pollution with the development of modern industry began to negatively affect public life. Organizations have been held responsible for the ecological problems that arise. This development has shown that organizations should be aware of the environment (De George 1987). During this period, organizations had to think about the environment in which they operate besides production due to the environmental problems they caused.

The 1960s were the period when organizations' tendencies on social issues emerged (De George 1987). In these years, the importance of participation in management, social pressures and laws and regulations that socially motivate organizations forced the organizations to think about the social consequences of their economic activities (Wellman 1975). As the concept of social responsibility began to be discussed, organizations became aware of the impact of their image in public on the success of the organization. With this awareness, it has been a necessity for organizations to operate with their image in mind, in other words, to manage their organizational image (McNamara 1999). With the influence of the developments in these periods, the concept of business ethics has started to attract the attention of organizations. There has been a great increase in research on business ethics (Freeman 2000), and in parallel with this development, business ethics has been seen as a management discipline (McNamara 1999). For the first time in the United States, the so-called ethics and business ethics courses have been included in the curriculum. Ethics committees have started to be established for the first time with the arrangements made in the continuation of this development (Kayacan 2005). As in previous years, the influence on the understanding and implementation of religious elements in this period has been considerably greater (Snoeyenbos and Hummel 1979). However, unlike the other period, the factors that brought business ethics to the agenda of the organization were not socially based, but rather social, consumer awareness, ecological problems and organizational image anxiety.

Business Ethics in the 1970s: In the 1970s, scandals and corruption caused by organizations became an expression of the ethical erosion of working life. These unethical practices of organizations have led societies to monitor the activities of the organization more closely. Organizations that started to have social concerns in this direction started to organize their activities within the framework of ethical principles and gave trainings in order to increase the sensitivity of their employees about business ethics (Brabeck 1984). In the 1970s, the images of organizations, in other words, were seen as the most important factor in the continuity of organizational life. During this period, the community expected organizations to be sensitive to social problems and to take an active role in improving the quality of life (Arslan and Berkman 2009). In the 1970s, organizations that understood the impact of the organisation's image on financial performance focused on the question of how to strengthen their image in public and develop an attitude towards social demands (De George, 1987). The most important factor that brought business ethics to the agenda of the organization during this period was the concern of how organizations appear by society. While these developments in working life, business ethics has started to develop rapidly as an academic field. In this period, philosophers who argued that ethical principles can also be applied to working life have made important contributions to the academic progress of business ethics within the framework of ethical theories and philosophical analyzes. The support of business ethics studies with philosophical knowledge has been an important factor both in forming the basis of business ethics and enriching its content (De George 1987). In addition, the participation of business professors, social scientists, theologians, philosophers, executives, employees, members of the press and media in the conferences and congresses on business ethics in this period highlighted the interdisciplinary characteristics of the field (Ferrell and Fraedrich 2015).

Business Ethics in the 1980s: During this period, business ethics became more on the agenda of organizations operating in the United States of America (USA). In this respect, it can be said that the penalties given as a result of lawsuits filed against unethical organizations in the country have a significant effect (Gandz and Hayes 1988). The transformation that began in the world economy in 1980 meant a new order for all organizations. These periods are known as the years when free market conditions began to emerge as a result of the state administrations giving up to interfere with the economic structure. This development has brought the concept of "competition" to the forefront. With the intensely felt competitive pressure, organizations began to adopt new management approaches and organizational mergers became one of these approaches. With the introduction of free market order, the international activities of the organizations have increased. Organizations that have started to operate in different cultures have begun to understand that cultural differences also differentiate management. In this sense, organizations have seen that a global approach to the concept of business ethics is necessary (Weiss 1994).

In the 1980s, organizations began to play important roles in society in parallel with the developments in the past years (De George 1986). The impact on economic life of large organizations with budgets competing with states has continued to increase. In this sense, it has become more important for the social welfare of organizations to act with a sense of responsibility in their activities. During this period, international organizations on business ethics were established for the first time. The most important of these is the European Business Ethics Organization, which was established in 1987 in the Netherlands (Arslan 2001). In addition to the Business Ethics Society, the Institute of Business and Professional Ethics, many applied ethics centers have been established in the world (Marcoux 2006). These organizations have sought to provide academic support to the organisation's ethical sensitivities.

The identification process for business ethics, which has become an academic field of study, continued in the 1980s. During this period, there was a significant increase in academic studies and the number of courses and courses given at universities related to business ethics (De George 1987). The accumulation of knowledge generated by the increasing number of work on business ethics has made an important contribution in adapting business ethics to organizational activities and guiding implementation (Stark 1993). With the beginning of the globalization process, organizations began to understand the impact of the competition phenomenon on success. Thus, organizations have taken their practices in a wider context as a necessity to operate in a volatile and complex environment. In summary, the most important factor that keeps business ethics on the agenda of organizations during this period was competition pressure.

Business Ethics in the 1990s: In these years, free trade and open competition have now begun to replace limited trade and controlled competition. More businesses enter into partnerships with their competitors, and host governments continue their activities with outside assistance. Governments have now become an integrated part of the global strategies of enterprises (Vallance and Elizabeth 1995). During this period, the dominance of the organizations in the economic structure gradually increased. In the 1990s, the restructuring practices that emerged in every sector, especially in the United States, in other words, the growth studies shrinking (Yeter 2009), the environmental pollution caused by organizations, the concept of the management of the differences arising from the transformation of the business world into a global structure and the practices of discrimination factors have been some of the factors that bring business ethics to the agenda of organizations (De George 1994). These developments have emphasized the fact that organizations should think about their activities on a global scale. In the period 1990, it is seen that organizations began to resort to unfair means to make a difference against their competitors. Various arrangements have been made to prevent this situation. The necessity of ethical treatment in competition rather than unfair competition has come to the fore. In other words, organizations understood that competition should be carried out ethically in the way they live (Enderle 1997). In the 1990s, with the rise of consciousness levels of societies, production processes began to be effective in the purchasing decisions of consumers as well as the quality of the products. As a matter of fact, nike was faced with a consumer boycott in North America and Europe because it employed the child workforce in those years (Arslan and Berkman 2009). The enforcement power of consumers against organizations, which began to gain certain rights in the 1960s, made it felt so intense during this period. Consumers have started to prefer organizations operating in this direction, taking into account not only the products that are the outputs of organizations but also the social responsibility levels of the organizations in their purchasing decisions.

Business Ethics in the 2000s: Since the 1990s, business ethics practices have started to become institutionalized. However, the incidents caused by unethical organization activities that are still happening today are indicative of the fact that organizations do not adequately adopt business ethics in their activities. For this reason, it can be said that there are not many organizations that carry out their activities within the framework of business ethics (Suhonen et al. 2011). In the 2000s, as in other periods, business ethics has been a topic that continues to be discussed in all segments of society, especially in the business world. However, debates on business ethics generally take place around the prevention of ethical erosion in the business world rather than developing the concept. In the 2000s, the issue of business ethics became an issue that was at the top of the agenda of every organization. During this period, organizations began to make efforts at the "institutional level" in order to carry out their activities within the framework of business ethics (González 2003). Organizations have sought to create ethical codes and share this information with the public in order to set certain standards in the way they do business (Arnold and Bustos 2005; Kline 2009).

Globalization creates a different range of activities requiring the application of business ethics (Boatright 2000). In this sense, the process of globalization transforms the meaning that is attached to business ethics. Organizations have now understood that being an ethical organization requires an effort to behave in accordance with the law (Hatch 1993). For this reason, the institutional initiatives of organizations for business ethics can be considered as an indicator of this understanding. When the historical development process of business ethics is examined, it is seen that certain factors bring up business ethics in each period. For example, in the 1960s, consumers who made business ethics important for organizations, ecological problems, while in the 1980s the competitive factor was effective. However, these ecological problems or other factors can not be said to lose their effect in the future. The dominance of factors that only increase disinterest in business ethics has changed relatively. For this reason, it would not be wrong to say that the importance of business ethics has not lost its popularity in any period, but its importance is emphasized more with the developments.

ORGANIZATION

The concept of the organization has survived to the present day by emphasizing different issues against various situations. The organization was to facilitate the achievement of the objectives that were both established by individuals in the organization, as well as the distribution of work and responsibilities within the organization and the distribution of employees within the organization. In 1958, March and Simon emphasized the interaction of employees in the organization. In 2003, Blau and Scott described the formal aspect of the organization. In 1964, Etzioni stated that organizations were an organization for a specific purpose. In these expressions of the concept of the organization, the purpose and formal structure of the organization is emphasized (Eisenberg and Riley 2001).

The concept of organization is to achieve the common goals of communities; The division of labor is defined as the coordination of activities carried out within the framework of responsibilities and enforcement power (Türkkan 2015). It is simply the name given to the community, which includes the activities that individuals manage by meeting in a common denominator for a specific purpose. Businesses, educational institutions, public institutions, sports clubs, army, associations, organizations and trade unions can be given as examples to organizations. Organizations are characterized by national global, new-old, large-small varieties. All of these communities meet in a common corporate denominator for a purpose and individuals are involved in this phenomenon by bringing their personality and social drive (Haman 2016).

Characteristics of Organizational Culture: Organizational culture has developed from the eighties onwards by adapting to time and the conditions created by time with different definitions (Frost et al. 1985). These definitions include; In organizational culture, individuals have shared value, the way they do and conduct business, differences in organizational personalities, and value judgments such as intra-organizational belief-slogan-story (Alvesson 2012). Organizational culture provides the whole of the attitudes, behaviors, information, values and norms in the organization, which ensures that individuals within the organization are together; determines the future of the organization through its aims, strategies and policies by regulating the life style of the organization, its relations with other organizations and its activities (Ott 1989). In terms of organizational culture characteristics; it may have been learned, but it may also be acquired by showing variability (Schein 1990).

While it is based on the judgment of sharing among individuals, it is also a developing situation based on the beliefs and values that develop in the mindset of these shares. In organizational culture, judgments that are repeated periodically or raised by individuals determine the characteristics of organizations (Ashkanasy et al. 2000). In 2004, Yamaguchi introduced the organization's culture; shared by most employees, expressing them as relevant norms, beliefs and values for the organization; In 2005, defined it as a way of doing things by Wilson, McCormack and Ives, and to include common beliefs and actions acquired through socialization and learning as an organization. In 2008, Casida defines the culture of the organization as deep underlying assumptions and beliefs, shared by members of the organization and working unconsciously.

Each organization has its own characteristics and the culture of the organization; which constitutes an organization's management system; it expresses fundamental values, beliefs and principles (Schein 1985). Messages from a business's culture; closely linked to the strategy and management practices of the organization and has an impact on individuals in the organization. While the culture system is quantitative because it cannot be easily seen or touched; they are all known to the employees of the organization (Allaire and Firsirotu 1984). The cultural laws of organizations outweigh any other law and are regarded as the most powerful message in many organizations (Denison and Neale 2017).

Relationship Between Organizational Culture and Business Ethics

It is a fact accepted by everyone that business ethics and organizational culture have great importance for management. However, in order to understand the two concepts and to determine the difference between them, the concepts must first be defined. Organizational culture, common values, norms and beliefs adopted by the members of the organization, the resulting symbols and symbolic movements, business ethics; It refers to the system of norms that direct the movements of both organizations and members of organizations within and outside the organization. As can be seen from here, organizational culture and business ethics are not synonymous with each other, but business ethics is an important part of the organizational culture. In other words, when the culture of the organization is taken as a whole, business ethics is one of the parts that make up this whole (Parker 1998). Ethical values and business ethics are the basis of organizational culture. That is, it defines who the organization is, what it believes in, and how it tries to achieve it; determines the principles of strategy, policy and behavior in this process; it shapes the beliefs, traditions and ways of working of the organization (Verstraeten 1998). Determination of values; it is mandatory to announce these values to the employees of the organization, the public and other relevant stakeholders for the following reasons (Arslan and Berkman 2009):

- Draws the basic framework for managers and employees regarding their behavior and decisions.
- The vision of the business, strategic approach and implementation support.
- Stance of the enterprise to the stakeholders the positive image of the enterprise increases the trust and respect of the company in all stakeholders. Thus, employees are motivated; the customer's sympathy for the product/service increases; the interest of investors increases; the approaches and audits of relevant public supervisory bodies take place positively; suppliers can see what is expected of them more clearly; communication with non-governmental organizations and the media can be more comfortable and smoother.

As a result, values form the basis of organizational culture. In fact, successful organizations, such as Peters and Waterman (1982) in their In Search of Excellence, have laid the foundation for value. Respecting others, contributing to the quality of life of the society and being fair and honest in all relationships are the examples of values. Issues such as product/service quality and safety, accuracy and honesty in pre- and post-sales applications, health and safety of the workplace and employees, investors' interests, energy saving, environmental protection are sub-headings of these values (Arslan and Berkman 2009).

Business ethics examines the problems that arise as a result of business activities in terms of ethical discipline (Trevino and Weaver 1994). These issues can be located inside or outside the business and can affect those who are directly involved in the business, as well as third parties who are not in any relationship. The company has responsibilities towards its employees, customers, the society as a whole and the natural environment in which they are located. How these responsibilities are perceived is related to organizational culture (Ouchi and Wilkins 1985).

In enterprises where some business values and rules are not shared by employees, forcing employees into a common culture can lead to dismissal strays from work and reaction behaviors. For this reason, in order to develop business ethics, it is necessary to create the organizational culture first. The creation of a group culture that constitutes a different element of behavior in the working life or in the social society, in other words, the subculture enables the development of business ethics and the determination of the values that constitute the majority of conflicting values; strengthens the sharing of ethical principles; it makes possible the positive development of self-regulation, criticism and discussion processes (Thiroux and Krasemann 1980).

Today, businesses are faced with the necessity to better understand different cultural expectations in order to gain legitimacy in social and social relations in the global market. Cultural differences in perceiving the ethical responsibilities of enterprises affect the management. Understanding the relationship between the perception of business ethics and culture contributes to the interpretation of ethical decisions and cultural trends (Khushf 1998).

The institutionalization of business ethics means that the subject is no longer official and external and placed in the daily business life. Institutionalizing business ethics will prevent negative impact on the reputation of both the business and the decision-maker (Ardichvili et al. 2009). In order for business ethics to be institutionalized within the enterprise, the organization must first have a culture of organization in the enterprise. The common philosophy between business managers and their employees as a result of the organizational culture allows these two groups to determine what they expect of each other and meet them in a common step (Sheridan 1992).

The process of institutionalizing business ethics refers to conscious and systematic efforts to make ethical values more effective in organizational culture and to constantly observe them. If managers succeed in integrating ethical values into the organizational culture, ethics will be assessed at the organizational level, not at the personal level, affecting the attitudes and behaviors of all employees. Despite its increasing importance, institutionalization in enterprises is not a phenomenon that can be realized all of a sudden. The institutionalization of business ethics in enterprises takes place only within a certain period of time (Lozano 2007).

Businesses that want to create an ethical environment in their business and make business ethics a part of the organizational culture should organize ethical programs covering all business employees for this purpose. These programs should be comprehensively prepared to ensure that employees are sensitive to ethical rules, to prevent scandals, and to establish ethical relations within the enterprise and with the external environment of the business. The programs to be created must also comply with the rules of business ethics, which guide the attitudes and behaviors that employees in the enterprise must follow and define responses to behavior contrary to business ethics. In addition, in these programs, the basic management functions of the enterprise in other words; During planning, organizing, directing, coordination and control activities, it should be emphasized that ethical behaviors should be given organizational priority (Brickley et al. 2002). Organizational participation is required in enterprises with organizational culture. Ensuring organizational participation is also achieved by employees' belief in the aims and values of the organization, efforts for the interests of the organization and efforts to become members of the organization. Therefore, participation in the objectives of the organisation will ensure compliance with ethical conduct (Nakano 2007).

SOLUTIONS AND RECOMMENDATIONS

Today, business ethics has become one of the most debated concepts with the acceptance of the view that businesses have a special ethical responsibility. The focus is on business policies, objectives, business ethics, internal business behavior and relationships, and the right and wrong behaviors encountered in business life. At the heart of business ethics are values such as responsibility, tolerance, honesty, keeping their promises, rights, justice, and response to injustices. The fact that such issues such as corruption, bribery, nepotism are frequently encountered in the business world has played an important role in the emergence of business ethics as a discipline and in increasing ethical sensitivity. In order to achieve high productivity in the business-employee relationship, it is very important to get rid of the negative features and to have the positive features on which business ethics is based. In order to survive in the competitive, complex, global order in today's business world, businesses and employees need to stick together and uphold the principles of business ethics. Spreading and ensuring compliance with all units of the business has become imperative for a business to be able to build trust in its employees, customers and society and to its reputation in the market.

Among the negative views of the employees in terms of organizational culture are the views of the employees, which are generally dissatisfied with the issue of wage policy, limited opportunities for promotion, and dissatisfaction with the regulation of social activities. Among the negative opinions of the employees in terms of business ethics, their managers should put themselves in their place while listening to others, to ensure that the primer participates in decisions about them, to act objectively in their evaluations, skeptical and ambivalent views on non-taking sides and openness to criticism. The relations between the employees' perceptions and attitudes about the situational and task organizational culture elements of the organization they work for and the principles of tolerance, honesty, responsibility, and justice, which are the principles of business ethics, and it is revealed that organizational culture is an important factor affecting business ethics are examined.

The most important element in the concept of business ethics is the discussion of the correct roles of organizations and employees in society in general (Sparks and Pan 2010). It is important that the associability of organizations for social admissibility and social approval meet the same denominator. Accordingly, organizations should try to provide a stable framework for what they look like by society. Awareness of business ethics affects customers' purchasing decisions and positively affects the success of the organization (Singhapakdi et al. 2001). This situation refers to organizational life. In other words, the concept of business ethics is effective in establishing a meaningful relationship between the approaches of organizations and their level of success. The quality of this relationship is also important to the concept of business ethics by organizations. It can be said that the concept of business ethics requires a long-term perspective and long-term investments in this direction. Moreover, the importance of management literature is increasing as it is the concept that attracts the most attention of researchers and gains more awareness of consumers and other interested people day by day. In the literature, the concepts of business ethics and business ethics are often used interchangeably and these use status concepts are reinforced by similar characteristics. Accordingly, concepts similar features in organizing relations in the business world, creating an atmosphere of trust, integrating, auditing, establishing and maintaining business peace, developing the concepts of honesty, respect and equality. Both concepts aim to ensure that working life is carried out in a certain order and that the parties concerned are protected from damage from their working relationships. According to professional ethics, the same professional employees, wherever they are in the world, must follow the rules of conduct established by professional ethics. For this purpose, today, those who do the same profession come together and form various social associations such as medical chambers and bar associations. The conformity of certain behavioral patterns in the relations with one another by those in the same profession also draws attention to the universal dimension of professional ethics (Bowie 2017). Thus, individuals engaged in the same professions should develop common behaviors regardless of the culture in which they live. It is observed that the concepts of professional ethics and business ethics differ from each other in this respect. While professional ethics point to universal elements independent of cultures for individuals doing the same profession, the possibility of a universal approach to business ethics is a controversial issue.

FUTURE RESEARCH DIRECTIONS

Individual ethics is about our individual moral duties and responsibilities, which we, as individuals, are obliged to fulfill, both to ourselves and to that outside of us. Our ideas and attitudes about ourselves create the concept of "self" or me. This is one of the most important points that makes man superior to other living organisms. Everyone has a sense of self, i.e. a belief about their personality. Ethical behavior is just images of the relationship between the self and the outside world. Thanks to the self-consciousness we have, we understand our own personality just like the personality of the person in front of us and see that we have some relationships with ourselves. Just as we need to be ethical with others, we feel the obligation to be ethical with ourselves. This feeling brings with it some ethical responsibilities to ourselves.

Organizational culture is critical to business ethics in terms of supporting behaviors in accordance with business ethics and guiding managers and employees to act in accordance with business ethics. Every organization has a culture. The impact of cultures on the behavior of managers and workers is too important to ignore. Businesses that want to develop a culture that cares about ethical values should try to satisfy a wider group of stakeholders in their strategic decisions and take care to maximize social returns as well as economic. The subjects in which the employees of the organization express a positive opinion in terms of organizational culture are generally concerned with the sense of integration within the organization, they are satisfied with the relations with their colleagues, they are integrated with the organizational objectives, they trust the management, they give importance to cooperation and solidarity, they adopt the norms and behaviors norms that should be followed in the organization. On the other hand, the subjects in which the employees of the organization give a positive opinion in terms of business ethics are generally that their managers avoid actions that will harm people, do not use illegal ways to realize their aims, do not accept exaggerated gifts, do not make corruption, do not discriminate their profession as a means of serving people. They aim to increase the image of the company in their activities and pay attention to working hours.

Ethical behavior and morality are primarily learned from the family. Then the behavior and values are learned again as they grow as part of the development of our system. Our values, and therefore our concepts of right and wrong, are highly influenced by the values of our family and our environment. A person who grows up in an agricultural society will have different values than the person who grew up in the urban environment. The values of a person whose father is a banker will differ from the values of a man who has a father who is a railway worker. These differences should be accepted without judgment. The behavior of the organization's founder and senior managers is a message, especially for other managers and employees. A high management that has been sensitive and exemplary in ethical behavior has taken the most critical step towards creating an ethical organizational culture. In other words, a senior management who can be a role model sends the message that ethical values and principles will be maintained. In this process, discourses, rituals, stories and symbols will reinforce ethical values and principles in communication within the enterprise. For example, jargons and emphasis on benevolence and honesty, rewards to ethical behavior, or emotional stories with some sacrifice motifs can be effective. An administration that has not tolerated unethical behavior and has elevated ethical behavior organizationally and systematically will create the ethical environment and culture of the organization.

CONCLUSION

The main issue of ethical debates is what makes human actions morally valuable or worthless. What makes action morally valuable is that it reveals good and creates good. Various ethical theories define the good in a different way. According to different ethical theories, it is good, pleasure, happiness, fulfillment of duty, truth, love. In General, at first glance, what is desired is good, and what is disgusted is bad. However, people do not agree on their wishes and often contradict each other. Here ethics focuses on the discussions about the process of determining the principles that will solve the conflicts of individuals and conflicts of individuals about the desired good (Saul 1981). Ethics does not produce morality, talks about morality (Morris 2004). The reason why ethics and morality are not identical is because of the fact that ethics is a moral philosophy, and morality is the subject of ethics. Ethical theories; In addition to the essence, origin and function of ethics in social life, it expresses opinions on the necessity of people to live together, norms and values of social life, relations between people and society, the purpose and meaning of individual life (De George 1986). In other words, ethics is the theory of right and wrong, and morality is its practice. It is more accurate to talk about moral principles rather than moral principles, not ethics, but moral behavior (Brinkmann 2001).

Morality determines the right, wrongs and how to behave in accordance with cultural values and ideals. The widely spread concept of morality contains unwritten standards of how to behave. Ethics, on the other hand, is based on more abstract concepts and tries to define what should be understood from these abstract concepts. The code of ethics is expected to include clear and written rules for a specific area. For example, although there are common principles in the fields of art ethics, political ethics, medical ethics, legal ethics, educational ethics, environmental ethics, bio ethics, media ethics, etc., these include more specific principles. These principles are developed based on universally accepted concepts, not according to the characteristics of the individuals who are expected to comply. Ethics also gives an important place to the concepts of justice, equality and compromise (Johnson 2018). There is a difference between morality and ethics in terms of breadth and narrowness, theory and practice. Morality, as a discipline, is a set of rules reflected in the practice of everyday life of ethics. Morality, in everyday life, while thinking about how individuals should live in fine details; ethics requires a more abstract and theoretical perspective (Morris 2004).

In general, many individuals prefer to be ethical. Individuals who participate in a strong organizational culture will easily internalize organizational content, which sends messages to be honest and respectful with their own personal beliefs. They will spontaneously turn to ethical behavior, because both the right tendency of the individual to act ethically and the organizational cultural messages they receive constantly provide this. Unfortunately, many individuals socialize in unethical organizational cultures. If it is normal to lie, an individual can behave the same way as long as you remain in that organization (Trevino and Nelson 2016). To be able to survive in a competitive environment, to carry out their activities without interruption, to prevent them from forming internal conflicts, to provide a safer work environment to their employees, to have a higher quality workforce, etc. to have an organizational culture; they should ensure that this culture is adopted by all employees, as well as informing employees of the principles of business ethics in writing, and whether or not they comply with these principles. In addition, organizing regular training seminars on business ethics and organizational culture in the business can be very effective for employees to better understand and protect the values of the organization.

REFERENCES

Allaire, Y., & Firsirotu, M. E. (1984). Theories of organizational culture. *Organization Studies*, *5*(3), 193–226. doi:10.1177/017084068400500301

Alvesson, M. (2012). *Understanding organizational culture.* Sage Publications Ltd., doi:10.4135/9781446280072

Anderson, E. (1995). *Value in ethics and economics*. Harvard University Press.

Appelbaum, S. H., Deguire, K. J., & Lay, M. (2005). The relationship of ethical climate to deviant workplace behaviour. *Corporate Governance: The International Journal of Business in Society, 5*(4), 43-55.

Ardichvili, A., Mitchell, J. A., & Jondle, D. (2009). Characteristics of ethical business cultures. *Journal of Business Ethics, 85*(4), 445–451. doi:10.100710551-008-9782-4

Arnold, D. G., & Bustos, K. (2005). Business, ethics, and global climate change. *Business & Professional Ethics Journal, 24*(1/2), 103–130. doi:10.5840/bpej2005241/26

Arslan, M. (2001). İş ve meslek ahlakı. Nobel yayın dağıtım. Istanbul, Turkey

Arslan, M., & Berkman, Ü. (2009). Dünyada ve Türkiye'de iş etiği ve etik yönetimi. *Türk Sanayici ve İş Adamları Derneği, Yayın No: TÜSİAD.*

Ashkanasy, N. M., Wilderom, C. P., & Peterson, M. F. (2000). *Handbook of organizational culture and climate*. Sage.

Audi, R. (2009). *Business ethics and ethical business.* Oxford University Press.

Audi, R. (2010). The place of ethical theory in business ethics. In The Oxford handbook of business ethics. OUP.

Baum, G. (2001). *Nationalism, religion, and ethics.* McGill-Queen's Press-MQUP.

Beauchamp, T. L., Bowie, N. E., & Arnold, D. G. (Eds.). (2004). Ethical theory and business. Academic Press.

Bennington, L. (2007). HR managers as ethics agents of the state. *Human resource management: Ethics and employment*, 137.

Bentham, J. (1989). *The Collected Works of Jeremy Bentham: Correspondence: Volume 9: January 1817 to June 1820.* Academic Press.

Blau, P. M., & Scott, W. R. (2003). *Formal organizations: A comparative approach*. Stanford University Press.

Boatright, J. R. (2000). Globalization and the Ethics of Business. *Business Ethics Quarterly*, *10*(1), 1–6. doi:10.2307/3857689

Booth, J. E., Heinz, R. S., & Howe, M. W. (2005). Ethical practice in a corporation: the Allina case. *The Ethics of Human Resources and Industrial Relations*, 229.

Bowie, N. E. (2017). *Business ethics: A Kantian perspective*. Cambridge University Press. doi:10.1017/9781316343210

Brabeck, M. (1984). Ethical characteristics of whistle blowers. *Journal of Research in Personality*, *18*(1), 41–53. doi:10.1016/0092-6566(84)90037-0

Bradburn, R. (2001). *Understanding business ethics*. Cengage Learning EMEA.

Braybrooke, D. (1983). *Ethics in the World of Business*. Rowman & Littlefield.

Brickley, J. A., Smith, C. W. Jr, & Zimmerman, J. L. (2002). Business ethics and organizational architecture. *Journal of Banking & Finance*, *26*(9), 1821–1835. doi:10.1016/S0378-4266(02)00193-0

Brinkmann, J. (2001). On business ethics and moralism. *Business Ethics (Oxford, England)*, *10*(4), 311–319. doi:10.1111/1467-8608.00247

Buchholz, R. A., & Rosenthal, S. B. (1998). *Business Ethics: The Pragmatic Path Beyond Principles to Process*. Prentice Hall.

Budd, J. W., & Scoville, J. G. (2005). Moral philosophy, business ethics, and the employment relationship. *The ethics of human resources and industrial relations*, 1-21.

Casida, J. J. (2008). Linking nursing unit's culture to organizational effectiveness: A measurement tool. *Nursing Economics*, *26*(2), 106. PMID:18524375

Chryssides, G. D., & Kaler, J. H. (1993). *An introduction to business ethics*. Cengage Learning EMEA.

Ciulla, J. B. (2011). Is business ethics getting better? A historical perspective. *Business Ethics Quarterly*, *21*(2), 335–343. doi:10.5840/beq201121219

Clegg, S., Kornberger, M., & Rhodes, C. (2007). Business ethics as practice. *British Journal of Management*, *18*(2), 107–122. doi:10.1111/j.1467-8551.2006.00493.x

Conroy, S. J., & Emerson, T. L. (2004). Business ethics and religion: Religiosity as a predictor of ethical awareness among students. *Journal of Business Ethics*, *50*(4), 383–396. doi:10.1023/B:BUSI.0000025040.41263.09

Cragg, W. (2000). Human rights and business ethics: Fashioning a new social contract. In *Business Challenging Business Ethics: New Instruments for Coping with Diversity in International Business* (pp. 205–214). Springer. doi:10.1007/978-94-011-4311-0_20

De George, R. T. (1986). Theological ethics and business ethics. *Journal of Business Ethics*, *5*(6), 421–432. doi:10.1007/BF00380748

De George, R. T. (1987). The status of business ethics: Past and future. *Journal of Business Ethics*, *6*(3), 201–211. doi:10.1007/BF00382865

De George, R. T. (1994). International business ethics. *Business Ethics Quarterly*, *4*(1), 1–9. doi:10.2307/3857554

De George, R. T. (2005). *A history of business ethics*. Paper delivered at the third biennial global business ethics conference. Santa Clara, CA.

De George, R. T. (2011). *Business ethics*. Pearson Education India.

Denison, D. R., & Neale, W. (1996). *Denison organizational culture survey*. Aviat.

DesJardins, J. R., & DesJardins, J. R. (2009). *An introduction to business ethics*. McGraw-Hill Higher Education.

Donaldson, T., & Dunfee, T. W. (1994). Toward a unified conception of business ethics: Integrative social contracts theory. *Academy of Management Review*, *19*(2), 252–284. doi:10.5465/amr.1994.9410210749

Drover, W., Franczak, J., & Beltramini, R. F. (2012). A 30-year historical examination of ethical concerns regarding business ethics: Who's concerned? *Journal of Business Ethics*, *111*(4), 431–438. doi:10.100710551-012-1214-9

Dunfee, T. W. (1991). Business ethics and extant social contracts. *Business Ethics Quarterly*, *1*(1), 23–51. doi:10.2307/3857591

Eisenberg, E. M., & Riley, P. (2001). Organizational culture. The new handbook of organizational communication: Advances in theory, research, and methods, 291-322.

Enderle, G. (1997). A worldwide survey of business ethics in the 1990s. *Journal of Business Ethics*, *16*(14), 1475–1483. doi:10.1023/A:1005866527497

Enderle, G. (Ed.). (1999). *International business ethics: Challenges and approaches*. University of Notre Dame Press.

Etzioni, A. (1964). Modern organizations. Academic Press.

Evans, R. (1991). Business ethics and changes in society. *Journal of Business Ethics*, *10*(11), 871–876. doi:10.1007/BF00383705

Ferrell, O. C., & Fraedrich, J. (2015). *Business ethics: Ethical decision making & cases*. Nelson Education.

Fisher, C., & Lovell, A. (2009). *Business ethics and values: Individual, corporate and international perspectives*. Pearson education.

Freeman, R. E. (2000). Business ethics at the millennium. *Business Ethics Quarterly*, *10*(1), 169–180. doi:10.2307/3857703

Frost, P. J., Moore, L. F., Louis, M. R. E., Lundberg, C. C., & Martin, J. E. (1985). *Organizational culture*. Sage Publications, Inc.

Gandz, J., & Hayes, N. (1988). Teaching business ethics. *Journal of Business Ethics*, *7*(9), 657–669. doi:10.1007/BF00382975

Garrett, T. M. (2004). Nirali Prakashan. *Business Ethics (Oxford, England)*, *24*.

Ghillyer, A. (2008). *Business ethics: A real world approach*. Academic Press.

Gini, A., & Marcoux, A. M. (Eds.). (2005). *Case studies in business ethics*. Pearson Prentice Hall.

González, A. M. (2003). Ethics in global business and in a plural society. *Journal of Business Ethics*, *44*(1), 23–36. doi:10.1023/A:1023230222707

Grace, D. (2006). For business ethics. *Australian Journal of Management*, *31*(2), 371–380. doi:10.1177/031289620603100211

Gupta, A. D. (Ed.). (2010). *Ethics, business and society: Managing responsibly*. SAGE Publications India.

Haman, Ö. (2016). *Örgüt Yapısı Ve Örgütsel İletişim Arasındaki İlişkinin Belirlenmesi Üzerine Bir Araştırma*. Yüksek Lisans Tezi, Nişantaşı Üniversitesi, Sosyal Bilimler Enstitüsü, İşletme Anabilim Dalı, İşletme Yönetimi Bilim Dalı.

Hartman, E. (1996). *Organizational ethics and the good life*. Oxford University Press on Demand.

Hatch, M. J. (1993). The dynamics of organizational culture. *Academy of Management Review*, *18*(4), 657–693. doi:10.5465/amr.1993.9402210154

Holme, C. (2008). Business ethics–part one: Does it matter? *Industrial and Commercial Training*, *40*(5), 248–252. doi:10.1108/00197850810886487

Hugman, R. (2012). *Culture, values and ethics in social work: Embracing diversity*. Routledge. doi:10.4324/9780203094907

Hurn, B. J. (2008). Ethics in international business. *Industrial and Commercial Training*, *40*(7), 347–354. doi:10.1108/00197850810912216

Johnson, C. E. (2018). *Organizational ethics: A practical approach*. Sage Publications.

Jones, C., Parker, M., & Ten Bos, R. (2005). *For business ethics*. Routledge. doi:10.4324/9780203458457

Kaler, J., & Chryssides, G. (1996). *Essentials of business ethics*. McGraw-Hill.

Kant, I. (2001). *Lectures on ethics* (Vol. 2). Cambridge University Press.

Kayacan, M. (2005). *Anonim şirketlerin sosyal sorumlulukları ve etik değerler. In İSMMMO I. Uluslararası Muhasebe Denetimi Sempozyumu*. Nisan.

Keller, A. C., Smith, K. T., & Smith, L. M. (2007). Do gender, educational level, religiosity, and work experience affect the ethical decision-making of US accountants? *Critical Perspectives on Accounting*, *18*(3), 299–314. doi:10.1016/j.cpa.2006.01.006

Khushf, G. (1998). The scope of organizational ethics. *HEC Forum*, *10*(2), 127–135. doi:10.1023/A:1008890902591 PMID:11645346

Kingsley Trezise, E. (1996). An introduction to business ethics for human resource management teaching and research. *Personnel Review*, *25*(6), 85–89. doi:10.1108/00483489610148554

Kirel, Ç. (2000). *Örgütlerde Etik Davranışlar*. Yönetimi ve Bir Uygulama Çalışması, Anadolu Üniversitesi Yayınları.

Kline, W. (2009). Business as an ethical standard. *Journal of Private Enterprise*, *24*(2), 35.

Koh, H. C., & El'fred, H. Y. (2004). Organisational ethics and employee satisfaction and commitment. *Management Decision*.

Kopperi, M. (1999). Business ethics in global economy. *EJBO. Electronic Journal of Business Ethics and Organization Studies*.

Lozano, J. (2007). Organizational ethics. In Ethics in the economy (Vol. 165, No. 186, pp. 165-186). Peter Lang Publishing Group in association with GSE Research.

March, J. G., & Simon, H. A. (1958). *Organizations*. Wiley.

Marcoux, A. M. (2006). The concept of business in business ethics. *Journal of Private Enterprise*, *21*(2), 50-67.

Mayer, D. (2001). Community, Business Ethics, and Global Capitalism. *American Business Law Journal*, *38*(2), 215–260. doi:10.1111/j.1744-1714.2001.tb00289.x

McNamara, C. (1999). *Complete Guide to Ethics Management: A Ethics Toolkit for Managers*. Academic Press.

Megone, C., & Robinson, S. J. (Eds.). (2002). *Case histories in business ethics*. Routledge. doi:10.4324/9780203361207

Morris, D. (2004). Defining a moral problem in business ethics. *Journal of Business Ethics*, *49*(4), 347–357. doi:10.1023/B:BUSI.0000020866.05705.d2

Nakano, C. (2007). The significance and limitations of corporate governance from the perspective of business ethics: Towards the creation of an ethical organizational culture. *Asian Business & Management*, *6*(2), 163–178. doi:10.1057/palgrave.abm.9200216

Ott, J. S. (1989). *The organizational culture perspective*. Dorsey Press.

Ouchi, W. G., & Wilkins, A. L. (1985). Organizational culture. *Annual Review of Sociology*, *11*(1), 457–483. doi:10.1146/annurev.so.11.080185.002325

Özgener, Ş. (2004). *İş Ahlakının Temelleri*. Yönetsel Bir Yaklaşım, Nobel Yayın Dağıtım.

Paliwal, M. (2006). *Business ethics*. New Age International.

Parker, M. (Ed.). (1998). *Ethics & organizations*. Sage. doi:10.4135/9781446280171

Parker, M. (2003). Business, ethics and business ethics: Critical theory and negative dialectics. *Studying management critically*, 197-219.

Payne, D., Raiborn, C., & Askvik, J. (1997). A global code of business ethics. *Journal of Business Ethics, 16*(16), 1727–1735. doi:10.1023/A:1005877726316

Peters, T. J., & Waterman, R. R. H. (1982). *Search of Excellence: Lessons from America's Best Run Companies* (Vol. 1). Harper & Row.

Petty, G. C., & Hill, R. B. (2005). Work ethic characteristics: Perceived work ethics of supervisors and workers. *Journal of Industrial Teacher Education, 42*(2), 5–20.

Porter, G. (2005). A "career" work ethic versus just a job. *Journal of European Industrial Training, 29*(4), 336–352. doi:10.1108/03090590510597160

Pratley, P. (1995). *The essence of business ethics* (Vol. 111). Prentice Hall.

Randall, D. M., & Gibson, A. M. (1990). Methodology in business ethics research: A review and critical assessment. *Journal of Business Ethics, 9*(6), 457–471. doi:10.1007/BF00382838

Ricoeur, P. (1973). Ethics and culture. *Philosophy Today, 17*(2), 153–165. doi:10.5840/philtoday197317235

Runes, D. D. (1964). Dictionary of Philosophy. Littlefields, Adams and Co.

Saul, G. K. (1981). Business ethics: Where are we going? *Academy of Management Review, 6*(2), 269–276. doi:10.5465/amr.1981.4287830

Schein, E. H. (1985). Defining organizational culture. *Classics of Organization Theory, 3*(1), 490-502.

Schein, E. H. (1990). Organizational culture. American Psychological Association.

Schwartz, M. (2007). The "business ethics" of management theory. *Journal of Management History, 13*(1), 43–54. doi:10.1108/17511340710715160

Shaw, W. H., & Barry, V. (2015). *Moral issues in business*. Cengage Learning.

Sheridan, J. E. (1992). Organizational culture and employee retention. *Academy of Management Journal, 35*(5), 1036–1056.

Sidgwick, H. (1877). Bentham and Benthamism in politics and ethics. *Fortnightly, 21*(125), 627–652.

Singhapakdi, A., Karande, K., Rao, C. P., & Vitell, S. J. (2001). How important are ethics and social responsibility?-A multinational study of marketing professionals. *European Journal of Marketing, 35*(1/2), 133–153. doi:10.1108/03090560110363382

Sintonen, T. M., & Takala, T. (2002). Racism and ethics in the globalized business world. *International Journal of Social Economics, 29*(11), 849–860. doi:10.1108/03068290210446267

Small, M. W. (1993). Ethics in business and administration: An international and historical perspective. *Journal of Business Ethics, 12*(4), 293–300. doi:10.1007/BF01666533

Smith, A. (1987). *The Essential Adam Smith*. WW Norton & Company.

Snoeyenbos, M., & Hummel, J. (1979). Business ethics. *Teaching Philosophy, 3*(2), 238–243. doi:10.5840/teachphil19793244

Sparks, J. R., & Pan, Y. (2010). Ethical judgments in business ethics research: Definition, and research agenda. *Journal of Business Ethics*, *91*(3), 405–418. doi:10.100710551-009-0092-2

Spash, C. L. (2002). *Greenhouse economics: Value and ethics* (Vol. 1). Taylor & Francis. doi:10.4324/9780203209059

Spence, L. (2000). What ethics in the employment interview. *Ethical issues in contemporary human resource management*, 43-58.

Stark, A. (1993). What's the matter with business ethics? *Harvard Business Review*, *71*(3), 38–40. PMID:10126154

Suhonen, R., Stolt, M., Virtanen, H., & Leino-Kilpi, H. (2011). Organizational ethics: A literature review. *Nursing Ethics*, *18*(3), 285–303. doi:10.1177/0969733011401123 PMID:21558106

Svensson, G., & Wood, G. (2003). The dynamics of business ethics: A function of time and culture–cases and models. *Management Decision*, *41*(4), 350–361. doi:10.1108/00251740310468195

Thiroux, J. P., & Krasemann, K. W. (1980). *Ethics: Theory and practice*. Glencoe Publishing Company.

Trevino, L. K., & Nelson, K. A. (2016). *Managing business ethics: Straight talk about how to do it right*. John Wiley & Sons.

Trevino, L. K., & Weaver, G. R. (1994). Business ETHICS/BUSINESS ethics: One field or two? *Business Ethics Quarterly*, *4*(2), 113–128. doi:10.2307/3857484

Türkkan, L. (2015). *Örgüt Kültürünün Örgütsel Bağlılığa Etkileri*. Yüksek Lisans Tezi, İstanbul Üniversitesi, Sosyal Bilimler Enstitüsü, Hastane ve Sağlık Kuruluşlarında Yönetim Bilim Dalı.

Ural, T. (2003). İşletme ve Pazarlama Etiği. Baskı, Ankara: Detay Yayıncılık.

Valentine, S., Godkin, L., Page, K., & Rittenburg, T. (2009). Gender and ethics: Ethical judgments, ethical intentions, and altruism among healthcare professionals. *Gender in Management*, *24*(2), 112–130. doi:10.1108/17542410910938808

Vallance, E., & Elizabeth, V. (1995). *Business ethics at work*. Cambridge University Press. doi:10.1017/CBO9781139166461

Velasquez, M. G., & Velazquez, M. (2002). *Business ethics: Concepts and cases* (Vol. 111). Prentice Hall.

Verstraeten, J. (1998). From business ethics to the vocation of business leaders to humanize the world of business. *Business Ethics (Oxford, England)*, *7*(2), 111–124. doi:10.1111/1467-8608.00096

Warren, R., & Tweedale, G. (2002). Business ethics and business history: Neglected dimensions in management education. *British Journal of Management*, *13*(3), 209–219. doi:10.1111/1467-8551.00238

Weiss, J. (2008). *Business ethics: A stakeholder and issues management approach*. Cengage Learning.

Weiss, J. W. (1994). Business ethics: A managerial, stakeholder approach (No. s 287). Belmont, CA: Wadsworth Publishing Company.

Wellman, C. (1975). *Morals and ethics*. Scott, Foresman.

Wilson, V. J., McCormack, B. G., & Ives, G. (2005). Understanding the workplace culture of a special care nursery. *Journal of Advanced Nursing*, *50*(1), 27–38. doi:10.1111/j.1365-2648.2004.03346.x PMID:15788063

Yamaguchi, S. (2004). Nursing culture of an operating theater in Italy. *Nursing & Health Sciences*, *6*(4), 261–269. doi:10.1111/j.1442-2018.2004.00199.x PMID:15507046

Yeter, D. (2009). İş Etiği Kavramı Ve İşletmeler Açısından Önemi. *İller Bankası Genel Md. UİB AB Ve Dış İl. MD. paribus. tr. googlepages. com/y_demir2. doc. Son Erişim Tarihi*, *2*, 2010.

ADDITIONAL READING

Baron, D. P., & Baron, D. P. (2003). *Business and its environment*. Prentice Hall.

Bird, F. B., & Bird, F. B. (1996). *The muted conscience: Moral silence and the practice of ethics in business*. Greenwood Publishing Group.

Gibson, K. (2007). *Ethics and business: An introduction*. Cambridge University Press. doi:10.1017/CBO9780511806148

Griseri, P., & Seppala, N. (2010). *Business ethics and corporate social responsibility*. Cengage Learning.

Hamington, M., & Sander-Staudt, M. (Eds.). (2011). *Applying care ethics to business* (Vol. 34). Springer Science & Business Media. doi:10.1007/978-90-481-9307-3_13

Hartman, L. P., DesJardins, J. R., & MacDonald, C. (2008). *Business ethics: Decision making for personal integrity and social responsibility*. McGraw-Hill.

Homann, K. (2007). *Globalisation and business ethics*. Ashgate Publishing, Ltd.

KEY TERMS AND DEFINITIONS

Business Ethics: It is expressed as being fair to people and acting in a way that creates good feelings about the business.

Ethics: A system of principles, values and standards that describes and defines how individuals behave correctly.

Moralisch: Moralisch are judgments (right or wrong, good or bad) about the actions directed according to the human attitudes and attitudes governed by beliefs and designs.

Morality: It is called a group of beliefs and designs that dominate the conscious life of an individual, a people, a social class and an age.

Organization: To achieve the common goals of the communities; The division of labor is defined as the coordination of activities carried out within the framework of responsibilities and enforcement power.

Pragmatism: It is called the current which unilaterally evaluates the truth and reality only with the results of the actions and only looks at them in terms of their benefit.

Religion: Religion is the order that unites and provides beliefs in the form of rules, institutions, customs and symbols.

Chapter 15
Meeting Customer Expectations in Islamic Tourism:
Effects of Islamic Business Ethics

Elif Baykal
Istanbul Medipol University, Turkey

ABSTRACT

Changing consumer behaviour and increasing consumer demands in many contemporary markets have affected the Muslim world as well. The ever-increasing Muslim population has become more affluent and more demanding with the rise of modernization, especially after the 1980s. Contemporary Muslims' increased demands for better recreation and vocation facilities necessitated the birth of a new sector is referred to as Islamic tourism in this chapter. Increase in the market penetration of Islamic tourism has brought about new anxieties regarding ethical issues in this newly emerging industry. In this chapter, Islamic tourism has been considered as an alternative tourism approach. It creates a niche market for Muslim tourist with religious sensitivities and Islamic tourism and ethics in this unique sector has been handled in detail.

INTRODUCTION

Tourism has a magical potential for generating revenue owing to the fact that it is a quick-growing sector and has broad economic, social, cultural and environmental consequences (Giaoutzi and Peter Nijkamp 2006). Worldwide, we can talk about an ever increasing geographic and psychological spread and diversification of touristic activities (Donyadide 2010). More and more people benefit from a widely diversified set of tourism products. Moreover, it has become a highly competitive industry making pressure on tourism professionals to differentiate their products in order to reach higher levels of profitability and increase their revenue (Donyadide 2010). That is why many countries display an inclination towards encouraging this unique industry wherein entrepreneurs can earn maximum output with the least possible R&D and technological investments.

DOI: 10.4018/978-1-7998-4117-3.ch015

Nowadays, continuous innovation and the ability to improve existing products, services and business processes have become very important to catch up with the latest developments (Baykal, 2019b). In this context, innovation should be conceived as the competency of modern organizations in developing novel ideas and products, with the aim of gaining new and advantageous positions in the market or strengthening their existing positions in their current markets. In an organizational sense, innovation can be regarded as a useful and challenging production of organizations with an innovation that has never been done before (Baykal 2019). In this point, halal tourism or in other words Islamic tourism can be considered as a new product of Muslim market including myriad tangible products and services with increasing number of tourists. In fact, global halal tourism market can be viewed as one of the most growing industries (Albayrak and Özkul 2014). It is a tourism type that is in alignment with Islamic principles (Tekin 2014). In halal tourism a holistic approach is embraced encompassing halal food, halal recreation, and even halal finance.

Actually, tourism encompasses all kinds of visits including travelling leisure, business and other purposes. Similarly, Islamic tourism can be explained as all kind of activities of Muslims that can be observed while travelling and staying in places apart from their usual environment that are mostly originating from or shaped by Islamic motivations and sensitivities. Before the development of Islamic tourism, many sensitive Muslims had difficulty in meeting their holiday needs and had to give up their holiday demands. However, with the development of Islamic tourism, many people with a high level of Islamic sensitivity began to take a comfortable holiday, but the emergence of this form of tourism has brought with it many ethical issues that need to be addressed.

BACKGROUND

Islamic Tourism

The increase in the welfare levels of Islamic countries and educational levels, and communication alternatives along with globalization increased the worldly expectations of Muslims. Among all these factors, actually, the most important factor affecting Muslims' changing demands is the increase in the income level of Muslims (Zulkharnain and Jamal 2012). Similarly, ascending religious sensitivity and the tendency to lead a more contemporary version of Islam paved the way for the demand for halal products and halal- concept hospitality. Halal-concept hospitality is another example of customer-oriented marketing in the tourism sector (Pamukçu and Armacı 2016). Since Islam is the second most believed religion in the world after Christianity (23% of the world's population) It is unthinkable that tourism enterprises ignore this market segment in their marketing activities. For this reason, most businesses design their existing businesses to serve such customers at the same time (Pamukçu and Armacı 2016). The enterprises that can serve in the concept of Islamic tourism are those which can be operated on the basis of Islamic principles from the above mentioned enterprises in terms of product-service content and presentation concepts. In other words, all businesses whose products, services and presentation concepts are in compliance with Islamic principles can be addressed to the tourist market (Pamukçu and Armacı 2016). Their target market are those people who wants to entertain and relax without committing any kind of sin that is prohibited in Islamic religion.

In Islamic tradition halal is a unique term that not only limited with the food but also affects many other aspects of Muslims' daily life (Alserhan 2010). The term halal refers to all kind of acts and objects that are permitted, with no restriction. In other words, it denotes 'permissible' according to Sharia law. In order to understand the term Islamic tourism, we should first clarify the meaning of the term 'Islamic' which denotes directly to the all things related to Islamic faith and its doctrines. However, the use of 'Islamic' and 'Halal' as if they have the same meaning cannot be conceived as true. In Arabic countries the term Halal connotates is anything that is permitted by the Shariah. It explains "lawful" or "permitted (Battour, Ismail and Battour 2010). And all Islamic deeds, laws and prohibitions stems from either from Islamic holy book Quran or the Sunnah of the Prophet which means his deeds and recommendations that directs Muslims life and manners of behavior. In fact, we can consider an activity as Islamic if it is viewed plausible by God and deserves rewarding. Anyway, in many Islamic countries, the ethical and moral alignment of Islam to social life is possible through implementation of Shariah law (Zakai 2008). Such legal frameworks also cover permitted and prohibited forms of tourism activities (Henderson 2009) thus creates a plausible framework for halal tourism. In marketing literature, the term Halal is used as an adjective for describing the level of compliance with Islamic laws and regulations. Thus, using the term 'Halal' rather than the term 'Islamic' in tourism industry is more meaningful. According to Zulküfli et al. (2011) halal businesses effects various sectors including; food, cosmetics, textile, finance, and even tourism. In Islamic perspective, the term "halal" explains all that is commanded in the teachings of Islamic religion and the basis for the behavior of all kinds of activities of Muslim individuals (Jaelani 2017). In this respect, Halal tourism is a tourism approach serving a new holiday style that is customized in alignment with the needs and demands of Muslim travelers."

Although the tourism industry today includes travels for a wide variety of purposes, recreational tourism specifically marine tourism during holidays is still the most common at conceptual and practical levels. The same inclination is valid in Islamic tourism. Actually, hospitality establishments are one of the most important factors that affect the demand of marine tourism in a positive or negative way. Factors such as facility features, room, location, capacity, activities and activities, personnel and service quality of accommodation establishments are some of the factors determining the expectations, perceptions and experiences of tourists. In addition to these features for accommodation in accordance with Islamic understanding, there are also distinguishing elements that indicate whether the tourist is an Islamic property. In Islamic tourism, the hotels and holiday villages should be designed in a way that aims to conform to Islamic understanding and should serve accommodation facilities that allow those who are sensitive to this holiday will be considered as "Islamic Hotel" (Doğan 2011).

One of the most important difference of halal tourism is the fact that this unique tourism style offers activities totally eliminating haram activities such as gambling, drinking and all other kinds of activities that are religiously forbidden. Moreover, in halal tourism we can come across visits to mosques and other Islamic related places, provision of halal meals and giving importance to time and place for routine daily (Albayrak and Özkul 2014). As Zulkifli claims, halal tourism is a niche market including halal hotels, halal transport halal restaurants, halal tours and halal finance institutions (Zulkifli et al. 2011). Halal tourism should be understood as any product or action that is allowed to use or engage in tourism industry, according to Islamic teachings. Hence, in order to be successful Halal tourism industry should be guided by Islamic teachings and principles (Boediman 2017). And more importantly, tourism industrialists should understand Muslim travel behavior and travel intentions in order to satisfy needs of Muslim tourists (Battour 2018). In other words, success in Islamic tourism lies in understanding halal perspective of Muslim consumers. Actually, nowadays halal tourism has become more and more popular all around the world owing to increasing Muslim population and higher levels of Islamic awareness. There are 57 countries around the world with a significant majority of their population or all Muslims.

The population of these countries are constantly increasing and their economic statuses are strengthening. Supporting this view, futurist Ian Yeoman, tourism management professor at the university, claims that Islam is the fastest-growing religion in the world, and perhaps by 2040, the Muslim population of the world would outstrip the Christian population (Tekin 2014). This highly crowded population will come about with its unique properties and requirements including recreation and tourism.

In fact, halal tourism, in other words Islamic tourism, should be considered as a kind of business model innovation. Actually, innovation is about developing novel ideas and products, creating niche markets or strengthening their existing positions. In an organizational sense, innovation can be regarded as a useful and challenging production of organizations with an innovation that has never been done before (Baykal 2019). Before the innovation of halal tourism, many Muslims were experiencing difficulties in finding proper alternatives for their holidays considering their Islamic sensitivities. In order to be able to comply with Islam, they either had to give up some enthusiasm for the holiday or had to restrict their holiday needs so that they would not sin.

There are various approaches regarding Islamic tourism, that is to say, halal tourism in Muslim countries. In fact, in some Muslim countries that have rather more conservative regimes are often inclined to pay less little attention to tourism but more moderate ones such as Malaysia, Turkey, and Morocco tend to give more attention to tourism. On the one hand, in theocratic regimes the tourism industry is substantially impacted quiet restricted (Seyfi and Hall 2019). In this paper, we will focus on Turkish version of halal tourism as a framework for understanding Islamic tourism and ethics relationship. Since halal tourism is noteworthy in Turkey with its noticeable successes especially in the last years. The increasing numbers of sincere Muslim population that has halal sensitivity and high economic levels of them affected tourism demands in Turkey like halal-restaurants and gender-separated pools and beaches (Tanrısever, Batman and Pamukçu 2016).

Etiics and Tourism

Ethics is defined as a branch of philosophy that deals with moral behavior in human beings. Morality in this description explains whether something is right or wrong (Abuznaid 2009). Ethics can be conceived as a set of rules and practices explaining right, good and authentic conduct among members of a society or profession (Fennell 2010). Today, although the word, ethics is widely used and has widespread application, it is observed that its meaning and purpose cannot be understood completely. One of the simplest definitions of ethics is the definition that explains it as a philosophy examining the values underlying human and individual relationships in the moral sense of right-wrong, good-bad (Özkalp and Kırel, 2011, p. 504). Ethics is the philosophy of morality in the shortest term. It refers to thinking, questioning, discussing, evaluating, grounding and raising awareness on good or bad value judgments in the field of morality.

On the one hand, changing working habits, fierce competition, greater demands for 'corporate citizenship', and demands of stakeholders for higher levels of openness created a pressure on companies regarding openness, accuracy and transparency (Baykal 2019b). Actually, business ethics or work ethics has become quite popular in recent years as a result of failures and ethical problems of companies like Enron and WorldCom (Rokhman 2010). As it is well known by many theorists and practitioner, after 2000s ethical issues has become more noteworthy owing to disappointing evens experienced in many corporate organizations. That is why, nowadays, lots of corporations have preferred to establish codes of ethics which members are expected to adhere to and contain the rules governing the conduct of members of a specific profession (Fleckenstein and Huebsch 1999).

The increase in the complexity and lack of trust in business life and in this environment, the necessity of organizations to have moral responsibilities, revealed the importance of ethical values and ethical behaviors of employees. Nowadays, products have become so technologically oriented and have geographically and psychologically wider effects. Thus, they have the capacity to effect wider markets simultaneously. That is why, the size of contemporary organizations and their worldwide effects can result in ethical failures which may have worldwide consequences. These considerable effects make business ethics significant. In fact, business ethics, based on integrity and honesty covers general moral rules also that are applied to business life, to avoid unfair competition, customer independence, social responsibility, public relations, including many different issues of operative moral philosophy.

Nowadays, most clients are also looking for increased quality and experience in their holiday. That is why, in this competitive environment tourism industrialist should struggle to differentiate their products and services considering market demands i.e. based on ethical issues. In this respect, a new issue, "ethics in tourism" has become a heated debate which claims that all parts involved in tourism industry to follow its codes and help the sustainable development of this industry (Donyadide 2010).Actually, recent developments in tourism literature showed that tourism can be viewed by modernist and post-modernist perspectives thanks to the depth of sociology (Cohen and Cohen 2019). This new approach is an important constituent of modern social life (Franklin & Crang 2001). Social realities, ethnic, cultural and religious sensitivities and requirements of communities has created the need for a more elaborate approach in analyzing tourism and ethical issues in tourism. In fact, the reason for the late recognition of the importance and relevance of tourism ethics stems from the Western attitude towards tourism, that prevailed previously, perceiving it as an unimportant and unserious managerial and social issue (Gibson 2008, p. 407). But, recent researches showed that even frivolous touristic activities can have important ethical consequences (Cohen and Cohen 2019). In fact, tourism should be understood as an activity that serves mutual understanding and respect between individuals and societies, emphasizing the importance of respecting the common ethical values of humanity and respecting religious and cultural differences (Akan 2007).

According to Reynolds (2000), the most common and most important ethical problems in the tourism sector are problems such as mobbing, physical and sexual harassment, conflicts of interest, injustice in salary and rewarding, bribery, and revealing trade secrets to the outside. Similarly, Coughlan (2001) also claims that unfortunately the tourism industry is highly open to unethical situations such as theft, misleading advertising, double booking in accounting processes, damage to the environment etc. owing to the structural characteristics of the tourism industry. Hence this unique industry necessitates some general and valid rules that are accepted by a wide range of stakeholders. Thus, Universal Principles of Tourism Ethics are accepted as a valid framework for a more ethical tourism industry. As a fundamental frame of reference for ethical tourism, GCET, Global Code of Ethics for Tourism, is a comprehensive rule set which is aimed to direct key stakeholders in tourism development. These principles have been adopted since 1999 by the General Assembly of the World Tourism Organization. Even though it is not legally binding for member countries, *code of ethics* creates the necessary baseline for voluntary implementation of ethical values in tourism industry (Sharma 2017). This widely accepted set of rules suggests that tourism is something universal and it should be both universal and ethical in order to be a sustainable industry. According to the seventh article it, tourism should be considered as a right for all individuals of the world. This is a fundamental right to rest and entertain, and all people must have the chance to recognize and benefit from the world tourism resources (Akan 2007).

The Code's 10 items are about different aspects tourism industry including economic, social and environmental aspects. For example, article 1 suggests that tourism should contribute to insight and respect among societies. In other words, tourism should be considered as a way for harmonizing communities. Anyway, this industry's nature is suitable for positioning its elements as a tool that unites people. Similarly, article 2 suggests that tourism should be considered as a vehicle for individual and collective fulfillment. It is one of the most plausible medium for entertainment and can create alternative ways for a more enjoyable life. Moreover, in the Article 3, it is suggested that tourism should be conceived as a factor of sustainable development for countries. If countries can utilize their touristic opportunities it will be a source for competitive advantage and it is a good way to close the current account deficit. Many countries enjoy the ease of solving their financial problems through utilizing their touristic facilities. It is a convenient and high value -added source of income. Furthermore, article 4 claims that tourism should be viewed as a user of the cultural heritage of communities and should contribute to the enhancement of this heritage. In fact, success and sustainability in tourism are closely related with high levels of awareness and appreciation of this heritage. According to code of ethics, appreciation of local affluences gives way to better utilization of tourism opportunities and is useful for citizens of the host country which can also be accepted as an ethical citizenship behavior. That is why, as it is mentioned in article 5, tourism is a beneficial activity for host countries and its citizens contributing to higher levels of welfare. In parallel with these items article 6 talks about the necessity of satisfying all kinds of stakeholders and it emphasizes the obligations of stakeholders in tourism development (Sharma 2017). Article 9 is also about ethical issues in tourism industry. It is related to the rights of tourism industry employees and investors. It is about occupational safety, employee rights, the need for training and development, and the transfer of expertise and experience. Moreover, according to this principle the significance of fair and accurate sharing of tourism profits among investors and stakeholders in the host countries is important for the development and sustainability of tourism (Akan 2007). According to this article rights of employees and employers in the tourism industry are both important and should be handled delicately. For a more ethical and fair working environment both employers and employees should feel their rights and benefits are looked after with good faith. The problems experienced in the institutionalization of investments in the tourism sector can create negative situations in which the employees in this industry are victimized, that is why, it is especially important to respect the rights of employees. Ensuring the legal assurance of the employees, adopting fair wage policies and reducing the benefits to the satisfactory points are among the sine qua non for a more ethical tourism industry (Sharma 2017).

Moreover, Donyadide (2010) claims that for an ethical tourism industry, stakeholders of tourism industry should be tolerant and respectful for different religions and worldviews and they should be respectful towards different culture. Similarly, Tourism activities should be conducted in harmony with lifestyles and customs of the host regions (Donyadide 2010). That is to say, tourism industrialists should be aware of the local culture and traditions and should avoid offending people regarding these issues. On the other hand, tourism industry has an obligation to ensure objective and honest information about destination, travel conditions, price and quality of their services (Donyadide 2010). Giving misleading information to customers, showing different than the quality of service, hiding some information in order to mislead customers are fundamental factors that negatively affect ethical behavior in the tourism industry. That is why the press, in all kinds of communication channels tourism companies ought to issue trustworthy and balanced information about their facilities (Donyadide 2010). On the one hand, it is expected from tourism professionals to contribute to the cultural and spiritual fulfillment of their clients and promote an atmosphere for them that is proper for practicing their practice their religions (Donyadide 2010). Both preserving the local culture and ensuring a plausible environment for the practice of guests' cultural and religious traditions are important phenomenon contributing to an ethical environment in tourism industry.

Islamic Ethics in Business

Quran, Islamic Holy Book, explains mankind as the vicegerents of Allah on Earth. In Islam the concept of Khilafa, that is to say, viceregency is the reason for human existence, giving way to ethical commitment. It is a kind of worship for Muslims (Ismail and Blaim 2012). Mankind is a kind of representative of Allah and should behave accordingly. Ethical behavior is a reflection and antecedent of this duty. In Islam, ethics are not bounded with business practices and relationships. According to Islam the business decision makers have a free choice in their daily conducts and religious norms and regulations ensures a framework for the most suitable ways of conduct. In Islam, ethics considers two important issues; first one's relationships with Allah, second one's relationships with other people. In this approach, a Muslim individual should have an ethical relationship with others by getting along well with others (Abuznaid 2009). According to Islamic teachings Muslims are responsible for all their activities and will be judged in the afterlife and their commitment to Islamic rules and obligations (Ismail and Blaim 2012). Shari'a creates the necessary rules and regulations for the Islamic way of life" (Dusuki 2008) that can be regarded as the rules that enhance business ethics in Muslims business practices. Supporting this view, Al-Ali and Al-Owaihan (2008) claimed that Islamic work ethics is an inclination that contributes to the involvement, engagement and participation of believers at work. As ethics in general, Islamic ethics explains what individuals should do or should not do in their daily transactions and activities.

In Islamic ethics, individuals should take into consideration some important topics. First of all, increasing efficiency and effectiveness in all kinds of processes is vital. Organizations should avoid lavishness and unwieldiness in order to attain their goals without incurring unnecessary costs and obtaining maximum benefit. On the other hand, having a holistic approach in all kind of processes and creating integrity are also important issues in Islamic business ethics. According to Islamic business ethics, in organizations interrelated subsystems should act with a harmony and all processes should be in alignment with Islamic religious norms and regulations. Furthermore, in Islamic business ethics cheating is strictly forbidden. Deceiving other people and harming them by giving inaccurate information is against Islamic business tradition. Individuals should be authentic and transparent in their daily conducts at work. Cheating and deceiving are among the most dangerous sins. Similarly avoiding corruption and abuse of power are very important issues. Moreover, in Islam bribery and misuse of power are considered sinful and unlawful. Power holders should be aware of the fact that their ways of conduct are closely scrutinized by God and taking this into consideration they should have transparent and fair conduct with their colleagues and subordinates. Furthermore, according to Islamic business ethics power holders should take better care of their subordinates and should consider their rights. Ethical leadership, morality on the side of leader and authentic behavior of the leader are important prerequisites in order to talk about Islamic work ethics. In Islamic culture paternalistic leadership is quiet pervasive wherein leaders mainly focus on their followers' welfare and behaves like a father figure with benevolence and compassion (Baykal, 2019,c).

On the one hand, Islamic work ethics give importance to consultation and cooperation among colleagues. According to Islamic tradition, one should give importance to other people's ideas and he should take into consideration their ideas and offers when trying to reach a decision. Above all, producing product and services with high quality is also an important requirement for ethical conduct in Islamic work ethics. It is considered unethical to produce low quality products or misdirecting customers by producing with a lower quality than promised. A Muslim should do his/her best whatever he is responsible of (Abuznaid 2009).

Islamic Business Ethics and Islamic Tourism

Islam considers hospitality as an important phenomenon and creates a suitable baseline that enhance societal bondage through hospitality (Omar and Islam 2019). In this point, hospitality refers to the kind of actions wherein hosts welcome and take care of guests or strangers or travelers. In fact, the most important antecedent of hospitality is ensuring healthy atmosphere and comfortable and safe accommodation for the guests. That is to say, the term „hospitality" reminds the act of building relationship between host and guest. Anyway, the term hospitality has its roots in the Latin term hospitaire that refers to receiving guest. Nevertheless, hospitality can be explained as a mechanism of display affection, love, respect, honor and kindness towards guests or tourists. With the guidance from al-Qur'an and Hadith, Islamic hospitality is about overprotecting the guest so that they can feel themselves comfortable and relaxed in any kind of situation. Islam emphasizes hospitality with the belief that Muslims can enhance and increase their societal, national as well as international bondage strongly (Omar and Islam 2019).

In Islam, the Islamic teachings, that is to say, Shariah laws have direct implications for tourism and hospitality industry. The Shariah laws forbid individuals to visit places wherein sins are committed such as alcohol consumption, ignorance of Islamic dress codes in beaches and gambling (Battour, Ismail and Battor 2010). For example, there is Quranic evidence regarding the ban of alcohol. The Holy Quran states: "They ask concerning wine and gambling, Say: In them is great sin, and some profit, for men; but the sin is greater than the profit" (Al-Quran, 2: 219). That is why, hotels have to be regulated by the Shariah in terms of forbidding alcohol (Battour, Ismail and Battor 2010). Moreover, swimming together with the opposite sex with inappropriate and revealing dress is strictly prohibited by the Shariah since it is conceived as one of the most important causes of Fitnah that leads to temptations with evil consequences including arousal of desires prohibited by Islamic teachings (Battour, Ismail and Battor 2010). Related literature confirms these assumptions. For example; in 2012, Dinar Standard Crescent Rating, a US-based company, conducted a survey examining various tourists' behavior on a global scale. According to the results of this global study: 'Halal standards of food production is of utmost importance in 67% of travel decisions of the tourists. In the second place, 'economic suitability comes to the forefront with a rate of 53%, and in the third place, Muslim-friendly practices are important with a rate of 49%. According to the results of the research, the criteria which are prominent in the preferences of the accommodation enterprises of the tourists encompasses clean rooms for accommodation and neat and properly designed environment wherein Islamic considerations are prominent, a sincere hospitality and service system, that is to say sensible and considerate hotel employees, Halal food facilities, including clean food, non-swine meat, convenient location for various activities, provides the required atmosphere for performing daily and weekly worships, a non-alcoholic atmosphere, halal entertainment and swimming facilities and sensitivity to privacy.

In fact, in Islamic teaching, we can come across some very important assumptions regarding Islamic hospitality. According to Islamic hospitability in order to make guest comfortable, spiritual, emotional Intellectual and physical needs of guest should be met (Boediman 2017). In order to meet their spiritual needs a proper place for worship halal food and halal entertainment facilities should be provided.

Moreover, for answering their emotional needs, a friendly, trustworthy and courteous accommodation atmosphere should be ensured. And for meeting intellectual need all the necessary information about the destination and accommodation place should be ensured. On the one hand, to meet guests' physical need a hygienic, clean and safe atmosphere should be provided. Above all, prayer is the most important pillar of Islam. It is among the five pillars of Islam and it is the most important manifestation of Shariah. The importance of prayer comes from the fact that it keeps a Muslim regularly thinking of and communicating with Allah. A Muslim cannot delay his/her prayer. Thus, provision of sufficient facilities and places to perform religious obligation is vital for Islamic tourism industrialists (Battour, Ismail and Battor 2010).

Tourism enterprises, which design their services only according to the halal concept, identify the customers with Islamic faith as the target market and position their enterprises as Halal Hotel Concept (Pamukçu and Armacı 2016). Properties expected from halal concept are: separation of recreation areas according to men and women, the sale, presentation and consumption of all kinds of alcoholic beverages are prohibited, that is why, in Islamic tourism basic prohibitions such as alcohol, display of feminine bodily beauties, abundance of everything including entertainment is characterizing Islamic ethics. Hence Islamic sensitivity effects design of Islamic hotels and the nature of their activities. They consider laws and prohibitions of Islamic tradition in their processes. For example, they have separate masjids for both men and women. The availability of separate pools for men and women is also very important and is a feature frequently highlighted in internet sites and in newspaper advertisements. On the other hand, ensuring proper environment for worship is also very important. The place where the hotel customers can worship, signs in the rooms showing the direction of Qibla, bathrooms, toilets and beds are all expected from an ethical Islamic hotel (Özdemir 2015, p. 10).

In Islamic tourism, privacy is a precursor to the ethical climate. In Islamic hotels they often offer a separate beach service for women, and swimming pools of women are isolated both in the entire facility and in a way that prevents them from being seen from the outside environment. In general, women's pools are covered with high walls or tarpaulins. Women can only use their swimsuits in the women's pool, and they can enter the beach or common pools with special clothes called as Islamic style ", which covers the whole body (Doğan, 2011).

Moreover, accurate information to customers should be provided in Islamic tourism in order to be considered ethical and Islamic. If there is misleading information in advertising, posters, brochures and similar tools, or if the customer is not provided or incomplete or the service given is not in the least equivalent qualifications, criminal sanctions can be applied in many Islamic countries. That is why, Islamic hotels should first provide their customers with clear information about some critical applications, rules and technical features in their websites, newspapers, TV and magazine advertisements. For example, it should be clearly stated that in Islamic hotels there are no alcoholic beverages, no foreign tourists, no nightclubs or disco, the pools are separate for women and men and even children of a certain age, and the beach conditions must be clearly stated. Otherwise, it must either meet the needs of its customers who make such requests or, if it does not, will be legally defective due to incorrect or incomplete information to the customer (Doğan, 2011). In order to ensure a more ethical advertisement policy in halal tourism, tourism companies should give importance to accuracy in the information they reveal in their advertorials. They should specifically avoid providing misleading information and it is not right for customers to hide information on sensitive issues. The advertisements should be accurate and ethical in its presentation and should not exaggerate a bit. In Islamic hotels, misleading information can sometimes be given about privacy conditions, such as exaggerated information about separate pools or beaches for women and men, which can raise customer expectations, which can create frustration for customers when they come to the hotel.

Extravagance is considered as a great source of waste in Islam. It explains transcending the limit in all cases. It is not possible for someone to be individual and socially virtuous if does not avoid lavishness. Because the culture of waste will ensure that the individual is a person who thinks himself, succumbs to his ego and focuses only on consumption. When it comes to waste, only food and beverage, property or financial opportunities should not be considered. In Islamic tradition, waste is not only an over-consumption of natural resources, but a disrespect rather ungratefulness to the Creator (Çınar 2018). With this logic, Islamic tourism needs to be designed by taking into account the principles of Islam regarding lavishness. According to Islam, luxury is forbidden and nothing should be avoided. According to Islamic teachings, Muslims should not overdo anything. In this sense, Islamic hotels are expected to do their best to make their guests comfortable, while at the same time avoiding waste and unnecessary luxury. One of the most criticized topics of Islamic tourism is that many Islamic hotels are not sensitive enough about luxury. Tourism industrialist should create comfortable and attractive accommodation centers without creating lavishness. sensitivity to modesty shows itself in all areas of hotel management, from restaurant services to furniture decoration. However, many Islamic hotels are condemned by lavishness. Sincere Muslims argue that the Islamic hotels are being treated too extravagantly and that this does not coincide with the Islamic business ethics.

Moreover, financial transparency and accuracy should also be conceived as important premises of ethical work climate in Islamic tourism. In this point, high quality financial communication is helpful in ensuring a more financially transparent and authentic business climate. According to the related literature, excellence in financial communication requires a holistic approach encompassing trust-based relationships that are built with main players of the financial market including investors and financial analysts (Whitehouse 2017). Actually,with the help of high quality financial communication, organizations can gain the trust of all their stakeholders (Daniel, Marioara, & Isabela 2017). Similarly, if tourism companies share accurate information and behave with great transparency, their shareholders would develop a positive perception about the company's ethical behavior. Achieving successful levels of financial communication necessitates the recreation of rules and norms of giving messages in congruent with financial corporate principles regarding accuracy, clarity, and transparency (Daniel, Marioara, & Isabela 2017). Without doubt, transparency and accuracy contribute to the perception of ethical behavior for an organization.

SOLUTIONS AND RECOMMENDATIONS

Islamic business ethics is strictly related to Islamic religious doctrines that are why in order to talk about ethics in Muslim business life, it should be known that, in most cases ethics in business life and religious ethics in Islamic daily routines are often intertwined. Flexibility and adaptability of Islamic business ethics are part of the dynamism of Shari'a (Ismaeel and Blaim 2012). Since, Islamic business ethics are characterized by Islamic Shari'a law, that is to say it has roots in basic rules and regulations of Islamic religion. In Islamic faith, a sincere Muslim should adhere to Islamic codes. Hence, ethical conduct should be guided by faith that in practice connotates following Islamic Shari'a law and engaging in what is halal (Abuznaid 2009). Hence, in order to talk about Islamic work ethics in Islamic tourism, ensuring alignment and compatibility with Islamic principles is a preliminary requirement.

FUTURE RESEARCH DIRECTIONS

In the future, this study can be supported by an empirical study wherein positive effects of Islamic business ethics on sustainability and success of Islamic tourism can be handled. Further studies can compare Islamic business ethics with western-style business ethics. And cultural differences can be examined regarding the differences in implication of business ethics on tourism industry. Moreover, business ethics and its positive effects can be examined in other related industries that are intensely affected from religious and cultural dynamics such as service media or education sectors.

CONCLUSION

According to the claims of some futurists, Islamic religion has the potential to encompass about 30% of the total world population by the year of 2025 (Horn 2000). It is one of the fastest growing religions on all over the World. And the increasing number of Muslims creates new demands and markets. Tourism industry is one of the most prominent developing markets among Muslim population. In Muslim countries, Tourism is a major market, however less than 10% of global tourism revenue goes to this market. It should still be conceived as a newborn market that needs to be elaborated delicately. Four countries including Morocco, Egypt, Turkey and Malaysia domain other Muslim countries in tourism with 17.5 million guests in 2004 countries, particularly Western tourist (Ghadami 2012). Most of these guests prefer marine tourism, culture tourism and religious tourism. Although not satisfying yet, increasing market shares in Islamic tourism, or in other words halal tourism brought about the need for a more ethical business environment.

In Islam, designating whether an act should be considered as a virtue or not determined in the Qur'an (Islamic law) and sunnah, literally meaning the beaten track of the Prophet, and it is mostly derived from his sayings (hadith) and actions (Abuznaid 2009), thus, as in the case in their private lives, when Muslims align their deeds and intentions with Islamic rules, they are considered as ethical professionals. And Islamic rules are flexible and adaptable to any era and meet the needs of people from many different cultures and professions. According to Sidani and al Ariss (2017), Islamic ethics encompasses some basic assumptions that most scholars agree upon. God's Unity (tawhid)—person's relationship with God, it is the most private assumption that reflects one's intimate relationship with God.

Justice ('adl)—removing inequity and injustice on all over the world, not only among Muslims, which is a very important assumption that is helpful i creating a fair and transparent work environment. Trusteeship (khilafah)—individuals are entrusted with blessings given to them with moderation (wasat or i'tidal)—eliminating excesses and deficiencies (Saeed et al., 2001); Beekun and Badawi (2005), hence Muslims are expected to behave accordingly and behave as prudent, reasonable, honest and determined professionals.

In Islamic tradition, tourism is an action utilized to obtain God's consent and to be physically, socially and psychologically healthy. Actually, when choosing the most plausible "vacation spot," the tourists assume that they should get the most for the money, they should be treated with dignity and respect (Fleckenstein & Huebsch 1999). For most customers of Islamic tourism, an ethical company is the one that serves the most considerate atmosphere that serves their Islamic sensitivities. In this point, the existence of facilities that attach importance to privacy such as separate pools and beaches for men and women and the provision of suitable facilities for worship are among the most important expectations from Islamic hotels that contributes to creation of a more ethical climate.

Moreover, in order to ensure a more ethical advertisement policy in halal tourism, tourism companies should give importance to accuracy in the information they reveal in their advertisements. They should specifically avoid providing misleading information and it is not right for customers to hide information on sensitive issues. The advertisements should be accurate and ethical in its presentation and should not exaggerate a bit. In Islamic hotels, misleading information can sometimes be given about privacy conditions, such as exaggerated information about separate pools or beaches for women and men, which can raise customer expectations, which can create frustration for customers when they come to the hotel.

On the other hand, as in the case with many other industries, in tourism industry companies should be sensitive regarding their financial responsibilities. They should consider financial benefits of all their stakeholders including their shareholders, managers, employees and customers. Being profitable and having higher market shares are not enough in Islamic tourism, companies should also be considerate, trustworthy and transparent in order to satisfy their stakeholders and be perceived as ethical organizations.

REFERENCES

Akyol, M., & Kılınç, Ö. (2014). Internet and Halal Tourism Marketing. *Electronic Turkish Studies, 9*(8).

Al-Qaradawi, Y. (2002). *Ethics and values in Islamic economics*. Resalah.

Ali, A. J., & Al-Owaihan, A. (2008). Islamic work ethic: A critical review. *Cross Cultural Management, 15*(1), 5–19. doi:10.1108/13527600810848791

Alserhan, B. A. (2010). Islamic Branding: A Conceptualization of Related Terms. *Journal of Brand Management, 18*(1), 34–49. doi:10.1057/bm.2010.18

Battour, M. M., Ismail, M. N., & Battor, M. (2010). Toward a halal tourism market. *Tourism Analysis, 15*(4), 461–470. doi:10.3727/108354210X12864727453304

Baykal, E. (2019d). Boosting Innovative Work Behavior in Organizations Through Absorptive Capacity. *Journal of Organizational Behavior Review, 1*(1), 16–33.

Baykal, E. (2019b). Transparency in Financial Communication: Effect of Authentic Leadership. In Handbook of Research on Global Issues in Financial Communication and Investment Decision Making (pp. 110-126). IGI Global.

Baykal, E. (2019c). Turkish Type Leadership: Sabri Ulker Example. *The Journal of Social Science, 3*(6), 425–438.

Baykal, E. (2019e). Transparency in Financial Communication: Effect of Authentic Leadership. In Handbook of Research on Global Issues in Financial Communication and Investment Decision Making (pp. 110-126). IGI Global.

Baykal, E. (2019a). Yenilikçi Örgüt İklimi: İşe Adanma Üzerine Etkisi. *Ekonomi İşletme ve Maliye Araştırmaları Dergisi, 1*(3), 266–279. doi:10.38009/ekimad.612630

Beekun, R. I. (1996). *Islamic business ethics*. International Institute of Islamic Thought.

Beekun, R. I., & Badawi, J. A. (2005). Balancing ethical responsibility among multiple organizational stakeholders: The Islamic perspective. *Journal of Business Ethics*, *60*(2), 131–145. doi:10.100710551-004-8204-5

Bhuiyan, Chanmhuri, Mohamad, & Rabiul. (2011). Potenatials of Islamic Tourism: A Case Study of Malaysia on East Coastal Economic Region. *Australian Journal of Basic and Applied Sciences, 5*(6), 1333-1340.

Boediman, E. P. (2017). Halal Lifestyle in Marketing Communication of Tourism and Hospitality. *International Journal of Economic Research*, *14*(4), 429–438.

Boğan, E., Batman, O., & Sarıışık, M. (2016). Helâl turizmin kavramsal çerçevesi ve Türkiye'deki uygulamalar üzerine bir değerlendirme. In *3rd International Congress of Tourism & Management Researches* (pp. 1425-1440). Academic Press.

Daniel, A. C., Marioara, A., & Isabela, D. (2017). Annual Financial Statements as a Financial Communication Support. Ovidius University Annals. *Economic Sciences Series*, *17*(1), 403–406.

Standard, D. (2012). *Global Muslim Lifestyle Travel Market 2012: Landscape and Consumer Need Study for Airlines, Destinastions, Hotels & Resorts Executive Summary*. https://www.dinarstandard.com/travel-study/

Doğan, M. (2011). Türkiye'de İslami Turizm'in Gelişimi: 2002–2009. *Political Economy, Crisis And Development (Politik İktisat, Kriz Ve Kalkınma),* 471-478.

Dusuki, A. W. (2008). What does Islam say about corporate social responsibility? *Review of Islamic Economics*, *12*(1), 5–28.

El Garah, W., Beekun, R. I., Habisch, A., Lenssen, G., Adaui, C. L., Ismaeel, M., & Blaim, K. (2012). Toward applied Islamic business ethics: Responsible halal business. *Journal of Management Development.*

Elaziz, M. F., & Kurt, A. (2017). Religiosity, consumerism and halal tourism: A study of seaside tourism organizations in Turkey. *Turizam: međunarodni znanstveno-stručni časopis, 65*(1), 115-128.

Fleckenstein, M. P., & Huebsch, P. (1999). Ethics in tourism-reality or hallucination. *Journal of Business Ethics*, *19*(1), 137–142. doi:10.1023/A:1006118526469

Franklin, A., & Crang, M. (2001). The trouble with tourism and travel theory? *Tourist Studies*, *1*(1), 5–22. doi:10.1177/146879760100100101

Giaoutzi, M., & Nijkamp, P. (2006). *Emerging trends in tourism development in an open world. Tourism and Regional Development: New Pathways*. Ashgate.

Gündoğdu, B., Ünal, A., & Kılınç, İ. (2018). *Üniversitelerin İş Etiği Algıları ve Uygulamaları: Batı Karadeniz Üniversiteler Birliği Üzerine Bir Araştırma*. Academic Press.

Henderson, J. C. (2003). Managing tourism and Islam in peninsular Malaysia. *Tourism Management*, *24*(4), 447–456. doi:10.1016/S0261-5177(02)00106-1

Horn, R. (2000). Border bordellos. Time Asia. İnceleme, "Uluslararası Sosyal Araştırmalar Dergisi", 29(7).

Ingbar, J. (2004). *From Organizational Ethics: Where Values and Culture Meet.* Institute for Global Ethics.

Jaelani, A. (2017). Halal tourism industry in Indonesia: Potential and prospects. *International Review of Management and Marketing*, *7*(3), 25–34.

Nurrachmi, R. (2019). *The Potential Industry of Islamic Tourism in ASEAN Countries*. Academic Press.

Okhovat, H. (2010, September). A Study on Religious Tourism Industry Management, Case Study: Islamic Republic of Iran. *International Journal of Academic Research*, *2*(5).

Omar, C. M. C. (2019). Exploring the Validation of Islamic Hospitality Management in Tourism Destination: Malaysian Perspectives. *Journal of Hospitality and Networks*, *1*(1), 34–43.

Özkalp, E., & Kırel, Ç. (2011). Örgütsel Davranış, Ekin Basım-Yayım, Ekim 2011. Bursa.

Pamukçu, H., & Sarıışık, M. (2017). Helal Turizm Kavramı ve Gelişimi Üzerine Genel Bir Değerlendirme. *Uluslararası İktisadi ve İdari Bilimler Dergisi*, *3*(1), 82–98.

Perran, A. K. A. N. (2007). Uygulama açısından iş etiği kuralları ve evrensel turizm etiği ilkeleri. Anatolia. *Turizm Araştırmaları Dergisi*, *18*(1), 7–20.

Reynolds, P. (2000). 'Profit and Principles: Business Ethics in Hotel Management Companies in Asia'. *Australian Journal of Hospitality Management*, *7*(1), 1–13.

Rokhman, W. (2010). The effect of Islamic work ethics on work outcomes. *EJBO. Electronic Journal of Business Ethics and Organization Studies*.

Saeed, M., Ahmed, Z. U., & Mukhtar, S. M. (2001). International marketing ethics from an Islamic perspective: A value-maximi- zation approach. *Journal of Business Ethics*, *32*(2), 127–142. doi:10.1023/A:1010718817155

Seyfi, S., & Hall, C. M. (2019). Deciphering Islamic theocracy and tourism: Conceptualization, context, and complexities. *International Journal of Tourism Research*, *21*(6), 735–746. doi:10.1002/jtr.2300

Tanrısever, C., Pamukçu, H., & Batman, O. (2016). New Tourism Trends in the World and Their Adaptations to Turkey. *Gümüshane University Electronic Journal of the Institute of Social Science/Gümüshane Üniversitesi Sosyal Bilimler Enstitüsü Elektronik Dergisi*, *7*(16).

Tekin, Ö. A. (2014). İslami Turizm: Dünyadaki ve Türkiye'deki Genel Durum Üzerine Bir İnceleme. *Journal of International Social Research*, *7*(29).

Tekin, Ö. A. (2014). *İslami Turizm*. Türkiye'deki ve Dünyadaki Genel Durumu Üzerine Bir.

Whitehouse, M. (2017). Financial analysts and their role in financial communication and investor relations. In A. V. Laskin (Ed.), *Handbook of financial communication and investor relations* (pp. 117–126). Wiley. doi:10.1002/9781119240822.ch10

Wuryasti, F. (2013). *Halal Tourism, a New Concept of Tourism Activities in Indonesia*. Available from: http://www.travel.detik.com/read/201 3/10/30/152010/2399509/1382/

Zulkharnain, A., & Jamal, S. A. (2012). Muslim Guest Perception of Value towards Syariah Concept Hotel. In Current Issues in Hospitality and Tourism, Research and Innovations (pp. 337-340). Taylor & Francis Group.

Zulküfli, W. S. W., Rahman, S. A., Awang, K. W., & Man, Y. B. C. (2011). Developing the Framework for Halal Friendly Tourism in Malaysia. *International Business Management*, *5*(6), 295–302.

ADDITIONAL READING

Al-Hamarneh, A., & Steiner, C. (2004). Islamic tourism: Rethinking the strategies of tourism development in the Arab world after September 11, 2001. *Comparative Studies of South Asia, Africa and the Middle East*, *24*(1), 173–182. doi:10.1215/1089201X-24-1-175

Battour, M., Ismail, M. N., Battor, M., & Awais, M. (2017). Islamic tourism: An empirical examination of travel motivation and satisfaction in Malaysia. *Current Issues in Tourism*, *20*(1), 50–67. doi:10.1080/13683500.2014.965665

Baykal, E. (2019). Spiritual Leadership in Collectivist Cultures: Turkey Example. In Science and Spirituality for a Sustainable World: Emerging Research and Opportunities (pp. 44-69). IGI Global.

Gabdrakhmanov, N. K., Biktimirov, N. M., Rozhko, M. V., & Mardanshina, R. M. (2016). Features of Islamic tourism. *Academy of Marketing Studies Journal*, *20*, 45.

Hall, C. M., Razzaq, N. H. A., & Prayag, G. (2020). Introduction to halal hospitality and Islamic tourism. The Routledge handbook of Halal hospitality and Islamic tourism, 1-18.

Prayag, G. (2020). Halal tourism: Looking into the future through the past. *Tourism Recreation Research*, 1–3.

Preko, A., Mohammed, I., Gyepi-Garbrah, T. F., & Allaberganov, A. (2020). Islamic tourism: Travel motivations, satisfaction and word of mouth, Ghana. *Journal of Islamic Marketing*, *12*(1), 124–144. doi:10.1108/JIMA-04-2019-0082

Suid, I. S., Nor, N. A. M., & Omar, H. (2017). A review on islamic tourism and the practical of islamic attributes of destination in tourism business. *International Journal of Academic Research in Business & Social Sciences*, *7*(12), 255–269.

Wahyono, Z., & Razak, M. A. A. (2020). Islamic Tourism in Southeast Asia: The Concept and its Implementation. *International Journal of Halal Research*, *2*(2), 90–105. doi:10.18517/ijhr.2.2.90-105.2020

KEY TERMS AND DEFINITIONS

Ethics: It is a branch of philosophy that "involves systematizing, defending, and recommending concepts of right and wrong behavior."

Halal: An Arabic word meaning lawful, used to describe meat from animals that have been slaughtered in accordance with Islamic law.

Halal Tourism: It is a subcategory of tourism which is geared towards Muslim families who abide by rules of Islam.

Islam: The word "Islam" means "submission to the will of God."

Islamic: It is a term denoting directly to all things related to Islamic faith and its doctrines.

Muslim: Muslims are those people who are monotheistic and worship one, all-knowing God, who in Arabic is known as Allah.

Tourism: It is a term encompassing all kinds of visits including travelling leisure, business and other purposes.

Chapter 16
Contrast in Ethics, Morality, Justice, and Behavior:
Some Reflections on Business and Islamic Ethics

Khaliq Ahmad
https://orcid.org/0000-0001-8129-5178
Qassim University, Saudi Arabia

Burhan Uluyol
https://orcid.org/0000-0002-9984-489X
Istanbul Sabahattin Zaim University, Turkey

Othman Altwijry
Qassim University, Saudi Arabia

ABSTRACT

The ethical dilemma begins due to greed, jealousy, and discontentment, and these have solutions in life. This chapter aims to analyze and apply the understanding of dichotomy of contrast in ethics, morality, justice, and behavior. Since ethics matters a lot in discharging corporate social responsibility in business and trade, the chapter will also examine the application of these ethics, morality, and behaviour from an Islamic perspective. The authors derive that "one must strengthen contentment, remembrance of death, the firm belief in Allah's mercy, generosity" while refraining from common contemporary business problems such as window dressing of financial results, fraud, deception, bribery to get contracts, and among others.

DOI: 10.4018/978-1-7998-4117-3.ch016

INTRODUCTION

This is an era of departure from 20th century laden with secularism, capitalism and so called market dynamism as a self-centered approach. However, concern for others, care, love and service to the humanity has been stressed in different ways in all faiths and religions. But do people follow what is taught to them? Also, to what extent people practice what they themselves preach? Islam is a faith based religion that also has prescription for values based on ethics, morality, justice and peace. An anecdote from personal experience of a friend and our daily life is for illustration as follows.

During a Friday prayer in a mosque inside a University campus in a predominantly Muslim nation the Imam gave a sermon on Islamic ethics and values. In that context, he referred to the unethical behavior of those (mostly students) who drove their motor cycles on the pavement as a short-cut to reach to their destination, that included the mosque. He stressed upon the Islamic values of empathy and care for others and warned that such action could also lead to accidents and injuries to those walking on and using the pavement especially handicapped ones. I was quite impressed with this timely sermon and so as soon as the prayer was over, I rushed out of the mosque to check the result. Regretfully, there was no change and it never happened even after months of prayers and sermons. This brings me to this assertion that moral values can't be taught it is rather caught?

This is not applicable to the act of worshipers alone. This is quite rampant in *Fiqh Muamalat* (Islamic transactions) by business owners and their agents' (managers) managing on their behalf. Despite listening sermons for fair deal (based on social justice) in conducting our trade and businesses and related services including Islamic banking and financing services, they are involved in too much profiteering. This affects the spirit perhaps not the letter due to being a *Shariah* compliant transactions. And therefore criticisms are quite often raised by the experts in terms of failing in their pursuits of poverty eradication due to failings in Islamic social financing.

Why people did not follow what was preached to them during the prayer especially during Friday's sermon or elsewhere in other religious preaching and gatherings? The possible explanation comes from social psychological studies on group norms and conformity behavior. Social norms play an important role in how people behave in social situations such as following or not following the traffic rules and shared values. Social psychologist Solomon Asch (1951) conducted a series of experiments on group norms and conformity and revealed that people were willing to ignore reality to conform to the rest of the group. Group or social norms, even if they are wrong, immoral and unethical, become the standard that guides people behavior. Simply put if most of the people are driving their motor cycles on the pavement then it becomes the norm to follow. Norms define the culture of the social unit, be it family, community, business and trade or nation.

Deutsch and Gerrard (1955) found that there are reasons due to which this happens. They termed it as two factors- normative conformity and informational conformity. The first requires compliance to remain in the group failing which he/she feel rejected whereby the second one happens due to ignorance and needed help from fellow members to guide him/her to be part of the chosen group. Businesses and corporate sectors are not exception to this. These establishments see and follow the trends and rampant trade practices in a society and at the market place.

On another note, there is no dearth of examples from the history where human beings have been subjected to extreme forms of brutalities in religious and faith based societies such as Buddhists that preaches non-violence, either at the individual or at the mass levels, such as the recent Rohingya crisis in Myanmar, Uighurs in East Turkistan. According to the media reports regarding mass deportation, burning houses, rapes and killing of innocent Rohingya and brainwashing and brutal approach of camping Chinese Muslims in Yinchuan are well documented and covered in many international forums and the crime against the humanity is done by the ruling class who are not illiterate rather they were well educated specialist who know things and had clear intent and purpose.

How and why do people engage in act of brutality and genocide on orders from people in power and control is rightly described in the famous Milligram study (1963) provides some possible answers. The author findings suggested that a person can go very far in obeying the group norms when it comes to harming people outside the group. His research interest was to know about the magnitude of harmful behavior due to bad influence in committing *zulm* (atrocities). Milgram discovered that as per the influence of the group leader authority a person may go this far to destroy and kill the innocent human beings and this according to author happens due to the socialization process and upbring of the person who is out to commit such a heinous act. Furthermore, Milligram (1974) entitled, "The Perils of obedience" explained the socio-legal and philosophical aspects of obedience that has enormous importance for behaving in such a way but did not elaborate about human conduct in the concrete situations.

Milgram (1974) again explained two behavioral patterns in the given socio-economic contents. Firstly, autonomous state of mind and the culprit is behaving directly and bear the full responsibility of his/her actions. Secondly, the person acted upon instruction by others as an agentic person. In such a situation the outcomes of the person is attributable to the person giving the instructions. In the later situations the actor has either of the two convictions; (a) the authority issuing instruction is fully qualified, and (b) any outcome is also responsibility of the instructing person who will eventually stand for his/her actions for what happens at the end of the day.

The objective of this paper is to analyze and apply the understanding of dichotomy of Contrast in Ethics, Morality, Justice and Behavior. Since ethics matters a lot in discharging CSR in business and trade, the paper will also cover the application of these ethics, morality and behavior from an Islamic perspective. This may help the future executives and managers to adhere the principles of justice and fairness in contemporary societies.

LITERATURE REVIEW

More recently, researchers in the area of organizational behavior have paid their attention on what is termed as justice perception and enactment. Sometimes justice needs to be seen as perception matters a lot in administration of justice not the justice alone per se. How an individual treat others matter whether it's just or unjust behavior. Recipient's situation varies from person to person and how the agent entrusted to serve justice enacted upon the desired outcome is a delicate matter. The literature shows multiplicity of aspects involved. There are group of factors at play, one group of factors lie within the individual actor such as his/her traits to define one's personality (Schuh, Zeng, Xin & Fernandez, 2017), his or her emotions (Cohen, Patner & Turan, 2012; Patient & Skarlicki, 2010) and finally ones' moral identity (Brebels, DeCremer, Van Dijke, Van Hiel, 2011).

Environment, Experience and Exposure

The socialization process in an environment is a factor that gave him/her exposure that could constitute a person's experiences and his/her restrained relationships in any social contract. Some studies are contextual and have focused on environment where one has grown and regulated his rather retorted their uncalled behavior such as motivation and will power to meet the standard of behavior, (e.g. Whiteside & Barclay, in press), compromised social class and status and eventual position in given social strata and power (Blader & Chen, 2012) and politics game (Monin, Noorderhaven, Vaara, & Kroon, 2013) which basically influence justice and its enactment. There are studies which examined justice enactment in relation to the enactor of the justice system themselves (Johnson, Lanaj, & Barnes, 2014). Authors such as Zapata, Olsen, & Martins, (2013) and Zhao, Chen, & Brockner, (2015) discussed how people conceptualise justice including the various form of justice system ranging from utilitarnism to rights and what is due to recipient of justice. All these studies indicate one thing common is the perception of fairness and justice and how the people go to accept and get motivated to behave justly.

Therefore, understanding human behavior is quite complex. It requires systematic analysis of all possible sources and factors. Inculcating moral and ethical values require creating, enforcing, reinforcing such norms and cleansing one's soul. It cannot be achieved simply through lectures and sermons. It is also important to understand that people may not hesitate harming others while obeying the orders of authority and power especially when they tend to believe that the accountability for their actions lie with those who order them. Therefore, Milligram was right when he stated that while law and philosophy behind it are important but they explain little about human behavior in a given situation.

Leadership Role

The personality of the leadership plays very significant role in demonstrating the ethical or otherwise his/her behavior. Following nine factors are necessary to see the role played by those in power who often exercise formal and informal authority to influence the outcome in ensuring the justice and fairness on the ground.

1. Leadership Style: There are hundreds of style and approaches to practice ones' leadership style. This includes traits, values, practices, process and socialization process. It also encompasses the consideration of the situations someone is facing while guiding and demanding followership.
2. Power Game and its influence on the style: Use of authority to influence some kind of followership is key components of any leadership. It defines behavior and adherence (Dewi & Tanjung, 2007; Popper & Zakkai, 1994).
3. Leader or Servant First: we can say that this method is akin to the principles of leadership from Islamic perspectives. This is according to a morality that leans toward "values, ethics, principles, virtues, good characters, God fearing, humanism and spirituality" (Beekun, 2006; Khaliq, 2009).
4. Whether Leader is Transactional or transformational: this leadership style focuses the exchange of rewards with subordinates for services rendered whereby transformational leadership style spearheads organizational change and transform group behavior by changing workplace culture, and it empowers people in the real life situation to transform an organization (James & Joseph, 2001; Muenjohn, 2008).
5. Trait Theories: This closely relates to human traits. Usually a leader should have "the visionary trait; they are expected to have a sense of direction and long-term vision for the organization or community" (Adnan, 2006; Beekun & Badawi, 1996).

6. Leadership vis-a-vis Situation: Situation changes and so is the need of a need for change in the style. One needs a best fit style of a leader's behavior to suit the situations that can guarantee followership (Adnan, 2006).
7. Islamic principles and Leadership: This is based on an idea of justice and fairness principles. The sources of such principles are Al-Quran and the Prophetic traditions. There are great emphasis on trust, *adl'*, *ihsan*, human welfare and goodness for all irrespective of race, religion and creed (Khaliq 2009; Khan 2007).
8. Sources of Knowledge on Leadership: There exist empirical researches in many Western and industrialized countries on leadership and these researches and theories serve as source of knowledge based on wisdom and reasoning. But there are theories which are outsourced through revelation and reason and for which the source of knowledge are Quran and Sunnah. (Adnan, 2006; Al-Buraey, 2006; Beekun, 2006; Khaliq 2009).

There is a range of choices from the above. However, the servant leadership is now being well researched in catholic communities especially from the Scandinavian region to see role of a leader as servant first and leader afterwards. This is what Islamic traditions have also encapsulated role of a leader is servant of a community (Syyidul Qaum Khadimuhum).

Religion and Practice

Beside ones' faith in a religion of choice, and belief system, it preaches certain values and practices. For example many religious beliefs come to play a role over here. Islamic belief system highly depends upon the Islamic worldview based upon *Tawhid* and accountability hereafter shapes the followers' behavior and ethical conduct. Since the religion is a matter of emotional attachment there is sometimes a trend to ignore reasoning and logic and hence there is a natural and systematic approach through which ones' sense of religious, social and moral experience to govern human conduct and behavior (DeGeorge, 1986).

Religion is again managing relationship either vertical or horizontal. Vertical relationship is with the Creator (Almighty) while horizontal is with fellow human. Thus it shapes one's behavior and harmonizes the relationship to strike peace deal and avoid its abuse. It can be used for social and human welfare. But this type of ethical behavior is sometimes synonymous with the idea of morality. Islamic prescription of morality is related to Prophet's behavior and teachings and it refers to special type of a code of ethics for followers of Islam and Muslims (Hanafi & Sallam, 1995).

The following section is about to describe the basis of Islamic ethics and ethos as part of morality and moral values since the Prophets of God are the torch bearers and source of guidance for mankind and overall of human development and personality conducts.

ISLAMIC ETHICS

Islam as a religion is based on Abrahamic traditions. During the last prophet who perfected by his noble practices and henceforth Islamic prescriptions of ethics, value chains and moral conducts have finality to them. These are based upon the revelation as Allah's message used to be descended through the last messenger, the Prophet (saw). Islam, as a religion, is sematic in its approach to ethics since it's aligned to Abrahamic religion offers "a complete and eternal code of ethics for humanity based on the *Tawhidic* paradigm".

Islamic perspective defines morality as: "the personal behavior based on Islamic revealed knowledge". Ethics therefore must be defined as "the accepted norms of social behavior; such norms being based on the Islamic revealed knowledge". Business ethics should be "the accepted norms of behavior in managing individual and business conducts; such norms being based on Islamic revealed knowledge".

The Uniqueness of Islamic Ethics

Theoretically speaking, Islamic philosophy of ethics and ethical conduct of followers of Islam as an Abrahamic religion has following five maxims:

1. *Unity of Command*: *Tawhidic* concept of "Oneness" religiously means the totality of power and existence in and outside of the Universe in an integrated manner.
2. *Equilibrium and Balance*: Again Islamic concept of "*adl*" means fairness a sense of balance among the components and various facets of human life,
3. *Freedom and Free will*: The creation of human beings is the best of the best creation by the God who is entrusted freedom as this is the *Fitrah* of the human, no coercion is welcome idea. However there is limits set by Allah, Who empowered human as His vicegerent of Allah (swt) on this earth.
4. *Benevolence or "Ihsan"*: This means an action carried out by a human being without any favour in the return. This charitable act for the welfare of mankind without expectation of any reward thereof rather just benefiting others, and
5. *Responsibility (Mas'uliyyah)*: "Accountability of human beings for all their actions" to their Creator in this word and hereafter.

In Islam, ethics dominates human social and economic conducts. The emphasis is on both the ends and the means. Therefore any economic activities must be morally, ethically and religiously legitimate. According to Naqvi (in Khalifa, 2001), who developed a model that is encompassing Islamic ethics based on similar previously mentioned axioms. These are four axioms: Unity of Command (*Tawhid*), Equilibrium and (*Adl*), Free Will Freedom and Free will (*Hurriyah*) and Responsibility (*Mas'uliyyah*). Khalifa (2001) who suggested another maxim: Steadfastness (*Istiquamah*) as a part of the model. The argument is based upon the verses from the holy Qur'an and the Sunnah of the Prophet Mohamad (saw).

Allah (swt) mentions in the Holy Qur'an: "Verily those who say: 'Our Lord is Allah,' and remain firm on that path [steadfast], on them shall be no fear, nor shall they grieve" (Suratul Kahf, 46:13). Also very popular and known jurist (Imam Muslim) relates that a companion of the messenger of Allah (saw) had come to him and asked: "O Messenger of Allah, tell me something about Islam, which I could not ask anyone about save you. He answered, 'Say: I believe in Allah,' and then be steadfast (in accordance with such a profession)".

The author, Khalifa (2001), subscribed and elaborated further that *Tawhid* and *Istiquamah*, these two axioms cover both "faith and intentions, actions and deeds". These eventually constituted "the realm of freedom, responsibility and accountability in Islam". *Tawhid* as Muslims' *Aqidah*, comprises three elements:

1. *Rububiyyah* (*Tawhid* implied as Lordship)

2. *Uluhiyyah* (*Tawhid* implied for Worship)
3. *Asma' wa Sifat* (*Tawhid* comprised the Names and Attributes of Allah).

The term *Rububiyyah* emphasized submission (*Taslim*), and it leads to *Tawakkul*. This is followed by patience (*Sabr*), and composure, or even delight (*Riza*) that leads to pleasure with the outcome as it comes from Almighty. But the *Tawhidic* concept implying *Uluhiyyah* focusses on sincerity (*Ikhlas*) that hereby means seeking the *Riza* (Pleasure of Allah in every act and intention).Finally, *Asma' wa Sifat* part of *Tawhidic* paradigm necessitates piety(*Taqwa*), and it leads to kindness and benefaction (Birr).

The author, Khalifa (2001) further believes that the attributes emphasized above are qualities need to be nourished and sustained since above mentioned ones are integral part of Islamic ethical system. Sustainability is hard to achieve, and that is why, remembrance of Allah (*Dhikrullah*) is highly counted and recommended in religious prescriptions. The Sustainability of the otherwise right intentions (*Istiquamah*) and human deeds is tough and the real challenge of reflecting and manifesting one's faith (*Iman*) in Muslim's life. *Istiquamah*, as an off shoot of *Tawhid* then are the two basic constituents of the author's model of an Islamic ethical system heavily used by the Islamic economists in the modern day academia.

Khalifa (2001) finally concluded that sincere intention seeking pleasure of Allah is the desired outcome of the real Unity of command concept inter-twined with the term *Tawhid*. One strong demonstration and human attributes noble concept in Islam is the quality is *Zuhd* (Free from materialistic temptation). The word *Zuhd* is meant as a Muslim controls the worldly temptations to be a master of his/her destiny by controlling himself/herself to not being enslaved by his/her desires and wants. This technically means that *Zuhd* keeps one away from *Dunia* (worldly life temptations) should be in line with *Akhirat* (Life hereafter). It contains that a Muslim who fears *mas'uliyyah* should not seek *Dunia* by his intentions, deeds and endeavors. His ultimate end and lifelong purpose of existence is much beyond this temporary life rather should be the home in the hereafter. It equally demonstrate that wealth and all economic activities and well-being are only the means. This eventually is not to be meant an end which is basically leading to the attainment of *Falah* by adhering to ethics from an Islamic perspective. Therefore as conclusion this combination of *Tawhid* and *Istiquamah* are the ways to achieve eternal life full of success (*Falah*).

ISLAMIC BUSINESS ETHICS

Beside the above explanations by scholars of ethics, Hanafy and Sallam (1995) are authors who are directly dealing with economic justice and fairness. This is implied in the modern day business affairs based on Islamic morality and ethics. These authors have categorized Islamic ethical principles into six categories for a conduct of modern day trade and business. These Islamic business ethics are as follows:

1. Truthfulness: The ultimate outcome is nothing but truth. Allah (swt) commands that all the followers of Islamic faith to be straight forward and truthful in their business and trade dealings and utterances (33:70). Therefore Islam as religion strongly condemns falsehood and deceits in any business form and trade dealings.
2. Trust: the accountability and trustworthiness, as the idea of having to appear in the life hereafter before Allah and to account for one's action. Islam prescribes that human life is an *amanah* and all its resources are entrusted upon human beings in worldly life is a trust by Allah. Followers of Islam to be conscious of this trust in every aspect of mundane life. A business and its dealing is

also a matter of trust between the members of a society and the individual. Trust is henceforth from Allah (swt).

3. Sincerity: Intentions and actions in every walk of life including commercial life matters as well. Islam attaches great importance to the sincerity of intentions. The performance of ones' duties to perfections implied therefore that individuals work with sincerity and devotion. Such ethical code conduct in economic activities results in more efficiency as well as a high rate of yields and economic total factors' productivity. Sincerity and purity of intentions also discouraged in Islam to eliminate manipulation or exploitation of others for personal reasons.
4. Brotherhood: All human beings are brothers to each other. The distinctions of race, creed and culture, feelings of color, tribal mindedness, caste and language are not valid criteria. This should not be used for judging the superiority or inferiority of an individual or members of any group. Everyone is entitled to an ethically right behavior irrespective of the above distinctions (caste, creed, race, or territory). Refer to *Qur'an* (*Suratul Hujurat*, verse 19).
5. Science and Knowledge: Religiously speaking it is mandated for Muslims to seeking knowledge even have to travel far and distant places such as China (earlier days China was considered far) and obtain excellence in performance by acquiring knowledge and wisdom. Research and development that breads innovation is highly encouraged in Islam.
6. Justice: Fairness in dealings is undoubtedly a prerequisite of business and trade. As it encompasses the entire gamut of human socio-economic life. The whole universe is based on the concept of justice, fairness and balance. Justice means that everyone should be treated as he/she deserves, without any undue negative consideration since human by nature is innocent unless proven otherwise which is including "fair treatment, equality and a sense of proportion and balance". Today's business dealings are full of problems for decisions related to "pricing, product quality, employee treatment, handling of environmental pollution".

Ethical Behavior

There are many factors that affect ethical behavior either positive or negative ways. These can be broadly classified among the following categories.

1. Socio-legal interpretation factors
2. Factors related to Organizations, and
3. Personal or Individual factors
    a. Moral development stages
    b. Personality traits
    c. Influences by family
    d. Influences by peers
    e. Influences by life experiences
    f. Situational factors influences
4. Legal interpretations in so called "secular societies are based upon contemporary and often transient values and standards whereby in an Islamic society, these values and standards are guided by the *Shariah*".
5. It is important that the ethical conduct is inevitable in the organization and it is the main element of organizational structure. Therefore, the many organizations call the code of ethics and it is dif-

ferent in different industries. Such codes or code of ethics may enhance ethical behavior among organizational participants, their use is sometimes inappropriate. An organization for example trading in *Khamr* or other *haram* products to be avoided by Muslims but many Muslims are working there as they found this is the only way earning living-hood. Put differently, the halal business organizations can cultivate the ethical and moral behaviors according to Islamic code of ethics for such Muslims seeking pleasure of Allah can work and earn livings.

6. Personal or Individual factors:
   a. In a *hadith* narrated by A'isha (rah), she narrated that: The Apostle of Allah (saw) said: "There are three (persons) whose actions are not recorded; a sleeping person till he awakes, an idiot till he is restored to reason, and a boy till he reaches puberty" (Abu Dawud, 4384). Thus the Prophet (SAW) said that individuals undergo two stages of moral development. The minor or pre-pubescent stage and the adulthood stage and both are important. These are three levels of the spiritual advancement and development for the human soul or *nafs* (Rizvi in Beekun, 1997): All are classified as *nafsul ammarah* (12:53), *nafsul lawwamah* (75:2) and *nafsul mutma'innah* (89:27).
      i. *Nafsul mutma'innah* is the highest level from all the three types. The soul achieves a highest level of satisfaction after the intellect (*aql*) has checked the evil tendencies.
      ii. One may be succumbing to the *nafsul ammarah*, if a person persists in behaving unethically. If any Muslim is behaving Islamically, he must be fighting and standing against the evil impulses of the *ammarah*, and responding to the directions of the *lawwamah,* and the *mutma'innah*.
      iii. Unethical behavior and the interaction among these three states of the soul ranging from ethical to evils is dependent upon one's level of *Taqwa* or piety.
      iv. How is the one's soul (*nafs*) is at a present point of time or whether he/she is winning or losing the battle against temptation by evil forces, he/she are likely to be more or less prone towards behaving ethically.

This is somehow closer to the understanding subscribed by the followers of Freudian motivation theory based on *id*, *ego* and *superego* as propounded by Sigmund Freud. This motivation theory narrates that unconscious psychological forces, such as hidden desires and motives, shape an one's behavior, like pulse buying patterns in the given market situation aroused out of full stimuli comprising of high pitch selling. Nevertheless theory by Sigmund Freud was subjected to outright criticism, who in addition to being a medical doctor, is synonymous with the field of psychoanalysis. Back to Islamic perspective based upon the religious and Islamic ethical principles, *nafsul ammarah* destroys the behavior of human beings. This is due to his/her temptations that force one who behaves like devils' agent. In case the person who, otherwise if control temptations, is defined vicegerents of Allah (saw).

b. Personality traits: This stresses on honesty and the person will likely behave differently from another who does not respect other people's right of property. In the nutshell a key personality variable which may affect the ethical behavior of an individual is his/her locus of personal restraints that regulate and control, whether internal or external one's behavior. If a person believes that he or she can control the events in his life, the person is capable to control with internal focus. Therefore, any person has to responsible for the consequences of his or her behavior. Meanwhile, a person also has external focus which will determine the fate or luck including the behavior of other people affect one's life. As a results, an individual believes that the external factors would cause him or her to behave either ethically or otherwise.

c. Influences by family: Family members who may institute as internal factors are more likely than externals put pressures to make ethical or unethical decisions. But the person will resist hurting others, even when ordered to do so by the seniors (Lefcourt, 1982). However people start to form ethical expectation as from the childhood. This is why the Prophet used to emphasize the importance of family nurturing: "*Command your children to pray when they become seven years old, and discipline them for it (prayer) when they become ten years old; and arrange their beds (to sleep) separately*". Our wishes that the children grow up as good citizens of a society, it needs to educate our children from the childhood. Like "children see, children do", we can cultivate the highly ethical or moral behaviours among the children if they observe their family members are behaving accordingly and if they are rewarded for ethical or punished for being untruthful or stealing from others etc. makes a great difference. Just as a reminder that any mixed gestures from parents are likely to result in unethical behaviour. A child who is told for example that stealing is bad; at the same time, he is given supplies "borrowed" from the parents' office at work falsely will acquire the same practice in life. Socialisation process- Family, School, Peers and mass Media have lifelong influences in whole life. As children attends school, they are influenced by the peers where they are either reinforce with family values or start differently behaving such way influenced by those with whom they interact daily.

d. Influences by peers on children when they grow and are admitted to school are established norm where they are influenced by the peers with whom they interact daily. Their family values as were practiced by these peers who come out of different family settings belong to different influence has positive or negative impact on him/her.

e. Influences by life experiences: both positive and negative events can affect the lives of people. These events can determine the ethical and moral behaviors of the individuals. Haley (1965) has shared Malcolm X's Hajj experience. They author exemplified this experience that had a major impact on his later years as a Muslim.

f. Situational factors influences: Why a person behaves unethically in certain situations or demonstrates a short cut in life? It is because they may see no way out. A director, for example, with a fictitious sales to cover losses within his area of responsibility. According to some researches in the history of Islam, debt is a major reason why individuals behave unethically. In a *hadith* narrated by A'isha (rah): Somebody said to (the Prophet), "Why do you so frequently seek refuge with Allah from being in debt?" The Prophet (saw) answered, "A person in debt tells lies whenever he speaks, and breaks promises whenever he makes (them)" (Sahih Al-Bukhari, 1795). Islam calls lenders to show leniency to the debtors due to the indebtedness is likely to lead to unethical conduct. It is also given as advice by the Prophet whereby the same time, debtors are urged to repay debts promptly.

SUMMARY AND CONCLUSION

One can conclude a root cause is the love for this never lasting world (dunya or materialistic temptations of life) is a key to the problems of unethical conducts. The Prophet of Allah (swt) says; "Nothing fills the stomach of a greedy person, except dust." As saying goes that "Human beings usually grow older, but two things grow younger in them; greed and long hopes". Same time, there are two things are never satisfied such as the seeking and knowledge and looking after the worldly life is a far gone conclusion. The greedy person wants to live longer and does not like to spend anything, even on him/her. One likes on others to contribute to the poor human. But his/her conducts and poorer ethical behaviour keeps them aloof.

Ethics or morality is the core foundation of Islamic worldview. Prophet Muhamad (saw) said: “On the Day of Judgement, nothing weightier than good conduct is placed in the Balance of every man on day of resurrection”. Quran also has warned believers as “do you think that most of them do hear or understand? They are nothing but as cattle; nay, they are straying farther off from the path”. (Qur'an: 25: 44). It is simply what makes a difference between two creatures; human versus animals: thus both the creatures eat and drink; have offspring; possess emotions; two of them move about with free-will and have senses that can differentiate between good and bad. However, we human beings are privileged to have thinking and reasoning attributes (intellect) so that we can act rationally and ethically.

SOLUTIONS AND RECOMMENDATIONS

Human greed, jealousy and discontentment are sources of today's ethical dilemma. It begins due to these aspects seeking solutions in life. There exists spiritual solution for “the lack of contentment, so one must strengthen contentment, remembrance of death, firm belief in Allah's mercy, generosity”. This may begin with act of charity by giving as much as we can and then keep doing it, do not compare yourself with others and no too much focus on worldly affairs. Also to bear in mind that these are sickness spiritually and ultimately punishes no one but the one suffering from it.

To seek bounties and blessings of Allah, work as an act of worship and entitlement to reward commensurate with effort. Reward in the hereafter too gives contentment. Allah enjoins comprehensive excellence [Ihsan] in everything (spiritual & worldly; personal & societal). Refrain from common contemporary business problems e.g. window dressing of financial results, fraud, deception, bribery to get contract etc.

FUTURE RESEARCH DIRECTIONS

In this research we just discuss the Islamic business ethics from theoretical perspective. We have examined the uniqueness, norms and codes of Islamic business ethics. However, the empirical aspects of Islamic business ethics are not discussed. Therefore, it is recommended that the future research may focus the empirical analysis of business ethics by employing behavioral models such Theory of Planned Behavior, Theory of Reasoned Action and among others.

REFERENCES

Adnan, A. (2006). *A Study of Islamic Leadership Theory and Practice in K-12 Islamic School in Michigan*. Retrieved January 30, 2009, from http://contentdm.lib.byu.edu/ETD/image/etd1273.pdf

Al-Buraey, M. A. (2006). *Islamic Principles in the Management of an Organization: A Focus on Leading*, Al-Dharan, Saudi Arabia: King Fahd University. Retrieved August 27, 2008, from http://ocw.kfupm.edu.sa/user062/MGT30100102/Islamic%20Princ%20Focus%20on%20Leadership%20malaysia2003c1.ppt

Arif Hassan. (2019). *Ethics*, Morality, Justice and Behavior. *International Journal of Human Development*.

Asch, S. E. (1951). Effects of group pressure on the modification and distortion of judgments. In H. Guestzkow (Ed.), *Group, leadership and men* (pp. 177–190). Carnegie Press.

Beekun, R. I., & Badawi, J. (1999). *Leadership: An Islamic Perspective*. Amana Publications.

Blader, S. L., & Chen, Y. R. (2012). Differentiating the effects of status and power: A justice perspective. *Journal of Personality and Social Psychology*, *102*(5), 994–1014. doi:10.1037/a0026651 PMID:22229456

Brebels, L., De Cremer, D., Van Dijke, M., & Van Hiel, A. (2011). Fairness as social responsibility: A moral self-regulation account of procedural justice enactment. *British Journal of Management*, *22*, 47–58. doi:10.1111/j.1467-8551.2010.00715.x

Cohen, T. R., Panter, A. T., & Turan, N. (2012). Guilt proneness and moral character. *Current Directions in Psychological Science*, *21*(5), 355–359. doi:10.1177/0963721412454874

Deutsch, M., & Gerard, H. B. (1955). A study of normative and informational social influences upon individual judgment. *Journal of Abnormal and Social Psychology*, *51*(3), 629–636. doi:10.1037/h0046408 PMID:13286010

Dewi, M. K., & Tanjung, H. (2007). *Discipline Without Punishment – Is It Possible?* Paper presented at International Conference, Management from Islamic Perspective, Kuala Lumpur.

James, C. S., & Joseph, C. S. (2001). The transformational – transactional leadership model in practice. *Leadership and Organization Development Journal*, *22*(8), 383–393. doi:10.1108/01437730110410107

Johnson, R. E., Lanaj, K., & Barnes, C. M. (2014). The good and bad of being fair: Effects of procedural and interpersonal justice behaviors on regulatory resources. *The Journal of Applied Psychology*, *99*(4), 99635–99650. doi:10.1037/a0035647 PMID:24446913

Khaliq, A. (2009). Leadership and Work Motivation from the Cross Cultural Perspective. *International Journal of Commerce and Management*, *29*(1), 72–84.

Milgram, S. (1963). Behavioral study of obedience. *Journal of Abnormal and Social Psychology*, *67*(4), 371–378. doi:10.1037/h0040525 PMID:14049516

Milgram, S. (1974). *Obedience to authority: An experimental view*. Harpercollins.

Monin, P., Noorderhaven, N., Vaara, E., & Kroon, D. (2013). Giving sense to and making sense of norms of justice in post-merger integration. *Academy of Management Journal*, *56*(1), 256–284. doi:10.5465/amj.2010.0727

Muenjohn, N. (2008). *Leadership Theories and Concepts: Its Past and Present*. Paper presented South-East Asia Regional Conference: Globalization, Innovation, and Leadership, Kuala Lumpur.

Patient, D. L., & Skarlicki, D. P. (2010). Increasing interpersonal and informational justice when communicating negative news: The role of the manager's empathic concern and moral development. *Journal of Management*, *36*(2), 555–578. doi:10.1177/0149206308328509

Rafik Essa Beekun. (1997). *Islamic Business Ethics*. The International Institute of Islamic Thought.

Schuh, S. C., Zheng, M. X., Xin, K. R., & Fernandez, J. A. (2017). The interpersonal benefits of leader mindfulness: A serial mediation model linking leader mindfulness, leader procedural justice enactment, and employee exhaustion and performance. *Journal of Business Ethics*, 1-19.

Whiteside, D. B., & Barclay, L. J. (in press). When wanting to be fair isn't enough: The effects of depletion and self-appraisal gaps on fair behavior. *Journal of Management*.

Zapata, C. P., Olsen, J. E., & Martins, L. L. (2013). Social exchange from the supervisor's perspective: Employee trustworthiness as a predictor of interpersonal and informational justice. *Organizational Behavior and Human Decision Processes*, *121*(1), 1–12. doi:10.1016/j.obhdp.2012.11.001

Zhao, G., Chen, Y., & Brockner, J. (2015). What influences managers' procedural fairness towards their subordinates? The role of subordinates' trustworthiness. *Journal of Experimental Social Psychology*, *59*, 96–112. doi:10.1016/j.jesp.2015.04.002

ADDITIONAL READING

Al-Aidaros, A. H., & Mohd Shamsudin, F. (2013). Ethics and ethical theories from an Islamic perspective. *International Journal of Islamic Thought*, *4*, 1–13. doi:10.24035/ijit.04.2013.001

Chapra, M. U. (2009). Ethics and economics: An Islamic perspective. *Islamic Economic Studies*, *16*(1), 1–26.

Cropanzano, R., & Stein, J. H. (2009). Organizational justice and behavioral ethics: Promises and prospects. *Business Ethics Quarterly*, *19*(2), 193–233. doi:10.5840/beq200919211

Folger, R., Cropanzano, R., & Goldman, B. (2005). What is the relationship between justice and morality. Handbook of organizational justice, 215, 215-245.

Harsanyi, J. C. (1977). Morality and the theory of rational behavior. *Social Research*, •••, 623–656.

Rupp, D. E., Wright, P. M., Aryee, S., & Luo, Y. (2015). Organizational justice, behavioral ethics, and corporate social responsibility: Finally the three shall merge. *Management and Organization Review*, *11*(1), 15–24. doi:10.1017/mor.2015.8

Zaroug, A. H. (1999). Ethics from an Islamic perspective: Basic issues. *The American Journal of Islamic Social Sciences*, *16*(3), 45–63. doi:10.35632/ajis.v16i3.2102

KEY TERMS AND DEFINITIONS

Adl': It is an Arabic word and it means justice.

Akhirat: It is an Arabic word and it means Life hereafter.

Dunia: It is an Arabic word and it means worldly life temptations.

Falah: It is an Arabic word and it means the ways to achieve eternal life full of success.

Fiqh Muamalat: It is an Arabic word and it means Islamic transactions.

Hurriyah: It is an Arabic word and it means Free will.

Ihsan: It is an Arabic word and it means benevolence and show respect and sympathy to others.

Mas'uliyyah: It is an Arabic word and it means Responsibility.

Shariah: It is an Islamic religious law that regulates not only religious rituals but also aspects of daily life of Muslims.

Tawhid: It is an Arabic word and it means Unity of Command.

Zuhd: It is an Arabic word and it means Free from materialistic temptation.

Zulm: It is an Arabic word and it means injustice and atrocities.

Chapter 17
Universal Principles of Tourism Ethics and the Future of Tourism and Hospitality Ethics

Sağbetullah Meriç
https://orcid.org/0000-0003-1949-623X
Van Yuzuncu Yil University, Turkey

ABSTRACT

Ethics refers to the norms and collectively formed criteria created by a group of people for specific purposes. Ethics is the set of principles that guide a person or group's behaviour, setting the standards of good or bad, right, or wrong. In this section, the issue of ethics in the tourism sector is examined. Secondary sources were used while preparing the chapter. First of all, the place and importance of the tourism sector today and the tourism sector's ethical problems have been stated. The ethical principles determined by the World Tourism Organization were examined by saying that all stakeholders in the tourism sector should adopt and share ethical principles. By stating the problems caused by ethical violations, suggestions for solutions are presented. The necessity of eliminating ethical issues was expressed with the recommendations that set the groundwork for future research in this field. Finally, the general scope and results of the chapter are presented and discussed.

INTRODUCTION

The existence of regulatory and supervisory rules for businesses increases the satisfaction of the employees by ensuring that the jobs in the business are carried out successfully. Employees' perceptions of justice within the organization are an important determinant of behavior within the organization (Güllüce, Bozkurt & Meriç, 2017: 40-45). The prerequisite for an institution's rules to be fair and effective is that the rules are required by all parties and the rules are accepted by all parties. For the rules to be understood and acceptable by the parties, all parties should be included in the process of creating the rules. These rules include acceptable and unacceptable (true and false) situations and behavior. Accept-

DOI: 10.4018/978-1-7998-4117-3.ch017

able rules, that is, behaviors that include how a person can behave correctly and appropriately in a given situation, are ethical rules.

Modern businesses today have quite different features than those established under the influence of the industrial revolution. Modern businesses have become socially responsible institutions by abandoning them only to make profit and continue their lives (Akova & Çalık, 2008: 2). Social responsibility behaviors of businesses as well as their product quality and prices affect customers. Social responsibility event refers to achieving positive results for nature and humanity by applying ethical principles and values. Therefore, the phenomenon of ethics has been an important tool for businesses to survive and regulate their relations with the environment. The purpose of businesses should not only be to make a profit, but to make a common benefit that all humanity can benefit.

Ethics is a set of criteria and values created by a group of people for specific purposes. For this reason, we can come across different views on the scope and definition of ethics (Fennell & Malloy, 1999: 930). One of these views argues that ethics is the behavior and practices accepted and applied by all circles. According to another view, ethics is a set of behaviors that benefit individuals and communities. In this view, it is the determinant of ethical behavior that the results of behavior benefit people and society. In this way, although there are different opinions and thoughts, there are some generally accepted definitions. According to one of them, ethics refers to the values that require people to take a conscientious responsibility for what they do and make them feel happy if they are adopted and applied.

The main subject of ethics is that everyone performs their duties in the best way in the interests of humanity and does not harm anyone. Accordingly, there are ethical principles that are accepted and applied in all sectors. Ethical values, which set out a certain order for behaviors and direct people to perform their behavior in accordance with this order, do not only benefit from a certain segment but also the benefit of all people.

Since tourism is an interdisciplinary field, it is in contact with many disciplines. Therefore, there are many segments in tourism that pursue different interests (Arslan Kalay, Şahin & Meriç, 2018: 67). The fact that these segments are directed only to their interests may negatively affect other segments and the tourism movement (Aslan & Kozak, 2006: 51). Ethical behavior is a necessity for tourism activities to gain as much income as possible. In order to ensure ethical behavior in tourism, ethical principles must be known and adopted by all individuals and organizations operating in the field of tourism. It is an important issue to examine the decisions and behaviors in tourism within the scope of ethics.

Tourism is a service-oriented sector. In order to benefit from the service, the interaction must be established between the service provider and the person receiving the service (Eralp, 1983: 35). This feature brings the tourism sector to an important position in ethics. Thanks to tourism, many people who do not know each other can interact with each other. Ethics also covers this interaction in the tourism sector due to its aspect that covers and evaluates human behavior. For this reason, it is an important requirement to address the behaviors that may exist or occur in tourism in terms of ethics.

In this section, it is aimed to discuss the ethical conformity of existing rules and policies in tourism by considering the concept of ethics within the scope of tourism. First of all, the definition and history of the concept of tourism and ethics are presented in the background. Later, ethical violations in the tourism sector were specified and solutions and suggestions were presented. Finally, the general scope of the department is discussed by determining the opinions and thoughts about the future of ethics in tourism and suggestions about future research opportunities are presented.

BACKGROUND

Tourism and Hospitality

Human beings have been traveling and changed places in the world since the day they existed. As the seasons changed and the animals migrated, people traveled to survive. Generally, these trips took place at short distances (Cook, Hsu & Marqua, 2014: 58-60). This displacement event, which is a troublesome occupation and occurs for many reasons, has become more effortless with the invention of the wheel. Thanks to the developing technology, people can travel thousands of kilometers within hours.

Tourism was expressed by philosophers of ancient Greek civilization as an attempt to travel and relocate concerning the concept of leisure. In Greek society, slaves and people from the substrate dealt with the needs of daily work, while Greek men dealt with increasing leisure time (Page, 2011). It is known that the tourism event in the modern sense has intensely and prominently appeared in Greek society. It is known that many people came to the region as tourists with the start of the Olympic games (Kozak, 2012). In addition to the ancient Greek civilization, it can be mentioned about the activities that can be considered within the scope of tourism in Roman and Egyptian civilizations.

Some travelers people also reveal the importance of the tourism event in the Middle Ages. Many discoveries were made during this period by traveling. With the increase in travel opportunities in the Middle Ages, tours involving especially cultural, educational and pleasure travels with the participation of individuals belonging to rich, aristocratic and privileged classes started. These tours, referred to as "Grand Tour", employed 5% of the population in the UK at that time (Lickorish & Kershaw, 1958). Today, according to World Tourism Organization data, 1.4 billion people have participated in international tourism activities. These people spent a total of $ 1.7 trillion (UNWTO, 2019). Considering the tourism mobility in the country, it can be said that the tourism event that emerged with the concept of leisure has turned into a big industry.

The tourism event that took place due to reasons such as trade, religion and health has reached its contemporary form during the Industrial Revolution period. Many changes took place in this period, when production increased and important discoveries in terms of human history were introduced. Thanks to radical changes in technology and human life, the tourism event has also improved. It is not a coincidence that the first package tour known in history was carried out in this period and the enterprises that are working on tourism started to be established in this period (Kozak, 2012: 40).

Most people know easily where to buy goods such as shoes, textbooks and computers. However, he does not know where to spend tourism and where to find tourism products. Because in the tourism industry, there are services, not goods. Services are often accompanied by products that qualify as goods (Cook, Hsu & Marqua, 2014: 80-82). These goods, which are concrete items that accompany or support the service offered in tourism activities, support the purchased service.

Tourism is a service sector focused on meeting the needs of travelers. The service can be expressed as a system where product items are created and delivered to the customer (Page, 2011). People need a wide range of services while traveling. These services can include car rental, plane tickets, accommodation, recreation activities, food and beverage, entrance tickets to the ruins, everything that needs to be done and seen (Cook, Hsu & Marqua, 2014). Tourism-oriented businesses that offer this service offer the same service to those who are excluded from those who need to travel and need to have fun and rest. People living in areas with these businesses may also request such services. Considering this aspect, it can be said that tourism businesses concern people who are not involved in the tourism event.

This situation of the tourism sector makes tourism the main factor of economic development, especially for growing and developing countries (Bozkurt & Altundaş, 2017).

Tourism is defined as all of the events and relations arising from the journey and accommodation of the individual, provided that they do not have any income-generating efforts and do not turn into permanent stay (Eralp, 1983). Tourism is a product of events and relationships that occur under certain conditions. As it is understood from the definition, although tourism is an activity that includes entertainment and recreational travel, it is an event that enables people to get to know different cultures and interact with people from different cultures. With technology, which contributes significantly to the virtual elimination of time and distance, the possibility of interaction with people has increased. Tourism, one of the largest industries in the world, leads millions of people who do not meet regularly to interact either formal or informal (Fennell & Malloy, 2007).

Tourism, which has been developing with different aspects since the day of human existence, continues to grow and develop today. Many factors are effective in the development of tourism. Some of the factors affecting the development of tourism (Kozak, 2012: 41) can be listed as follows;

Increasing Free Time: The decrease in the time people have to work provides people the opportunity to do activities outside of work. People who had to work 14 hours in the past continue their lives by working under 8 hours today. This decrease in working time played an important role in the development of tourism.

Right to Paid Vacation: Today, thanks to the right to paid vacation in many countries, people working eleven months of the year can allocate one month of the year to them without any interruption in the fee they receive. This situation causes people to have free time without financial difficulties. Considering that most of the people spend their free time in the tourism event today, the right to a paid holiday is an important development affecting the development of tourism.

Technological Developments: Developing technology provides many convenience to people, especially transportation. With the ships developed thanks to technology, large masses can participate in tourism much cheaper. Developments in travel vehicles such as trains and planes have contributed to the transportation being quite fun and comfortable. Also, people who do their jobs in a shorter time with technology increase their spare time.

Increasing Income Level: For people to participate in tourism activities, they need to make more than the income to meet the compulsory requirements. It is difficult for people to participate in tourism activities if they do not have the economic power to evaluate leisure time. It used most of the income generated in the past to meet people's physiological needs. Today, physiological needs correspond to a very small proportion of human income. Tourism activity has led to the emergence of a new trend that has emerged as a healthy individual for a longer period of time. Geriatrics and third age tourism can be given as an example.

Urbanization and Population Growth: With the development of urbanization and industry, cities have become difficult to live for people. Factories built near the cities pollute the air and the area in the cities is gradually decreasing. This situation motivates people to participate in tourism activities. The business advantages arising from urbanization are an important factor in increasing the population.

Prolonging Human Life: The health sector has developed in parallel with many developments from past to present. Many diseases that are fatal to humans can be treated today. This situation has also led to the prolongation of the average human life. The fact that people are healthy after retirement makes it easier for them to participate in tourism. Pensioners who have enough free time and income make up a large part of the tourism event.

Social Security: Due to the fact that the person experiences a situation that will prevent him from working permanently or temporarily, his wage cut will make life difficult for his family and himself. Social security is a whole of institutions that are in contact with each other and aim to secure the future of people. People lacking social security practices, which are mandatory in many countries, cannot participate in tourism activities.

Freedom of Travel: Considering traveling as a right and not restricting people to use these rights is called freedom of travel. Visa applications, which are applied mutually or unilaterally between countries, are one of the most important obstacles restricting freedom of travel. With the agreements made between countries, many people in the world have the opportunity to travel freely to different parts of the world.

Tourism Awareness: The decision to participate in tourism is usually due to the individual impulses of the people. The opportunities offered by tourism such as entertainment, rest, innovation, adventure, interaction make people want to participate in tourism. These features have made tourism a necessity by removing it from being a luxury for people. This awareness causes people to participate more in tourism activities.

Increasing the Level of Education: Education is an important determinant in decisions about participation in tourism because of the values and information that education adds to people. The fact that the education level of the people participating in tourism is higher than those who do not, indicates that the increase in the education level positively affects tourism. Besides, many people travel from where they live to learn languages and study in better schools.

Tourism and Hospitality Ethics

Freedom has always been an important component of tourism. People can leave their routine jobs and go anywhere they want. People who act as they wish can get rid of the moral cloak that regulates their behavior in daily life thanks to tourism (Fennell & Malloy, 2007). However, these people do not want to be treated well and to be deceived by the businesses they serve. The ethical issue caused controversy in tourism due to this dilemma (Jamal, 2019). However, the ethical principles that regulate this situation aim to protect the interests of the vacationer in the people who are on holiday. It does not include the rules that restrict the freedom and ethics of ethical people, but the rules that support people to have fun and be happy without harming others.

Since laws are based on ethical rules, legal rules and ethical principles generally overlap. Ethical behavior is not just about legal responsibilities. A legally non-criminal behavior may not be ethical either. Ethical behavior may not be included in the law (Yarcan, 2007: 34). Most researches on tourism focuses on the causes and consequences of human behavior. This shows that business ethics in tourism is a subject that attracts researchers' attention. The business ethics that emerge with the application of ethics to business life includes opinions that suggest that the issues concerning business life should be evaluated within the framework of general ethical rules (Akan, 2007). Business ethics includes decision-makers following the right processes in allocating justice for employees (Güllüce, Özer & Erkılıç, 2015), that is, organizational justice.

Business ethics examines the appropriateness of behavior between employees and managers in terms of ethical principles (Somers, 2001). In addition, it covers the compliance of the relationship between the purchaser and the service segment with ethical rules (Sentürk, Bayirli & Güclü, 2016). Business ethics addresses all business-related activities. The fact that standardization is difficult in the services within the scope of tourism and that this process is carried out by human hands makes business ethics important. Especially, compliance of the personnel who are in direct contact with the customer increases the service quality and customer satisfaction (Bektaş, 2008).

In the 1980s, environmental concerns increased, especially in western societies, as environmental issues such as the destruction of rainforests, the destruction of the ozone layer and global warming became media issues. In the mid-1990s, 35% of respondents in a UK study stated that the environmental problem is the most important. This rate is higher than the unemployment and inflation problem (Holden, 2003). Negative environmental impacts such as the evacuation of wetlands, which increased with the development of tourism, the destruction of coral reefs, the shrinking of forests and the increase of pollution, put pressure on stakeholders to question their positions according to nature. Considering the success of the tourism industry depends on the use of the natural environment, questioning the ethical values of the stakeholders towards nature has become an important issue.

Environmental ethics, which questions how human behavior towards nature should be, has increased its importance due to environmental damage that has increased in recent years (Smith & Duffy, 2004). This problem, which arises from the massive nature of tourism activities, poses great risks especially in underdeveloped and developing countries (Aslan & Kozak, 2006: 49). According to environmental ethics, animals should be given the right to uninterrupted freedom of existence. Animals should have the same rights as humans. This ethic applies the principle of individual rights not only to animals but also to all ontological entities and objects (such as trees, plants, and rocks) (Nash, 1989). Based on environmental ethics, there is an obligation to respect the environment and nature. Human behavior can become compatible with environmental ethics by adopting the idea that human beings are part of nature, rather than being an entity consuming nature.

Rules are the principles required to minimize the negative effects of a developing sector. The ethical principles that will be universally accepted and applied in tourism, and which were introduced by the United Nations World Tourism Organization in the field of ethics in tourism, constitute the most important rules in this respect (Ruiz-Lozano, De-los-Ríos-Berjillos & Millán-Lara, 2018). The Universal Principles of Tourism Ethics are rapidly increasing and have been put in place in order to provide the most benefit

Table 1. Universal Principles of Tourism Ethics

Principles	
Article 1	Tourism serves mutual understanding and respect between individuals and communities.
Article 2	Tourism is a personal and collective rest, entertainment, development and recognition of other cultures
Article 3	Tourism serves sustainable development
Article 4	Tourism benefits from the common cultural heritage of humanity and contribute to its development.
Article 5	Tourism serves the benefit of host countries and people of the country
Article 6	In the host country, individuals and organizations that play a role in the production and delivery of the service, directly and indirectly, have responsibilities in the development of tourism
Article 7	Tourism is a right for all citizens of the world. This right should be evaluated within the framework of a basic right as a right to rest and have fun.
Article 8	The principle of freedom must be adhered to in the circulation of tourists.
Article 9	The rights of investors and employees should be recognized and protected in the tourism sector
Article 10	Stakeholders in the private and public sectors within the tourism sector should adopt the Universal Principles of Tourism Ethics and contribute to their effective implementation.

*Source: (*UNWTO, 2019*)*

for the destination of developing tourism activities and to minimize the negative environmental and cultural impacts that these activities may cause (Akan, 2007: 7). These principles are essential for achieving the benefits of tourism in an egalitarian and sustainable way, which extends to all segments of society.

The first article of the universal principles of tourism emphasizes the importance of adhering to the common ethical values of humanity and respecting cultural and religious differences. Tourism activities should take place within the traditions and rules of the host country and be in accordance with the law. The host country should also respect the lifestyle, taste, and expectations of the tourists.

The second article reveals the role that tourism plays in developing, resting, having fun and connecting with other cultures. It is an important opportunity when it is realized for tourism purposes. Besides, tourism should contribute to the individual rights of disabled, elderly and ethical minorities.

The third article states that development should be sustainable and tourism serves this. This article underlines the importance of protecting natural balance and diversity in the environment.

The subject of the fourth article is the protection and development of cultural assets. This article emphasizes the importance of using some of the revenues from touristic areas for the protection and development of these assets. These works do not belong to a certain segment, they are the common heritage of human beings.

The fifth article emphasizes that tourism is an activity that serves the benefit of the people of the host country. This article emphasizes the necessity of reflecting the social, economic and cultural benefits obtained from tourism to the local people. Indirect and direct employment, especially created by tourism, is an important benefit for the host country.

The sixth article emphasizes the responsibilities of individuals and organizations that directly and indirectly play a role in the production and delivery of tourism products. Managers are required to provide the infrastructure in the field opened to tourism and to convey information about these fields in a realistic way. It is also the responsibility of local administrators to ensure the safety of life and property of the people in the region visited. Media outlets should be realistic and impartial in news on tourism.

The seventh article emphasizes that tourism is not only an activity of certain people but that the right to tourism is a common right of all citizens of the world. This article states that the factors that will prevent tourism should be eliminated and the opportunity of tourism should be offered equally to all the people of the world.

The eighth article emphasizes the need to adhere to the principle of freedom in touristic trips. With this article, it is stated that tourists should not be treated differently than the citizens of the country in matters such as accommodation, sightseeing, eating and drinking.

Article nine concerns the rights of tourism workers and investors. Employees on issues such as social rights, job security, and in-service training; Investors have the right to a fair share of tourism income like other stakeholders in the country.

The tenth article contains the necessary arrangements to ensure that the Universal Principles of Tourism Ethics are announced to stakeholders and implemented. The public and private sectors must be in cooperation for the correct and effective implementation of these articles. It is emphasized in the article that all tourism-related groups in the world should get to know non-governmental organizations and international organizations such as the World Tourism Organization.

SOLUTIONS AND RECOMMENDATIONS

Private industries such as tourism have their problems and concerns. Therefore, unique ethical issues in tourism should be evaluated with a special perspective. The tourism industry is constantly developing new products to meet ever-growing consumer demands (Walle, 1995: 264). It is very important for the tourism industry that is in a rapid change to come up with solutions that will adapt to this change.

The tourism industry, which has grown rapidly in recent years, seems to continue to grow under the influence of technological developments. Although there are advantages and benefits of technological developments, some tourism businesses are extremely reluctant to apply technological innovations. Fear of losing human relationships, which has been the basis of hospitality, has been expressed as the main reason for this reluctance (Cook, Hsu & Marqua, 2014). For tourism to develop by benefiting societies and people, ethical rules and principles should be developed accordingly. In addition, it should be noted that the changes that will occur in tourism are technologies and practices that will direct people to act ethically. In cases where the control is left to people through inspections at certain intervals, people behave ethically, even with the fear of punishment. Since this behavior is repeated for a long time, it becomes a habit over time. As a result, people show this behavior not because of fear of punishment, but because they are used to behaving ethically. For example, when a person who is afraid of getting punished performs the behavior of not throwing garbage in the environment for a long time, he makes it a habit and does not throw garbage in the environment.

In a study examining 362 business ethics articles published between 1995 and 2014, it is stated that there has been a significant increase in the number of business ethics articles in recent years. There are many studies in the literature such as sexual harassment, fraud, employee privacy and theft of company property (Köseoglu, Sehitoglu, Ross & Parnell, 2016: 1603). However, since many ethical concepts are heavily influenced by culture, these studies need to be increased.

States intervene directly or indirectly by supporting organizations that promote tourism. Tourism becomes an activity that harms the environment and society if it is not controlled (Page, 2011). The most important stage of ensuring control is to establish laws and policies that will ensure that these activities operate in accordance with ethical principles. By ensuring that these laws and policies comply with universal ethical principles and punish those who do not comply with these rules, it can be ensured that tourism activities are carried out in accordance with the ethical concept.

Since the tourism sector has a complex structure, it is quite difficult to create homogeneous environmental ethics for the sector (Selvi & Şahin, 2012). However, government policy has a major impact in establishing the main framework for stakeholder interaction with the environment. For example, the first of the government policies within the scope of tourism in Spain was initiated by General Franco in the 1950s. The environment has been neglected and damaged due to these policies built on growth at all costs. Due to this policy, there has been a serious decrease in the number of tourists in Spain as it has lost its environmental quality (Holden, 2003). Therefore, when determining tourism policies, it should be questioned by the government whether these policies comply with environmental ethics.

In order to ensure environmental ethics, it is necessary to have information not only about tourists but also about all stakeholders in tourism activities. It is necessary to know that a clean and healthy environment is necessary to advance tourism (Mason & Mowforth, 1996). Protecting the environment and nature is not something to be boasted, it is a must for people to do. If this obligation is not applied, the rights of plants, animals, and people using this area like us are usurped.

FUTURE RESEARCH DIRECTIONS

Since the ethics in the tourism sector generally cover different geographies and stakeholders, it contains more general principles than the rules of profession or business (Akan, 2007). Therefore, it is not easy to determine ethical principles and adopt everywhere the tourism event takes place. Therefore, the issue of ethics in tourism is a subject that is seriously handled by researchers. This subject is thought to be a source of many kinds of research in the future. The subject of ethics is important for tourism research as well as for tourism. According to the research on this subject (Frechtling, 2017), in many tourism ethics searches published in the field of social sciences, the universal principles of tourism have not been followed.

The most important segment for ethical behavior to become widespread and adopted in the tourism sector is those who have decision-making and sanction power (white-collar) in the sector. Although the state intervenes on some issues, the tourism sector is a sector dominated by the capitalist system. When the researches on ethics in tourism are examined, no research on the ethical behavior of the white-collar has been found. Therefore, researchers have the opportunity to address important questions about the ethical behavior of white-collar workers.

Another research opportunity related to tourism and ethics is related to geography. Most of the studies published in this area have been prepared by researchers in developed countries. Most articles cover examples of work done in tourism businesses in developed countries. Since there are more problems with ethics in underdeveloped or developing countries, researchers should make researches in these areas and offer solutions. These searches will contribute to the identification of ethical problems and the elimination of ethical problems. In addition, researchers in underdeveloped or developing countries use research from developed countries to identify and solve their problems. Researching with the same measurement techniques in different research areas may prevent the research from reaching the targeted results.

Most of the business ethics researches conducted in the tourism sector has been done within the scope of quantitative methods. These studies provide information on ethical violations. However, since behaviors will only fall within the scope of ethics and will be insufficient, there is a need for studies that reveal the reasons for unethical behavior.

Ethnographic studies involving scenario management can be carried out to develop conceptual frameworks through research in this area. Researchers researching business ethics should consider creating research models to include these methods.

CONCLUSION

The ethical values of academics, consultants, planners, politicians, and other developers should be taken into account to encourage research, planning, and development that fit tomorrow's needs. In other words, for the future to be more livable, there is a need to question and understand the ethical and moral status of all people, especially those with powers and sanctions. It is a simple necessity to lead a moral life. Otherwise, the world will turn into an environment of confusion where strong people exploit weak people and people are unhappy.

It is inevitable to experience ethical problems in the tourism sector where the service delivery, which is knitted and abstract with human behavior, takes place. Although it is not possible to eliminate this problem, it should be reduced as much as possible. The principles set by businesses in the tourism sector or specific stakeholders are insufficient in solving ethical problems. Global comprehensive ethical principles are needed to reduce existing problems as much as possible. Tourism universal ethical principles determined by the World Tourism Organization are clear and inclusive. However, these principles are not known to many people. These principles should be known and applied to all segments related to tourism. Compliance of local principles and rules applied within the scope of tourism with these universal rules should be investigated.

Tourism policy and planning decisions affect nature and human society. Therefore, it is unacceptable to make these decisions only for political reasons. In addition, the determination of these decisions only by government officials independent of the stakeholders in the tourism sector prevents the decisions from delivering the desired results. Decisions serve only short-term interests, without an ethical stance that is clearly articulated and expressed. For this reason, the results of the decisions taken by the state should be discussed and opinions should be taken from stakeholders. For the decisions to be decisions that will benefit all parties, the advantages of these decisions and the negative situations that they will create should be discussed.

Businesses are part of their environment. While in this social environment, the distance of businesses from ethical behavior can undermine the image of the business and the trust of people in the business. Getting a bad image of the business in the social system reduces trust in the business. The loss of trust has negative consequences for businesses. For this reason, ethics is necessary for the business environment as well as in personal relationships. Adopting the ethical principles of businesses and offering products and services within the framework of these principles will be to the advantage of businesses and increase customer satisfaction. In addition, when the behaviors of the tourism employees and the compliance of these behaviors with ethical principles are questioned, significant developments will be achieved within the scope of business ethics.

The fact that increasing travel and entertainment activities with tourism have consequences that destroy the environment causes criticisms towards the tourism industry. Although tourism activities are largely dependent on the environment, it is ironic that tourism causes environmental pollution. Some non-governmental organizations accuse tourism stakeholders of not acting ethically towards the environment, while others see tourism as an opportunity for environmental protection. In recent years, the non-governmental organization, which defends that tourism is a harmful activity for the environment and wants to ban this activity due to its harm, has almost decreased. Thanks to the economic opportunities obtained from tourism, initiatives to support environmental protection can be financed. Besides, the threat of a partial area to be used within the scope of tourism and other development activities such as agriculture can be eliminated. When tourism is planned and implemented in a way that does not harm the environment, it can provide important advantages in protecting the environment and repairing existing damages.

REFERENCES

Akan, P. (2007). Codes of Ethics in Business and the Implementation of Global Code of Ethics for Tourism. *Anatolia: Turizm Araştırmaları Dergisi*, *18*(1), 7–20.

Akova, O., & Çalık, İ. (2013). A Research on Ethical Perceptions of Students Receiving Tourism Education. *The Social Science Journal*, *1*, 1–13.

Arslan Kalay, H., Şahin, S., & Meriç, S. (2018). The Effect Of Usıng Archıtectural Heritage In Tourism OnThe Conservation Of Archıtectural Heritage: A Study On The Students Of Art Hıstory And Tourısm. Ordu University Journal of Social Science Research, 8(1), 65-78.

Aslan, A., & Kozak, M. (2006). Tourism Development and Ethical Problems: A Research on University Students. *Ege Academic Review*, *6*(1), 49–61.

Bektaş, Ç. (2008). Contrıbutıons of Ethıc Codes to Hotel Industry and A Fıeld Investıgatıon. *Journal of Commerce & Tourism Education Faculty*, *1*(1), 1–17.

Bozkurt, Ö., & Altuntaş, Y. (2017). SWOT Analysis for Determining the Tourism Potentials of Certain Regions: Akçakoca, *Research Journal of Politics. Economics and Management*, *5*(1), 89–99.

Cook, R., Hsu, C. H., & Marqua, J. (2014). *Tourism: The Business of Hospitality and Travel*. Pearson Press.

Eralp, Z. (1983). *General Tourism*. Basin-Yayin Yüksek Okulu Press.

Fennel, A., & Malloy, D. C. (1999). Measuring the Ethical Nature of Tourism Operators. *Annals of Tourism Research*, *26*(4), 928–943. doi:10.1016/S0160-7383(99)00032-8

Fennell, D. A., & Malloy, D. (2007). *Codes of Ethics in Tourism: Practice, theory, synthesis*. Channel View Publications. doi:10.21832/9781845410629

Frechtling, D. C. (2018). On the Ethics of Tourism Research. *Journal of Travel Research*, *57*(8), 1054–1067. doi:10.1177/0047287517729756

Güllüce, A. Ç., Bozkurt, Ö., & Meriç, S. (2017). Relationship Between Organizational Justice Actions and Internal Entrepreneurship: A Research on Hotel Managers. *Girişimcilik ve Kalkınma Dergisi*, *12*(2), 39–50.

Güllüce, A. Ç., Özer, S., & Erkiliç, E. (2015). The Effect of Organizational Justice Perception on Organizational Commitment among Healthcare Sector Employees. *Journal of Business and Management*, *4*(3), 16–25.

Holden, A. (2003). In Need of New Environmental Ethics for Tourism? *Annals of Tourism Research*, *30*(1), 94–108. doi:10.1016/S0160-7383(02)00030-0

Jamal, T. (2019). *Justice and Ethics in Tourism*. Routledge Press. doi:10.4324/9781315162942

Köseoglu, M. A., Sehitoglu, Y., Ross, G., & Parnell, J. A. (2016). The Evolution of Business Ethics Research in the Realm of Tourism and Hospitality: A Bibliometric Analysis. *International Journal of Contemporary Hospitality Management*, *28*(8), 1598–1621. doi:10.1108/IJCHM-04-2015-0188

Kozak, M. (2012). *General Tourism Information*. Anadolu Üniversitesi Press.

Lickorish, L. J., & Kershaw, A. G. (1958). *The Travel Trade*. Practical Press.

Mason, P., & Mowforth, M. (1996). Codes of Conduct in Tourism. *Progress in tourism and hospitality research*, *2*(2), 151-167.

Nash, R. (1989). *The Rights of Nature: A History of Environmental Ethics*. The University of Wisconsin Press.

Page, S. (2011). *Tourism Management*. Routledge Press. doi:10.4324/9780080879369

Ruiz-Lozano, M., De-los-Ríos-Berjillos, A., & Millán-Lara, S. (2018). Spanish Hotel Chains Alignment with the Global Code of Ethics for Tourism. *Journal of Cleaner Production, 199*, 205–213. doi:10.1016/j.jclepro.2018.07.133

Selvi, M. S., & Şahin, S. (2012). Sustainable Tourism: Local Government Perspective The Example of Western Blacksea Region. *International Journal of Social and Economic Sciences, 2*(2), 23–36.

Sentürk, F. K., Bayirli, M., & Güçlü, C. (2016). The Effects of Islamic Work Ethics and the Ethics of Love of Money over the Perception of Smuggling:The Example of the Alanya Tradesmen. *Turkish Journal of Business Ethics, 9*(2), 208–220.

Smith, M., & Duffy, R. (2004). *The Ethics of Tourism Development*. Routledge Press.

Somers, M. J. (2001). Ethical Codes of Conduct and Organizational Context: A Study of the Relationship Between Codes of Conduct, Employee Behavior and Organizational Values. *Journal of Business Ethics, 30*(2), 185–195. doi:10.1023/A:1006457810654

UNWTO. (2019). *World Tourism Organization*. https://www.e-unwto.org/doi/pdf/10.18111/9789284421152

Walle, A. H. (1995). Business Ethics and Tourism: From Micro to Macro Perspectives. *Tourism Management, 16*(4), 263–268. doi:10.1016/0261-5177(95)00015-G

Yarcan, Ş. (2007). A Conceptual Evaluation of Code of Ethics for Professional Tourist Guides. *Anatolia: Turizm Araştırmaları Dergisi, 18*(1), 33–44.

ADDITIONAL READING

Akan, P. (2007). Codes of Ethics in Business and the Implementation of Global Code of Ethics for Tourism. *Anatolia: Turizm Araştırmaları Dergisi, 18*(1), 7–20.

Fennell, D. A., & Malloy, D. (2007). *Codes of Ethics in Tourism: Practice, theory, synthesis*. Channel View Publications. doi:10.21832/9781845410629

Frechtling, D. C. (2018). On the Ethics of Tourism Research. *Journal of Travel Research, 57*(8), 1054–1067. doi:10.1177/0047287517729756

Holden, A. (2003). In Need of New Environmental Ethics for Tourism? *Annals of Tourism Research, 30*(1), 94–108. doi:10.1016/S0160-7383(02)00030-0

Köseoglu, M. A., Sehitoglu, Y., Ross, G., & Parnell, J. A. (2016). The Evolution of Business Ethics Research in the Realm of Tourism and Hospitality: A Bibliometric Analysis. *International Journal of Contemporary Hospitality Management, 28*(8), 1598–1621. doi:10.1108/IJCHM-04-2015-0188

Page, S. (2011). *Tourism Management*. Routledge Press. doi:10.4324/9780080879369

Sentürk, F. K., Bayirli, M., & Güçlü, C. (2016). The Effects of Islamic Work Ethics and the Ethics of Love of Money over the Perception of Smuggling:The Example of the Alanya Tradesmen. *Turkish Journal of Business Ethics, 9*(2), 208–220.

Walle, A. H. (1995). Business Ethics and Tourism: From Micro to Macro Perspectives. *Tourism Management, 16*(4), 263–268. doi:10.1016/0261-5177(95)00015-G

KEY TERMS AND DEFINITIONS

Business Ethics: The type of ethics in a particular profession, covering ethical behavior that everyone who does this profession must comply with.

Environmental Ethics: The type of ethics aimed at realizing human behavior against nature within the framework of ethical principles.

Tourism Businesses: Enterprises serving in the tourism sector in areas such as accommodation, entertainment, food, beverage, and travel.

Universal Principles of Tourism Ethics: Principles, which were announced by the United Nations World Tourism Organization in 2001, aimed at achieving appropriate behavior in tourism.

Compilation of References

ACOG Commıttee Opınıon Summary. (2019). Ethical Considerations for the Care of Patients With Obesity. *American College of Obstetricians and Gynecologists, 133*(1), 231-232.

Adkins, A. W. H. (1984, February). The connection between Aristotle's ethics and politics. *Political Theory, 12*(1), 29–49. doi:10.1177/0090591784012001003

Adnan, A. (2006). *A Study of Islamic Leadership Theory and Practice in K-12 Islamic School in Michigan*. Retrieved January 30, 2009, from http://contentdm.lib.byu.edu/ETD/image/etd1273.pdf

Aghazadeh, S.-M. (2003). The Future of Human resource Management. *Work Study, 52*(4), 201–207. doi:10.1108/00438020310479045

Ahammad, T. (2017). Personnel Management to Human Resource Management (HRM): How HRM Functions? *Journal of Modern Accounting and Auditing, 13*(9), 412–420.

Ahn, D., & Shin, D. H. (2013). Is the social use of media for seeking connectedness or for avoiding social isolation? Mechanisms underlying media use and subjective well-being. *Computers in Human Behavior, 29*(6), 2453–2462. doi:10.1016/j.chb.2012.12.022

Akan, P. (2007). Codes of Ethics in Business and the Implementation of Global Code of Ethics for Tourism. *Anatolia: Turizm Araştırmaları Dergisi, 18*(1), 7–20.

Akarsu, B. (1982). *Ahlak Öğretileri*. Remzi Kitapevi.

Akdemir, B. (2005). Küresel rekabet ortamında insan kaynakları yönetiminin değişen rolü: Stratejik insan kaynakları yönetimi. *Sosyal Siyaset Konferansları Dergisi*, (49).

Akdoğan, A., & Demirtaş, Ö. (2014). The Impact of Ethical Leadership Behaviors on Ethical Climate: The Mediating Role of Organizational Political Perceptions. *Afyon Kocatepe University Journal of Economics and Administrative Sciences, 16*(1), 107–124. doi:10.5578/jeas.7819

Akgeyik, T. (2009). *Türkiye'de iş etiği: İnsan kaynakları yönetimi boyutuyla*. ITO Yayınları.

Akman, G. (2019). *Fazla kilolu ve obez bireylerde beden algısı, benlik saygısı ve algılanan sosyal desteğin yaşam doyumuyla ilişkisinin incelenmesi* (Unpublished Master's thesis). University of Maltepe, Istanbul, Turkey.

Akmeşe, E. (2017). Evrensel Bir Simgenin Sıradışı Öyküsü: Chevolution. In H. Kuruoğlu & A. F. Parsa (Eds.), *Belgesel Filmde Zamanın Ruhu: Belgesel Filmde Değişen Anlam ve Anlatım* (pp. 117–137). Detay Publications.

Akmeşe, Z. (2017). *Televizyonda Çerçeveleme ve Toplumsal Bakış Açısının Oluşturulması*. Yayınlanmamış Doktora Tezi, İstanbul Üniversitesi Sosyal Bilimler Enstitüsü.

Akova, O., & Çalık, İ. (2013). A Research on Ethical Perceptions of Students Receiving Tourism Education. *The Social Science Journal*, *1*, 1–13.

Aksoy, B., & Koçancı, M. (2019). Labor Relations and Ethics: A Study on the Mainstream Human Resources Management Perspective. *Suleyman Demirel University Journal of Faculty of Economics & Administrative Sciences*, *24*(2).

Akyol, M., & Kılınç, Ö. (2014). Internet and Halal Tourism Marketing. *Electronic Turkish Studies, 9*(8).

Alalwan, A. A. (2018). Investigating the impact of social media advertising features on customer purchase intention. *International Journal of Information Management*, *42*, 65–77. doi:10.1016/j.ijinfomgt.2018.06.001

Al-Buraey, M. A. (2006). *Islamic Principles in the Management of an Organization: A Focus on Leading*, Al-Dharan, Saudi Arabia: King Fahd University. Retrieved August 27, 2008, from http://ocw.kfupm.edu.sa/user062/MGT30100102/Islamic%20Princ%20Focus%20on%20Leadership%20malaysia2003c1.ppt

Aldrich, T., & Hackley, B. (2010). The impact of obesity on gynecologic cancer screening: An integrative literature review. *Journal of Midwifery & Women's Health*, *55*(4), 344–356.

Ali, A. J., & Al-Owaihan, A. (2008). Islamic work ethic: A critical review. *Cross Cultural Management*, *15*(1), 5–19. doi:10.1108/13527600810848791

Alkan, A., Erdem, R., & Çelik, R. (2016). Discrimative attitudes and behaviour in field of healthcare: A conceptual review. *Hacettepe Sağlık İdaresi Dergisi*, *19*(3), 365–390.

Allaire, Y., & Firsirotu, M. E. (1984). Theories of organizational culture. *Organization Studies*, *5*(3), 193–226. doi:10.1177/017084068400500301

Al-Qaradawi, Y. (2002). *Ethics and values in Islamic economics*. Resalah.

Alserhan, B. A. (2010). Islamic Branding: A Conceptualization of Related Terms. *Journal of Brand Management*, *18*(1), 34–49. doi:10.1057/bm.2010.18

Altun Uğraş, G., Yüksel, S., Erer, M. T. I., Kettaş, E., & Randa, S. (2017). Are Nurses Willing to Provide Care to Obese Surgical Patients? *Bariatric Surgical Practice and Patient Care*, *12*(3), 116–122.

Altun, S., & Ercan, A. (2015). *Üniversite öğrencilerinin obeziteye ilişkin önyargılarının belirlenmesi (Yayınlanmamış Yüksek Lisans Tezi)*. Başkent Üniversitesi Sağlık Bilimleri Fakültesi Beslenme ve Diyetetik Anabilim Dalı.

Altun, S., & Ercan, A. (2016). Obezite Önyargısı. *Turkiye Klinikleri J NutrDiet-Special Topics*, *2*(1), 17–22.

Alvesson, M. (2012). *Understanding organizational culture*. Sage Publications Ltd., doi:10.4135/9781446280072

Amy, N. K., Aalborg, A., Lyons, P., & Keranen, L. (2006). Barriers to routine gynecological cancer screening for White and African-American obese women. *International Journal of Obesity*, *30*(1), 147.

Anchorage Declaration. (n.d.). Retrieved from https://unfccc.int/resource/docs/2009/smsn/ngo/168.pdf

Anderson, E. (1995). *Value in ethics and economics*. Harvard University Press.

Anghel, L. D., Grigore, G. F., & Roşca, M. (2011). Cause-related marketing, part of Corporate Social Responsibility and its influence upon consumers' attitudes. *Amfiteatru Economic*, *13*, 29, 72–85.

Anic, I.-D., Škare, V., & Milaković, I. K. (2019). The determinants and effects of online privacy concerns in the context of ecommerce. *Electronic Commerce Research and Applications*, *36*, 1–10. doi:10.1016/j.elerap.2019.100868

Appelbaum, S. H., Deguire, K. J., & Lay, M. (2005). The relationship of ethical climate to deviant workplace behaviour. *Corporate Governance: The International Journal of Business in Society, 5*(4), 43-55.

Araslı, O. (1979). *Özel yaşamın gizliliği hakkı ve TC Anayasasında düzenlenişi.* Yayınlanmamış Doçentlik Tezi.

Ardichvili, A., Mitchell, J. A., & Jondle, D. (2009). Characteristics of ethical business cultures. *Journal of Business Ethics, 85*(4), 445–451. doi:10.100710551-008-9782-4

Arif Hassan. (2019). *Ethics*, Morality, Justice and Behavior. *International Journal of Human Development.*

Aristotle. (1984). The Complete Works of Aristotle (vols. 1-2, J. Barnes, Ed.). Princeton University Press.

Armağan, A. (2013). *Obez kadınlarda benlik saygısının beden imgesi üzerindeki etkisinde biçimleyici değişkenlerin incelenmesi* (Unpublished doctoral dissertation). University of Okan, İstanbul, Turkey.

Armstrong, D. (1983, January). The fabrication of nurse-patient relationships. *Social Science & Medicine, 17*(8), 457–460. doi:10.1016/0277-9536(83)90051-5 PMID:6879243

Armstrong, G., Stewart, A., Denize, S., & Kotler, P. (2015). *Principles of Marketing*. Pearson.

Armstrong, M. (2006). *A handbook of human resource management practice.* Kogan Page Publishers.

Arnold, D. G., & Bustos, K. (2005). Business, ethics, and global climate change. *Business & Professional Ethics Journal, 24*(1/2), 103–130. doi:10.5840/bpej2005241/26

Arslan Kalay, H., Şahin, S., & Meriç, S. (2018). The Effect Of Usıng Archıtectural Heritage In Tourism OnThe Conservation Of Archıtectural Heritage: A Study On The Students Of Art Hıstory And Tourısm. Ordu University Journal of Social Science Research, 8(1), 65-78.

Arslan, M. (2001). *İş ve meslek ahlakı.* Nobel yayın dağıtım.

Arslan, M. (2001). İş ve meslek ahlakı. Nobel yayın dağıtım. Istanbul, Turkey

Arslan, M. (2005). *Is ve Meslek Ahlaki* [Business and Professional Ethics]. Ankara: Siyasal yayinevi.

Arslan, M., & Berkman, Ü. (2009). Dünyada ve Türkiye'de iş etiği ve etik yönetimi. *Türk Sanayici ve İş Adamları Derneği, Yayın No: TÜSİAD.*

Arslan, M., & Berkman, Ü. (2009). *Turkey and the World Business Ethics and Ethical Management.* TUSIAD.

Arslantaş, S., Binark, M., Dikmen, E. Ş., & Fidaner, I. B. (2012). *Türkiye'de Dijital Gözetim: T.C. Kimlik Numarasından E-Kimlik Kartlarına Yurttaşın Sayısal Bedenlenişi.* https://ekitap.alternatifbilisim.org/turkiyede-dijitalgozetim.html

Asch, S. E. (1951). Effects of group pressure on the modification and distortion of judgments. In H. Guestzkow (Ed.), *Group, leadership and men* (pp. 177–190). Carnegie Press.

Ashkanasy, N. M., Wilderom, C. P., & Peterson, M. F. (2000). *Handbook of organizational culture and climate.* Sage.

Aslan, A., & Kozak, M. (2006). Tourism Development and Ethical Problems: A Research on University Students. *Ege Academic Review, 6*(1), 49–61.

Aswathappa, K. E. M. A. L. (2005). *Human resource and personnel management.* Tata McGraw-Hill Education.

Attfield, R. (1983). *The Ethics of Environmental Concern.* Basil Blackwell.

Attfield, R. (1994). Reasoning about the Environment. In R. Attfield (Ed.), *Environmental Philosophy: Principles and Prospects* (pp. 135–149). Avebury and Brookfield.

Attfield, R. (2003). *Environmental Ethics: An Overview for the Twenty-First Century*. Polity Press.

Audi, R. (2010). The place of ethical theory in business ethics. In The Oxford handbook of business ethics. OUP.

Audi, R. (2009). *Business ethics and ethical business*. Oxford University Press.

Australian Academy of Science. (n.d.). *Population and Environment: A Global Challenge*. Retrieved from https://www.science.org.au/curious/earth-environment/population-environment

Awad, N. F., & Krishnan, M. S. (2006). The Personalization Privacy Paradox: An Empirical Evaluation of Information Transparency and the Willingness to Be Profiled Online for Personalization. *Management Information Systems Quarterly*, *30*(1), 13–28. doi:10.2307/25148715

Ay, C., Kartal, B., & Nardalı, S. (2010). *Pazarlamada Etik Yaklaşımlar*. Beta.

Aydın, E. (2006). Tıp Etiği. Güneş kitapevi, 10-11.

Aydın, E., & Ersoy, N. (1994). The Prıncıple of Justıce In Medıcal Ethıcs. *Turkiye Klinikleri J Med Ethics*, *2*(2), 61–63.

Aydoğan, F. (2004). *Medya ve Popüler Kültür*. Mediacat.

Back, M. D., Stopfer, J. M., Vazire, S., Gaddis, S., Schmukle, S. C., Egloff, B., & Gosling, S. D. (2010). Facebook profiles reflect actual personality, not self-idealization. *Psychological Science*, *21*(3), 372–374. doi:10.1177/0956797609360756 PMID:20424071

Bailey, M., Christensen, J. E., Danis, C. M., Ellis, J. B., Erickson, T. D., Farrell, R. G., & Kellogg, W. A. (2014). *U.S. Patent No. 8,903,847*. Washington, DC: U.S. Patent and Trademark Office.

Balaban, D. C. (2017). Online Communication and PR in Romanian Public Administration. The Case Study of Public Institutions From Transylvania. *Transylvanian Review of Administrative Sciences*, *47*(1), 20–34.

Bansal, G., Zahedi, F. M., & Gefen, D. (2016). Do context and personality matter? Trust and privacy concerns in disclosing private information online. *Information & Management*, *53*(1), 1–21. doi:10.1016/j.im.2015.08.001

Barabasi, A. L. (2003). *Linked: How Everything Is Connected To Everything Else And What It Means*. Penguin Group.

Barr, J., & Cunneen, J. (2001). Understanding the bariatric client and providing a safe hospital environment. *Clinical Nurse Specialist CNS*, *15*(5), 219–223.

Barrett, E. (1999). Justice in the workplace? Normative ethics and the critique of human resource management. *Personnel Review*, *28*(4), 307–318. doi:10.1108/00483489910273633

Barry, C. A., Stevenson, F. A., Britten, N., Barber, N., & Bradley, C. P. (2001). Giving voice to the lifeworld. More humane, more effective medical care? A qualitative study of doctor-patient communication in general practice. *Social Science & Medicine*, *53*(4), 487–505. doi:10.1016/S0277-9536(00)00351-8 PMID:11459399

Barry, C. T., Doucette, H., Loflin, D. C., Rivera-Hudson, N., & Herrington, L. L. (2017). "Let me take a selfie": Associations between self-photography, narcissism, and self-esteem. *Psychology of Popular Media Culture*, *6*(1), 48–60. doi:10.1037/ppm0000089

Bartels, K. K., Harrick, E., Martell, K., & Strickland, D. (1998). The relationship between ethical climate and ethical problems within human resource management. *Journal of Business Ethics*, *17*(7), 799–804. doi:10.1023/A:1005817401688

Baruah, T. D. (2012). Effectiveness of Social Media as a tool of communication and its potential for technology enabled connections: A micro-level study. *International Journal of Scientific and Research Publications*, *2*(5), 1–10.

Başarır, Ç. (2006). *İnsan kaynakları yönetiminde etiksel ilkelerin yerleştirilmesi ve bir uygulama* (Master's thesis). Balıkesir Üniversitesi Sosyal Bilimler Enstitüsü.

Başpınar, N., & Çakıroğlu, D. (2011). *Professional Ethics*. Nobel Publications.

Battour, M. M., Ismail, M. N., & Battor, M. (2010). Toward a halal tourism market. *Tourism Analysis*, *15*(4), 461–470. doi:10.3727/108354210X12864727453304

Baum, G. (2001). *Nationalism, religion, and ethics*. McGill-Queen's Press-MQUP.

Baykal, E. (2019b). Transparency in Financial Communication: Effect of Authentic Leadership. In Handbook of Research on Global Issues in Financial Communication and Investment Decision Making (pp. 110-126). IGI Global.

Baykal, E. (2019e). Transparency in Financial Communication: Effect of Authentic Leadership. In Handbook of Research on Global Issues in Financial Communication and Investment Decision Making (pp. 110-126). IGI Global.

Baykal, E. (2019a). Yenilikçi Örgüt İklimi: İşe Adanma Üzerine Etkisi. *Ekonomi İşletme ve Maliye Araştırmaları Dergisi*, *1*(3), 266–279. doi:10.38009/ekimad.612630

Baykal, E. (2019c). Turkish Type Leadership: Sabri Ulker Example. *The Journal of Social Science*, *3*(6), 425–438.

Baykal, E. (2019d). Boosting Innovative Work Behavior in Organizations Through Absorptive Capacity. *Journal of Organizational Behavior Review*, *1*(1), 16–33.

Baym, N. K., Zhang, Y. B., & Lin, M. C. (2004). Social interactions across media: Interpersonal communication on the internet, telephone and face-to-face. *New Media & Society*, *6*(3), 299–318. doi:10.1177/1461444804041438

Bayraktutan, G., Binark, M., Çomu, T., Doğu, B., İslamoğlu, G., & Telli Aydemir, A. (2012). *Social media in political communication and the use of twitter in the 2011 general elections in Turkey*. Academic Press.

Bayram, S. (2005). *İşletme Etiği ve Çalışanların Şirketlerinde Mevcut Olan Etik Uygulamalara İlişkin Algılarının Ölçümüne Yönelik Bir Araştırma*. Yayınlanmış Yüksek Lisans Tezi, Uludağ Üniversitesi SBE.

Beardwell, J., & Claydon, T. (Eds.). (2007). *Human resource management: A contemporary approach*. Pearson Education.

Beauchamp, T. L., Bowie, N. E., & Arnold, D. G. (Eds.). (2004). Ethical theory and business. Academic Press.

Becker, W. (2010). Ethics in human resources: An exercise involving new employees. *Journal of Human Resources Education*, *4*(4).

Beekun, R. I. (1996). *Islamic business ethics*. International Institute of Islamic Thought.

Beekun, R. I., & Badawi, J. (1999). *Leadership: An Islamic Perspective*. Amana Publications.

Beekun, R. I., & Badawi, J. A. (2005). Balancing ethical responsibility among multiple organizational stakeholders: The Islamic perspective. *Journal of Business Ethics*, *60*(2), 131–145. doi:10.100710551-004-8204-5

Bektaş, Ç. (2008). Contrıbutıons of Ethıc Codes to Hotel Industry and A Fıeld Investıgatıon. *Journal of Commerce & Tourism Education Faculty*, *1*(1), 1–17.

Bektaş, Ç., & Köseoğlu, M. A. (2008). Iş etiği ve iş etiğinin yayilim süreci. *Süleyman Demirel Üniversitesi İktisadi ve İdari Bilimler Fakültesi Dergisi*, *13*(1), 145–158.

Benligiray, S., Geylan, A., & Duman, E. (2010). İnsan Kaynakları Yönetiminin Stratejik Olarak Yönlendirilmesinin Finansal Performansı Etkileyip Etkilemediğinin Analizi. *Anadolu Üniversitesi Sosyal Bilimler Dergisi*, *10*(1), 61–84.

Bennington, L. (2007). HR managers as ethics agents of the state. *Human resource management: Ethics and employment*, 137.

Bentham, J. (1989). *The Collected Works of Jeremy Bentham: Correspondence: Volume 9: January 1817 to June 1820*. Academic Press.

Berdah, C. (2010). Obesity and psychiatric disorders. *Annales Médico-Psychologiques*, *168*(3), 184–190.

Berenbeim, R. E. (2010). Utilizing HR and ethics and compliance collaboration to promote an ethical culture. *Employment Relations Today*, *37*(1), 17–26. doi:10.1002/ert.20281

Berger, S., Denner, M.-S., & Röglinger, M. (2008). *The Nature of Digital Technologies – Development of a Multi-layer Taxonomy*. www.researchgate.net/publication/330881198

Berkman, Ü., & Arslan, M. (2009). Dünyada ve Türkiye'de iş etiği ve etik yönetimi. İstanbul: Tüsiad Yayınları.

Beschorner, T. (2006). Ethical theory and business practices: The case of discourse ethics. *Journal of Business Ethics*, *66*(1), 127–139. doi:10.100710551-006-9049-x

Besler, T., Bilici, S., Buzgan, T., Çakır, B., Çakır, B., Çom, S., Güler, S., Hasbay, A., & Irmak, H. (2010). Türkiye obezite (şişmanlık) ile mücadele ve kontrol programı (2010-2014). In TC Sağlık Bakanlığı Temel Sağlık Hizmetleri Genel Müdürlüğü. Ankara: Kuban Matbaacılık. Sağlık Bakanlığı Yayın.

Beutell, J. M. (2020). *Children's rights and social media: An analysis of TikTok's Terms of Service through the lens of a young user*. Academic Press.

Bhuiyan, Chanmhuri, Mohamad, & Rabiul. (2011). Potenatials of Islamic Tourism: A Case Study of Malaysia on East Coastal Economic Region. *Australian Journal of Basic and Applied Sciences, 5*(6), 1333-1340.

Binark, M., & Çomu, T. (2012). *Sosyal Medyanın Nefret Söylemi İçin Kullanılması İfade Özgürlüğü Değildir!* Retrieved from https://yenimedya.wordpress.com/2012/01/20/sosyalmedyanin-nefret-soylemi-icin-kullanilmasi-ifade-ozgurlugu-degildir/

Binark, M., & Bayraktutan, G. (2013). *Ayın karanlık yüzü: yeni medya ve etik*. Kalkedon.

Bivins, T. (2004). *Mixed media*. Lawrence Erlbaum Associates.

Blackwell, D., Leaman, C., Tramposch, R., Osborne, C., & Liss, M. (2017). Extraversion, neuroticism, attachment style and fear of missing out as predictors of social media use and addiction. *Personality and Individual Differences*, *116*, 69–72. doi:10.1016/j.paid.2017.04.039

Blader, S. L., & Chen, Y. R. (2012). Differentiating the effects of status and power: A justice perspective. *Journal of Personality and Social Psychology*, *102*(5), 994–1014. doi:10.1037/a0026651 PMID:22229456

Blau, P. M., & Scott, W. R. (2003). *Formal organizations: A comparative approach*. Stanford University Press.

Bloom, P. N., Milne, G. R., & Adler, R. (1994). Avoiding Misuse of New Information Technologies: Legal and Societal Considerations. *Journal of Marketing*, *58*(1), 98–110. doi:10.1177/002224299405800108

Boatright, J. R. (2000). Globalization and the Ethics of Business. *Business Ethics Quarterly*, *10*(1), 1–6. doi:10.2307/3857689

Boddington, P. (2017). *Towards a Code of Ethics for Artificial Intelligence*. Springer. doi:10.1007/978-3-319-60648-4

Boediman, E. P. (2017). Halal Lifestyle in Marketing Communication of Tourism and Hospitality. *International Journal of Economic Research*, *14*(4), 429–438.

Boğan, E., Batman, O., & Sarıışık, M. (2016). Helâl turizmin kavramsal çerçevesi ve Türkiye'deki uygulamalar üzerine bir değerlendirme. In *3rd International Congress of Tourism & Management Researches* (pp. 1425-1440). Academic Press.

Bolat, T., & Seymen, O. A. (2003). An Evaluation on the Effects of "Transformational Leadership Style" on the Establishment of Business Ethics in Organizations. Balıkesir University Journal of Social Sciences Institute, 6(9), 59-85.

Bonini, T., & Morello, G. (2014). No news from abroad: A comparative content analysis of news issues on Italian radio. *European Journal of Communication*, *29*(2), 218–229. doi:10.1177/0267323114520782

Bonnefon, J.-F., Shariffs, A., & Rahwan, I. (2015). *Autonomous vehicles need experimental ethics: Are we ready for utilitarian cars?* arXiv:1510.03346.

Bonnefon, J.-F., Shariffs, A., & Rahwan, I. (2016). The social dilemma of autonomous vehicles. *Science*, *352*(6293), 1573–1576. doi:10.1126cience.aaf2654 PMID:27339987

Booth, J. E., Heinz, R. S., & Howe, M. W. (2005). Ethical practice in a corporation: the Allina case. *The Ethics of Human Resources and Industrial Relations*, 229.

Bowie, N. E. (2017). *Business ethics: A Kantian perspective*. Cambridge University Press. doi:10.1017/9781316343210

Bowie, N. E., & Beauchamp, T. L. (Eds.). (2001). *Ethical theory and business*. Prentice Hall.

Bowie, N. E., & Jamal, K. (2006). Privacy Rights on the Internet: Self-Regulation or Government Regulation? *Business Ethics Quarterly*, *16*(3), 323–342. doi:10.5840/beq200616340

Boxall, P., & Purcell, J. (2007). Strategic management and human resources: the pursuit of productivity, flexibility, and legitimacy. *Human resource management: Ethics and employment*, 66.

Bozkurt, Ö., & Altuntaş, Y. (2017). SWOT Analysis for Determining the Tourism Potentials of Certain Regions: Akçakoca, *Research Journal of Politics. Economics and Management*, *5*(1), 89–99.

Bozkurt, V. (2012). *Değişen Dünyada Sosyoloji Temel Kavramlar Kuramlar*. Ekin Publications.

Brabeck, M. (1984). Ethical characteristics of whistle blowers. *Journal of Research in Personality*, *18*(1), 41–53. doi:10.1016/0092-6566(84)90037-0

Bradburn, R. (2001). *Understanding business ethics*. Cengage Learning EMEA.

Brand, M., Young, K. S., Laier, C., Wölfling, K., & Potenza, M. N. (2016). Integrating psychological and neurobiological considerations regarding the development and maintenance of specific internet-use disorders: An Interaction of Person-Affect-Cognition-Execution (I-PACE) model. *Neuroscience and Biobehavioral Reviews*, *71*, 252–266. doi:10.1016/j.neubiorev.2016.08.033 PMID:27590829

Brant, T. (2016). Who should your self-driving car save in a crash? You or pedestrians? *PC Magazine*. https://uk.pcmag.com/electronics/82490/who-should-your-self-driving-car-save-in-a-crash-you-or-pede

Brantner, C., & Herczeg, P. (2013). The life of a new generation: Content, values and mainstream media perception of transcultural ethnic media – An Austrian case. *Communications*, *38*(2), 211–235. doi:10.1515/commun-2013-0012

Braybrooke, D. (1983). *Ethics in the World of Business*. Rowman & Littlefield.

Brebels, L., De Cremer, D., Van Dijke, M., & Van Hiel, A. (2011). Fairness as social responsibility: A moral self-regulation account of procedural justice enactment. *British Journal of Management*, *22*, 47–58. doi:10.1111/j.1467-8551.2010.00715.x

Brickley, J. A., Smith, C. W. Jr, & Zimmerman, J. L. (2002). Business ethics and organizational architecture. *Journal of Banking & Finance*, *26*(9), 1821–1835. doi:10.1016/S0378-4266(02)00193-0

Brinkmann, J. (2001). On business ethics and moralism. *Business Ethics (Oxford, England), 10*(4), 311–319. doi:10.1111/1467-8608.00247

Brinkmann, J. (2002). Business and marketing Ethics: Concepts, Approaches and Typologies. *Journal of Business Ethics, 41*(1/2), 159–177. doi:10.1023/A:1021318710382

Buari, F. (2001). *Public relations or trust strategy*. Aspect Press.

Buchholz, R. A., & Rosenthal, S. B. (1998). *Business Ethics: The Pragmatic Path Beyond Principles to Process*. Prentice Hall.

Büchi, M., Just, N., & Latzer, M. (2016). Caring Is Not Enough: The Importance of Internet Skills for Online Privacy Protection. *Information Communication and Society*, 1261–1278.

Buckler, S. (2017). Corporate Social Responsibility, CSR, Sustainability, Ethics & Governance. In Imagined Communities Incorporated: Corporate Social Responsibility and Value Creation in a Globalised World. Springer International Publishing. doi:10.1007/978-3-319-35083-7_1

Buckley, M. R., Beu, D. S., Frink, D. D., Howard, J. L., Berkson, H., Mobbs, T. A., & Ferris, G. R. (2001). Ethical issues in human resources systems. *Human Resource Management Review, 11*(1-2), 11–29. doi:10.1016/S1053-4822(00)00038-3

Budd, J. W., & Scoville, J. G. (2005). Moral philosophy, business ethics, and the employment relationship. *The ethics of human resources and industrial relations*, 1-21.

Budd, J. W., & Scoville, J. G. (Eds.). (2005). *The ethics of human resources and industrial relations*. Cornell University Press.

Buhler, P. (2002). *Human resources management: All the information you need to manage your staff and meet your business objectives*. Simon and Schuster.

Bulunmaz, B. (2016). Gelişen Teknolojiyle Birlikte Değişen Pazarlama Yöntemleri ve Dijital Pazarlama. *TRT Akademi Dergisi, 1*(2), 347–365.

Burke, R. J., & Ng, E. (2006). The changing nature of work and organizations: Implications for human resource management. *Human Resource Management Review, 16*(2), 86–94. doi:10.1016/j.hrmr.2006.03.006

Burnay, J., Billieux, J., Blairy, S., & Larøi, F. (2015). Which psychological factors influence Internet addiction? Evidence through an integrative model. *Computers in Human Behavior, 43*, 28–34. doi:10.1016/j.chb.2014.10.039

Butt, S., & Phillips, J. G. (2008). Personality and self reported mobile phone use. *Computers in Human Behavior, 24*(2), 346–360. doi:10.1016/j.chb.2007.01.019

Cadwalladr, C., & Graham-Harrison, E. (2018). *Revealed: 50 million Facebook profiles harvested for Cambridge Analytica in major data breach*. https://www.theguardian.com/news/2018/mar/17/all

Çakmak, F. (2019). Medyada Dezenformasyon Sorunsalı ve Manipülasyonun Gücü: İdeoloji ve Söylem Açısından Seçim Haberlerinin Siyasal İletişimdeki Rolü. *Selçuk İletişim Dergisi., 12*(2), 1127–1154.

Callicott, J. B. (1998). The Conceptual Foundations of the Land Ethic. In M. E. Zimmerman, J. B. Callicott, G. Sessions, K. J. Warren, & J. Clark (Eds.), *Environmental Philosophy: From Animal Rights to Radical Ecology* (pp. 101–123). Prentice Hall.

Calvert, S. L. (2008). Children as consumers: Advertising and marketing. *The Future of Children*, ▪▪▪, 205–234.

Caplan, S. E. (2003). Preference for online social interaction: A theory of problematic Internet use and psychosocial well-being. *Communication Research, 30*(6), 625–648. doi:10.1177/0093650203257842

Carson, R. (1962). *Silent Spring*. Houghton Mifflin.

Cascio, W. F. (2015). *Managing human resources*. McGraw-Hill.

Casida, J. J. (2008). Linking nursing unit's culture to organizational effectiveness: A measurement tool. *Nursing Economics, 26*(2), 106. PMID:18524375

Castells, M. (2004). *The network society A cross-cultural perspective*. Edward Elgar.

Caudill, E. M., & Murphy, P. E. (2000). Consumer Online Privacy: Legal and Ethical Issues. *Journal of Public Policy & Marketing, 19*(1), 7-19.

CBS News. (2017). *Dutch man charged in Amanda Todd case allegedly targeted 2nd Canadian child*. https://www.cbc.ca/news/canada/british-columbia/amanda-todd-aydin-coban-1.3951334

Çelik, S. (2010). The relationship of business ethics practices with employee satisfaction. *Business Ethics Journal, 3*(5), 21-40.

Center for Sustainable Systems. (n.d.). *Environmental Justice Factsheet*. Retrieved from http://css.umich.edu/factsheets/environmental-justice-factsheet

Cespedes, F. V., & Smith, J. H. (1993). Database Marketing: New Rules for Policy and Practice. *Sloan Management Review, 34*(4), 7–22.

Çetintaş, H. B. (2016). Examination of Corporate Ethical Principles in the Context of Code of Ethics: Istanbul Chamber of Industry, Turkey's Top 500 Industrial Enterprises Case. *Electronic Turkish Studies, 11*(8).

Chaffey, D., & Ellis-Chadwick, F. (2016). *Digital Marketing Strategy ımplementation and Practice*. Pearson Education.

Chaffey, D., & Smith, P. (2012). *Emarketing Excellence, Planning and optimising your digital marketing*. Taylor & Francis.

Changing childhood in a globalizing world. (2005). InProut, A. (Ed.), *The future of childhood*. Routledge. Retrieved from https://learn.illinois.edu/pluginfile.php/5030953/mod_resource/content/1/Prout%2C%20A.%20%28Ed.%29%20%282005%29.%20Changing%20childhood%20in%20a%20globalizing%20world.pdf

Chang, Y. T., Yu, H., & Lu, H. P. (2015). Persuasive messages, popularity cohesion, and message diffusion in social media marketing. *Journal of Business Research, 68*(4), 777–782.

Chaturvedi, B. K., & Chandra, S. (2015). *Corruption and economic growth*. Academic Foundation.

Chen, G. M. (2011). Tweet this: A uses and gratifications perspective on how active Twitter use gratifies a need to connect with others. *Computers in Human Behavior, 27*(2), 755–762. doi:10.1016/j.chb.2010.10.023

Chertow, M. R. (2008). The IPAT Equation and Its Variants. *Journal of Industrial Ecology, 4*(4), 13–29. doi:10.1162/10881980052541927

Chiang, H. S., & Hsiao, K. L. (2015). YouTube stickiness: The needs, personal, and environmental perspective. *Internet Research, 25*(1), 85–106. doi:10.1108/IntR-11-2013-0236

Choi, Y. Y., & Seo, Y. S. (2015). The relationship between insecure adult attachment and smartphone addiction: the mediation effect of impulsiveness moderated by social support. *Korean J Psychol Couns Psychother, 27*(3), 749e72.

Chomsky, N., Herman, E. S., & Peterson, D. (2004). *Medyanın Kamuoyu İmalatı, Medyanın Tekelleşmesi Kitlelerin Yönlendirilişi ve Zorunlu İtaat* (A. Köymen, Trans.). Chiviyazıları Publications.

Chukwunonso, F. (2009). *The development of human resource management from a historical perspective and its implications for the human resource manager*. Academic Press.

Çiftçioğlu, B. A. (2011). Business and Ethics. Beta Publications.

Cihan, G. (2019). Obez bireylerde damgalanma hissi ile yeme davranışları ve depresyon arasındaki ilişkinin incelenmesi (Yayınlanmamış Yüksek Lisans Tezi). Başkent Üniversitesi Sağlık Bilimleri Enstitüsü, Fakültesi Beslenme ve Diyetetik Anabilim Dalı, Ankara, Turkey.

Ciulla, J. B. (2011). Is business ethics getting better? A historical perspective. *Business Ethics Quarterly*, *21*(2), 335–343. doi:10.5840/beq201121219

Clegg, S., Kornberger, M., & Rhodes, C. (2007). Business ethics as practice. *British Journal of Management*, *18*(2), 107–122. doi:10.1111/j.1467-8551.2006.00493.x

Clough, G. (1982). Environmental Effects on Animals Used in Biomedical Research. *Biological Reviews of the Cambridge Philosophical Society*, *57*(3), 487–523. doi:10.1111/j.1469-185X.1982.tb00705.x PMID:6982731

Clow, K. E., & Baack, D. (2012). *Integrated Advertising, Promotion and Marketing Communication*. Pearson.

Cnn.com. (2019). Retrieved from https://edition.cnn.com/2019/02/28/tech/tiktok-ftc-fine-children/index.html

Çobanoğlu, N. (2007). *Tıp Etiği. In Özerklik*. İlke Yayınevi.

Cochrane, U. K. (2018). *Facebook video genom Twitter*. Academic Press.

Coen, J. E. (1995). Population growth and earth's human carrying capacity. *Science*, *269*(5222), 341–346. doi:10.1126cience.7618100 PMID:7618100

Cohen, T. R., Panter, A. T., & Turan, N. (2012). Guilt proneness and moral character. *Current Directions in Psychological Science*, *21*(5), 355–359. doi:10.1177/0963721412454874

Cole, G. A. (2002). *Personnel and human resource management*. Cengage Learning EMEA.

COMEST. (n.d.). *The Precautionary Principle*. Retrieved from https://unesdoc.unesco.org/ark:/48223/pf0000139578

Computer Ethics. (2020). Retrieved from https://www.cmpe.boun.edu.tr/~say/c150/intro/lit10.html

Conley, C., & Tosti-Kharas, J. (2014). Crowdsourcing content analysis for managerial research. *Management Decision*, *52*(4), 675–688. doi:10.1108/MD-03-2012-0156

Conroy, S. J., & Emerson, T. L. (2004). Business ethics and religion: Religiosity as a predictor of ethical awareness among students. *Journal of Business Ethics*, *50*(4), 383–396. doi:10.1023/B:BUSI.0000025040.41263.09

Cooke, M., & Buckley, N. (2008). Web 2.0, social networks and the future of market research. *International Journal of Market Research*, *50*(2), 267–292. doi:10.1177/147078530805000208

Cook, R., Hsu, C. H., & Marqua, J. (2014). *Tourism: The Business of Hospitality and Travel*. Pearson Press.

Coombs, A. A. T., & King, R. K. (2005). Workplace discrimination: Experiences of practicing physicians. *Journal of the National Medical Association*, *97*(4), 467. PMID:15868767

Correa, T., Hinsley, A. W., & De Zuniga, H. G. (2010). Who interacts on the Web?: The intersection of users' personality and social media use. *Computers in Human Behavior*, *26*(2), 247–253. doi:10.1016/j.chb.2009.09.003

Costa, R., & Menichini, T. (2013). A multidimensional approach for CSR assessment: The importance of the stakeholder perception. *Expert Systems with Applications*, *40*(1), 150–161. doi:10.1016/j.eswa.2012.07.028

Cragg, W. (2000). Human rights and business ethics: Fashioning a new social contract. In *Business Challenging Business Ethics: New Instruments for Coping with Diversity in International Business* (pp. 205–214). Springer. doi:10.1007/978-94-011-4311-0_20

Csigene-Nagypäl, N. (2014). Corporate Social Responsibility of Hungarian SMEs with good environmental practices. *Journal for East European Management Studies*, *19*(3), 327–347. doi:10.5771/0949-6181-2014-3-327

Culnan, M. J. (2000). Protecting Privacy Online: Is Self-Regulation Working? *Journal of Public Policy & Marketing*, *19*(1), 20–26. doi:10.1509/jppm.19.1.20.16944

Culnan, M. J., & Armstrong, P. K. (1999). Information privacy concerns, procedural fairness, and impersonal trust: An empirical investigation. *Organization Science*, *10*(1), 104–115. doi:10.1287/orsc.10.1.104

Cymbala, A., & Owczarczuk, M. (2011). Variables importance in questionnaire data on advertising. *Expert Systems with Applications*, *38*, 14218–14224. doi:10.1016/j.eswa.2011.04.234

Daniel, A. C., Marioara, A., & Isabela, D. (2017). Annual Financial Statements as a Financial Communication Support. Ovidius University Annals. *Economic Sciences Series*, *17*(1), 403–406.

Danley, J., Harrick, E., Schaefer, D., Strickland, D., & Sullivan, G. (1996). HR's view of ethics in the work place: Are the barbarians at the gate? *Journal of Business Ethics*, *15*(3), 273–285. doi:10.1007/BF00382953

Day, S., Coomes, R. C., McGrath-Lone, L., Schoenborn, C., & Ward, H. (2017). Stratfied, precision or personalized medicine? Cancer services in the 'rea world' of a London hospital. *Sociology of Health & Illness*, *39*(1), 143–158. doi:10.1111/1467-9566.12457 PMID:27460935

De George, R. T. (2005). *A history of business ethics.* Paper delivered at the third biennial global business ethics conference, Santa Clara, CA.

De George, R. T. (2005). *A history of business ethics.* Paper delivered at the third biennial global business ethics conference. Santa Clara, CA.

De George, R. T. (1986). Theological ethics and business ethics. *Journal of Business Ethics*, *5*(6), 421–432. doi:10.1007/BF00380748

De George, R. T. (1987). The status of business ethics: Past and future. *Journal of Business Ethics*, *6*(3), 201–211. doi:10.1007/BF00382865

De George, R. T. (1994). International business ethics. *Business Ethics Quarterly*, *4*(1), 1–9. doi:10.2307/3857554

De George, R. T. (2011). *Business ethics.* Pearson Education India.

De Jans, S., Van de Sompel, D., Hudders, L., & Cauberghe, V. (2019). Advertising targeting young children: An overview of 10 years of research (2006–2016). *International Journal of Advertising*, *38*(2), 173–206. doi:10.1080/02650487.2017.1411056

De Veirman, M., Hudders, L., & Nelson, M. R. (2019). What Is Influencer Marketing and How Does It Target Children? A Review and Direction for Future Research. *Frontiers in Psychology*, *10*, 2685. doi:10.3389/fpsyg.2019.02685 PMID:31849783

Deckop, J. R. (Ed.). (2006). *Human resource management ethics*. IAP.

Deloitte. (2017). *Global Powers of Luxury Goods 2017*. Retrieved from https://www2.deloitte.com/content/dam/Deloitte/global/Documents/consumer-industrial-products/gx-cip-global-powers-luxury-2017.pdf

Denison, D. R., & Neale, W. (1996). *Denison organizational culture survey*. Aviat.

Derr, P., & McNamara, E. (2003). *Case studies in environmental ethics*. Rowman & Littlefield Publishers.

DesJardins, J. R., & DesJardins, J. R. (2009). *An introduction to business ethics*. McGraw-Hill Higher Education.

Deutsch, M., & Gerard, H. B. (1955). A study of normative and informational social influences upon individual judgment. *Journal of Abnormal and Social Psychology*, *51*(3), 629–636. doi:10.1037/h0046408 PMID:13286010

Deuze, M., & Dimoudi, C. (2002). Online journalists in the Netherlands: Towards a profile of a new profession. *Journalism*, *3*(1), 85–100. doi:10.1177/146488490200300103

Devall, B., & Sessions, G. (1985). *Deep Ecology: Living as if Nature Mattered*. Peregrine Smith Books.

Dewi, M. K., & Tanjung, H. (2007). *Discipline Without Punishment – Is It Possible?* Paper presented at International Conference, Management from Islamic Perspective, Kuala Lumpur.

Diamond, I., & Orenstein, G. F. (Eds.). (1990). *Reweaving the World: The Emergence of Ecofeminism*. Sierra Club Books.

Diedrichs, P. C., & Barlow, F. K. (2011). How to lose weight bias fast! Evaluating a brief anti-weight bias intervention. *British Journal of Health Psychology*, *16*, 846–861.

Dinu, V. (2008, Feb. 23). The ethical dimension of business. *Amfiteatru Economic*, 7-8.

Dinu, V. (2011). Corporate Social Responsibility– opportunity for reconciliation between economic interests and social and environmental interests. *Amfiteatru Economic*, 13(29), 6–7.

Doğan, M. (2011). Türkiye'de İslami Turizm'in Gelişimi: 2002–2009. *Political Economy, Crisis And Development (Politik İktisat, Kriz Ve Kalkınma)*, 471-478.

Doğan, N. (2007). *İş Etiğinin Kurumsal Sosyal Sorumluluktaki Rolü ve Bir Araştırma*. Yayınlanmış Yüksek Lisans Tezi, Marmara Üniversitesi SBE.

Dominick, J. R. (1999). Who do you think you are? Personal home pages and self-presentation on the World Wide Web. *Journalism & Mass Communication Quarterly*, *76*(4), 646–658. doi:10.1177/107769909907600403

Donaldson, T., & Dunfee, T. W. (1994). Toward a unified conception of business ethics: Integrative social contracts theory. *Academy of Management Review*, *19*(2), 252–284. doi:10.5465/amr.1994.9410210749

Doyal, L., & Gough, I. (1991). *A Theory of Human Need*. Macmillan. doi:10.1007/978-1-349-21500-3

Drover, W., Franczak, J., & Beltramini, R. F. (2012). A 30-year historical examination of ethical concerns regarding business ethics: Who's concerned? *Journal of Business Ethics*, *111*(4), 431–438. doi:10.100710551-012-1214-9

Duffy, D. J. (2016). Problems, challenges and promises: Perspectives on precision medicine. *Briefings in Bioinformatics*, *17*(3), 494–504. doi:10.1093/bib/bbv060 PMID:26249224

Dunfee, T. W. (1991). Business ethics and extant social contracts. *Business Ethics Quarterly*, *1*(1), 23–51. doi:10.2307/3857591

Durso, L. E., & Latner, J. D. (2008). Understanding self-directedstigma: Development of theweightbiasinternalization-scale. *Obesity (Silver Spring, Md.)*, *16*(Suppl. 2), 580–586.

Dusuki, A. W. (2008). What does Islam say about corporate social responsibility? *Review of Islamic Economics, 12*(1), 5–28.

Dutton, W. H. (2017). Fostering a cyber security mindset. *Internet Policy Review, 6*(1). https://policyreview.info/node/443/pdf

Easterlin, R. A. (1974). Does economic growth improve the human lot? Some empirical evidence. In P. A. David & M. W. Reder (Eds.), *Nations and Households in Economic Growth* (pp. 89–125). Academic Press. doi:10.1016/B978-0-12-205050-3.50008-7

Ecevit Alpar, Ş., Bahçecik, N., & Karabacak, Ü. (2013). *Çağdaş Hemşirelikte Etik*. İstanbul Tıp Kitapevi.

Eggenschwiler, J. (2017). Accountability challenges confronting cyberspace governance. *Internet Policy Review, 6*(3). https://policyreview.info/node/712/pdf

Eisenberg, E. M., & Riley, P. (2001). Organizational culture. The new handbook of organizational communication: Advances in theory, research, and methods, 291-322.

El Garah, W., Beekun, R. I., Habisch, A., Lenssen, G., Adaui, C. L., Ismaeel, M., & Blaim, K. (2012). Toward applied Islamic business ethics: Responsible halal business. *Journal of Management Development*.

Elaziz, M. F., & Kurt, A. (2017). Religiosity, consumerism and halal tourism: A study of seaside tourism organizations in Turkey. *Turizam: međunarodni znanstveno-stručni časopis, 65*(1), 115-128.

Emanuel, E. J., & Emanuel, L. L. (1992). Four Models of the Physician-Patient Relationship. *Journal of the American Medical Association, 267*(16), 2221–2226. doi:10.1001/jama.1992.03480160079038 PMID:1556799

Emirza, E., İştahlı, S. B., & İştahlı, Y. H. (2012). Sosyal Medyanın Çok Katlı Pazarlamada Kullanımı ve İş Tatmini Üzerindeki Etkileri. *Yönetim ve Ekonomi Araştırmaları Dergisi, 10*(18), 20–32.

Enderle, G. (1997). A worldwide survey of business ethics in the 1990s. *Journal of Business Ethics, 16*(14), 1475–1483. doi:10.1023/A:1005866527497

Enderle, G. (Ed.). (1999). *International business ethics: Challenges and approaches*. University of Notre Dame Press.

Eralp, Z. (1983). *General Tourism*. Basin-Yayin Yüksek Okulu Press.

Eroğlu, E., & Atabek, N. (2006). İletişim Fakültesi Öğrencilerinin "Etik Eğitimi"ne ve "İletişim Etiği"ne İlişkin Görüşlerinin Değerlendirilmesi. *Küresel İletişim Dergisi, 2*. https://dergipark.org.tr/tr/download/article-file/177843

Erol, G. (2009). Medya ve Etik: İnternet Haberciliğinde Nesnellik, Etik İlkelerın İçeriği, Uygulamadaki Sorunlar Ve Televizyon Haberciliği İle Karşılaştırılması. *Fırat Üniversitesi Medya ve Etik Sempozyumu*, 7-9.

Ersoy, N., & Aydın, E. (1994). Autonomy And The Prıncıple of Respect For Autonomy In Medıcal Ethıcs. *Turkiye Klinikleri J Med Ethics, 2*(2), 71–74.

Eteokleous, P. P., Leonidou, L. C., & Katsikeas, C. S. (2015). Corporate Social Responsibility in international marketing: Review, assessment, and future research. *International Marketing Review, 33*(4), 580–624. doi:10.1108/IMR-04-2014-0120

Etzioni, A. (1964). Modern organizations. Academic Press.

European Advertising Standards (EASE). (2019). *Children*. Retrieved from: https://www.easa-alliance.org/issues/children

European Union. (2012). *European Strategy for a Better Internet for Children*. Retrieved from https://ec.europa.eu/digital-single-market/en/european-strategy-better-internet-children

Europol's European Cybercrime Centre. (2015, Oct.). Virtual Global Taskforce Child Sexual Exploitation Environmental Scan 2015. EC3-Europol, 12.

Evans, D. (2008). Social media marketing: An hour a day. Academic Press.

Evans, R. (1991). Business ethics and changes in society. *Journal of Business Ethics*, *10*(11), 871–876. doi:10.1007/BF00383705

Fanning, A. L., & O'Neill, D. W. (2019). The Wellbeing–Consumption paradox: Happiness, health, income, and carbon emissions in growing versus non-growing economies. *Journal of Cleaner Production*, *212*, 810–821. doi:10.1016/j.jclepro.2018.11.223

Farris, P. F., Bendle, N. T., Pfeifer, P. E., & Reibstein, D. J. (2010). *Marketing Metrics:The definitive guide to measuring marketing performance*. Pearson Education.

Fennel, A., & Malloy, D. C. (1999). Measuring the Ethical Nature of Tourism Operators. *Annals of Tourism Research*, *26*(4), 928–943. doi:10.1016/S0160-7383(99)00032-8

Fennell, D. A., & Malloy, D. (2007). *Codes of Ethics in Tourism: Practice, theory, synthesis*. Channel View Publications. doi:10.21832/9781845410629

Ferrell, O. C., Gresham, L. G., & Fraedrich, J. (1989). *A Synthesis of Ethical Decision Models for Marketing*. Academic Press.

Ferrell, O. C., & Fraedrich, J. (2015). *Business ethics: Ethical decision making & cases*. Nelson Education.

Ferrell, O., & Gresham, L. (1985). A Contingency Framework for Understanding Ethical Decision Making in Marketing. *Understanding Ethical Decision Making in Marketing*, *49*, 87–96.

Ferris, G. R., Buckley, M. R., & Fedor, D. B. (Eds.). (2002). *Human resources management: Perspectives, context, functions, and outcomes*. Prentice Hall.

Fichtner, L. (2018). What kind of cyber security? Theorising cyber security and mapping approaches. *Internet Policy Review*, *7*(2). https://policyreview.info/node/788/pdf

Fifka, M. S. (2013). Corporate responsibility reporting and its determinants in comparative perspective–a review of the empirical literature and a meta-analysis. *Business Strategy and the Environment*, *22*(1), 1–35. doi:10.1002/bse.729

Fisher, C., & Lovell, A. (2009). *Business ethics and values: Individual, corporate and international perspectives*. Pearson education.

Fleckenstein, M. P., & Huebsch, P. (1999). Ethics in tourism-reality or hallucination. *Journal of Business Ethics*, *19*(1), 137–142. doi:10.1023/A:1006118526469

Folkvord, F., Bevelander, K. E., Rozendaal, E., & Hermans, R. (2019). Children's bonding with popular YouTube vloggers and their attitudes toward brand and product endorsements in vlogs: An explorative study. *Young Consumers*, *20*(2). Advance online publication. doi:10.1108/YC-12-2018-0896

Foregger, S. K. (2008). *Uses and gratifications of Facebook. com*. Michigan State University. Department of Communication.

Forhan, M., & Salas, X. R. (2013). Inequities in healthcare: A review of bias and discrimination in obesity treatment. *Canadian Journal of Diabetes*, *37*(3), 205–209.

Franklin, A., & Crang, M. (2001). The trouble with tourism and travel theory? *Tourist Studies*, *1*(1), 5–22. doi:10.1177/146879760100100101

Fraser, N., & Honneth, A. (2003). *Redistribution or Recognition: A Political-Philosophical Exchange*. Verso. doi:10.1002/9780470756119.ch54

Frazer, E. (2006). Max Weber on ethics and politics. *Politics and Ethics Review*, *2*(1), 19-37.

Frechtling, D. C. (2018). On the Ethics of Tourism Research. *Journal of Travel Research*, *57*(8), 1054–1067. doi:10.1177/0047287517729756

Freeman, R. E. (2000). Business ethics at the millennium. *Business Ethics Quarterly*, *10*(1), 169–180. doi:10.2307/3857703

Frigo, G. (2018). *The energy ethic and strong sustainability: Outlining key principles for a moral compass. In Strongly Sustainable Societies: Organising Human Activities on a Hot and Full Earth*. Routledge.

Frost, P. J., Moore, L. F., Louis, M. R. E., Lundberg, C. C., & Martin, J. E. (1985). *Organizational culture*. Sage Publications, Inc.

Fruh, S. M., Nadglowski, J., Hall, H. R., Davis, S. L., Crook, E. D., & Zlomke, K. (2016). Obesity stigma and bias. *The Journal for Nurse Practitioners*, *12*(7), 425–432.

Fusilier, M. R., Aby, C. D. Jr, Worley, J. K., & Elliott, S. (1996). Perceived seriousness of business ethics issues. *Business & Professional Ethics Journal*, *15*(1), 67–78. doi:10.5840/bpej19961519

Galas, D. J., & Hood, L. (2009). Systems Biology and Emerging Technologies Will Catalyze the Transition from Reactive Medicine to Predictive, Personalized, Preventive and Participatory (P4) Medicine. *Interdisciplinary Bo Central*, *1*(2), 1–4. doi:10.4051/ibc.2009.2.0006

Galbraith-Emami, S., & Lobstein, T. (2013). The impact of initiatives to limit the advertising of food and beverage products to children: A systematic review. *Obesity Reviews*, *14*(12), 960–974. doi:10.1111/obr.12060 PMID:23845093

Gandz, J., & Hayes, N. (1988). Teaching business ethics. *Journal of Business Ethics*, *7*(9), 657–669. doi:10.1007/BF00382975

Gangone, A.-D., & Gănescu, M.-C. (2014). Corporate Social Responsibility in emerging and developing economies in Central and Eastern Europe – a measurement model from the stakeholder theory perspective. *Economic Research-Ekonomska Istraživanja*, *27*(1), 539–558. doi:10.1080/1331677X.2014.967535

Gao, W., Liu, Z., & Li, J. (2017). How does social presence influence SNS addiction? A belongingness theory perspective. *Computers in Human Behavior*, *77*, 347–355. doi:10.1016/j.chb.2017.09.002

Garcia, J. T., Amankwah, E. K., & Hernandez, R. G. (2016). Assessment of weight bias among pediatric nurses and clinical support staff toward obese patients and their caregivers. *Journal of Pediatric Nursing*, *31*(4), e244–e251.

Gardiner, S. M., & Thompson, A. (Eds.). (2017). *The Oxford Handbook of Environmental Ethics*. Oxford University Press.

Garrett, T. M. (2004). Nirali Prakashan. *Business Ethics (Oxford, England)*, *24*.

Gasper, D. (2015). *Human security: From definitions to investigating a discourse. In Routledge Handbook of Human Security*. Routledge.

Gefen, D., Karahanna, E., & Straub, D. W. (2003). Trust and TAM in Online Shopping: An Integrated Model. *Management Information Systems Quarterly*, *27*(1), 51–90. doi:10.2307/30036519

Geller, G., & Watkins, P. A. (2018). Addressing medical students' negative bias toward patients with obesity through ethics education. *AMA Journal of Ethics, 20*(10), 948–959.

Gert, B., & Gert, J. (2017). The Definition of morality. In *The Stanford Encyclopaedia of Philosophy.* https://plato.stanford.edu/archives/fall2017/entries/morality-definition/

Ghaemi, S. N. (2017). Biomedical Reductionist, Humanist, and Biopsychosocial Models in Medicine. In T. Schramme & S. Edwards (Eds.), *Handbook of the Philosophy of Medicine* (Vol. 2, pp. 772–791). Springer. doi:10.1007/978-94-017-8688-1_38

Ghillyer, A. (2008). *Business ethics: A real world approach.* Academic Press.

Giannetti, B. F., Agostinho, F., Almeida, C. M. V. B., & Huisingh, D. (2015). A review of limitations of GDP and alternative indices to monitor human wellbeing and to manage eco-system functionality. *Journal of Cleaner Production, 87,* 11–25. doi:10.1016/j.jclepro.2014.10.051

Giaoutzi, M., & Nijkamp, P. (2006). *Emerging trends in tourism development in an open world. Tourism and Regional Development: New Pathways.* Ashgate.

Gibbs, J. L., Ellison, N. B., & Heino, R. D. (2006). Self-presentation in online personals: The role of anticipated future interaction, self-disclosure, and perceived success in Internet dating. *Communication Research, 33*(2), 152–177. doi:10.1177/0093650205285368

Giddens, A. (2008). *Ulus Devlet ve Şiddet.* Kalkedon Yayınları.

Gilligan, C. (1982). *In a different voice.* Harvard University Press.

Gini, A., & Marcoux, A. M. (Eds.). (2005). *Case studies in business ethics.* Pearson Prentice Hall.

Girardin, B. (2012). *Ethics in politics.* Retrieved from https://www.globethics.net/documents/4289936/13403252/FocusSeries_05_EthicsinPolitics_Benoit_text.pdf/bf28841e-216a-41ff-bca6-4ef9d4022d86

Girgin, A. (2000). *Yazılı Basında Haber ve Habercilik Etik'i.* İnkılap Kitapevi.

Global Footprint Network. (n.d.). Retrieved from https://data.footprintnetwork.org/#/

Godin, S. (1999). *Permission Marketing: Turning Strangers Into Friends And Friends Into Customers.* Simon&Schuster.

González, A. M. (2003). Ethics in global business and in a plural society. *Journal of Business Ethics, 44*(1), 23–36. doi:10.1023/A:1023230222707

Gospel, H. (2009). Human Resources Management: A Historical. The SAGE handbook of human resource management, 12.

Grace, D. (2006). For business ethics. *Australian Journal of Management, 31*(2), 371–380. doi:10.1177/031289620603100211

Grand-Clement, S. (2017). *Digital learning: Education and skills in the digital age.* RAND Europe. Retrieved from https://www.rand.org/content/dam/rand/pubs/conf_proceedings/CF300/CF369/RAND_CF369.pdf

Greener, J., Douglas, F., & van Teijlingen, E. (2010). More of the same? Conflicting perspectives of obesity causation and intervention amongst overweight people, health professionals and policy makers. *Social Science & Medicine, 70*(7), 1042–1049.

Greenwood, M., & De Cieri, H. (2007). Stakeholder theory and the ethics of HRM. *Human resource management: Ethics and employment,* 119.

Greenwood, M. (2007). Ethics and human resource management. *Human Resource Management in Australia, 3e,* 562–587.

Greenwood, M. R. (2002). Ethics and HRM: A review and conceptual analysis. *Journal of Business Ethics*, *36*(3), 261–278. doi:10.1023/A:1014090411946

Gruzd, A., Jacobson, J., Wellman, B., & Mai, P. (2016). *Understanding communities in an age of social media: the good, the bad, and the complicated.* Academic Press.

Gruzd, A., Haythornthwaite, C., Paulin, D., Gilbert, S., & Del Valle, M. E. (2018). Uses and gratifications factors for social media use in teaching: Instructors' perspectives. *New Media & Society*, *20*(2), 475–494. doi:10.1177/1461444816662933

Guest, D. E. (2007). HRM and performance: can partnership address the ethical dilemmas? *Human Resource Management: Ethics and Employment*, 52.

Güllüce, A. Ç., Bozkurt, Ö., & Meriç, S. (2017). Relationship Between Organizational Justice Actions and Internal Entrepreneurship: A Research on Hotel Managers. *Girişimcilik ve Kalkınma Dergisi*, *12*(2), 39–50.

Güllüce, A. Ç., Özer, S., & Erkiliç, E. (2015). The Effect of Organizational Justice Perception on Organizational Commitment among Healthcare Sector Employees. *Journal of Business and Management*, *4*(3), 16–25.

Gulyás, A. (2009). Corporate Social Responsibility in the British media industries – preliminary findings. *Media Culture & Society*, *31*(4), 657–668. doi:10.1177/0163443709335281

Gündoğdu, B., Ünal, A., & Kılınç, İ. (2018). *Üniversitelerin İş Etiği Algıları ve Uygulamaları: Batı Karadeniz Üniversiteler Birliği Üzerine Bir Araştırma.* Academic Press.

Gupta, A. D. (Ed.). (2010). *Ethics, business and society: Managing responsibly.* SAGE Publications India.

Halman, L., & Van Ingen, E. (2015). Secularization and Changing Moral Views European Trends in Church Attendance and Views on Homosexuality, Divorce, Abortion, and Euthanasia. *European Sociological Review*, *0*(5), 0, 1–12. doi:10.1093/esr/jcv064

Halmi, A. (1996). *Kvalitativna metodologija u društvenim znanostima.* Samobor: A.G. Matoš d.d.

Haman, Ö. (2016). *Örgüt Yapısı Ve Örgütsel İletişim Arasındaki İlişkinin Belirlenmesi Üzerine Bir Araştırma.* Yüksek Lisans Tezi, Nişantaşı Üniversitesi, Sosyal Bilimler Enstitüsü, İşletme Anabilim Dalı, İşletme Yönetimi Bilim Dalı.

Hand, W. B., Robinson, J. C., & Creel, E. (2013). Ethical Issues Surrounding Weight Bias and Stigma in Healthcare. *Online Journal of Health Ethics / the University of Mississippi Medical Center*, *8*(2).

Hardin, G. (1968). The Tragedy of the Commons. *Science*, *162*(3859), 1243–1248. doi:10.1126cience.162.3859.1243 PMID:5699198

Hartman, E. (1996). *Organizational ethics and the good life.* Oxford University Press on Demand.

Hartman, L. P. (1998). *Perspectives in business ethics.* Irwin/McGraw-Hill.

Hartman, L. P., DesJardins, J. R., & MacDonald, C. (2008). *Business ethics: Decision making for personal integrity and social responsibility.* McGraw-Hill.

Hartog, M., & Winstanley, D. (2002). Ethics and human resource management: Professional development and practice. *Business & Professional Ethics Journal*, *21*(2), 3–9. doi:10.5840/bpej200221210

Hart, P., & Dinev, T. (2006). An Extended Privacy Calculus Model for E-Commerce Transactions. *Information Systems Research*, 61–80.

Hartz, R., & Steger, T. (2010). Heroes, villains and 'honourable merchants': Narrative change in the German media discourse on corporate governance. *Organization*, *17*(6), 767–785. doi:10.1177/1350508410372558

Harvey, B. (Ed.). (1994). *Business ethics: A European approach*. Prentice Hall.

Hasta Hakları Yönetmeliği. (1998). *Resmi Gazete*. Retrieved from https://www.mevzuat.gov.tr/Metin.Aspx?MevzuatKod=7.5.4847&sourceXmlSearch=&MevzuatIliski=0

Hatch, M. J. (1993). The dynamics of organizational culture. *Academy of Management Review*, *18*(4), 657–693. doi:10.5465/amr.1993.9402210154

Heim, K. M. (1991). *Human resources management: Ethics in personnel*. Allerton Park Institute.

Held, V. (2007). *The Ethics of Care: Personal, Political, and Global*. Oxford University Press. doi:10.1093/oxfordhb/9780195325911.003.0020

Henderson, J. C. (2003). Managing tourism and Islam in peninsular Malaysia. *Tourism Management*, *24*(4), 447–456. doi:10.1016/S0261-5177(02)00106-1

Heyd, T. (2010). Climate Change, Individual Responsibilities and Cultural Frameworks. *Human Ecology Review*, *17*(2), 86–95.

Hite, R. E., & Eck, R. (1987). Advertising to Children-Attitudes of Business vs Consumers. *Journal of Advertising Research*, *27*(5), 40–53.

Hofmann, B. (2011). *Ethical aspects of bariatric treatment of adult obesity*. Institute of Health Economics.

Holden, A. (2003). In Need of New Environmental Ethics for Tourism? *Annals of Tourism Research*, *30*(1), 94–108. doi:10.1016/S0160-7383(02)00030-0

Holland, B. (2014). *Allocating the Earth: A Distributional Framework for Protecting Capabilities in Environmental Law and Policy*. Oxford University Press. doi:10.1093/acprof:oso/9780199692071.001.0001

Holme, C. (2008). Business ethics–part one: Does it matter? *Industrial and Commercial Training*, *40*(5), 248–252. doi:10.1108/00197850810886487

Hong, F. Y., Huang, D. H., Lin, H. Y., & Chiu, S. L. (2014). Analysis of the psychological traits, Facebook usage, and Facebook addiction model of Taiwanese university students. *Telematics and Informatics*, *31*(4), 597–606. doi:10.1016/j.tele.2014.01.001

Honneth, A. (1995). *The Struggle for Recognition: The Grammar of Social Conflicts*. Polity.

Honneth, A. (2007). *Disrespect: The Normative Foundations of Critical Theory*. Polity.

HoofnagleC. J. (2001). Privacy Practices Below the Lowest Common Denominator: The Federal Trade Commission's Initial Application of Unfair and Deceptive Trade Practices Authority to Protect Consumer Privacy (1997-2000). https://papers.ssrn.com/sol3/papers.cfn?abstractid=507582

Horn, R. (2000). Border bordellos. Time Asia. İnceleme, "Uluslararası Sosyal Araştırmalar Dergisi", 29(7).

Howard, P. N. (2011). *Castells and the media*. Polity Press.

Hudders, L., Cauberghe, V., & Panic, K. (2016). How advertising literacy training affect children's responses to television commercials versus advergames. *International Journal of Advertising*, *35*(6), 909–931. doi:10.1080/02650487.2015.1090045

Hughes, J. (2015). Transhumanist Position on Human Germline Genetic Modification. *Institute for Ethics and Emerging Technologies*. https://ieet.org/index.php/IEET/more/hughes20150320

Hugman, R. (2012). *Culture, values and ethics in social work: Embracing diversity*. Routledge. doi:10.4324/9780203094907

Hunt, S. D., & Vitell, S. J. (2006). The general theory of marketing ethics: A revision and three questions. *Journal of Macromarketing*, *26*(2), 143–153. doi:10.1177/0276146706290923

Hunt, S., & Vitell, S. (1986). The General Theory of Marketing Ethics. *Journal of Macromarketing*, 1–12.

Hurn, B. J. (2008). Ethics in international business. *Industrial and Commercial Training*, *40*(7), 347–354. doi:10.1108/00197850810912216

Hussein, M. (2009). Hiring and firing with ethics. *Human Resource Management International Digest*, *17*(4), 37–40. doi:10.1108/09670730910963334

Ienca, M., & Andorno, R. (2017). Towards new human rights in the age of neuroscience and neurotechnology. *Life Sciences, Society and Policy*, *13*(1), 5–22. doi:10.118640504-017-0050-1 PMID:28444626

Indonesia elections: Prabowo alleges fraud, says he will reject official vote tally. (2019, May 16). *The Straits Times*. Retrieved from https://www.straitstimes.com/asia/se-asia/indonesia-elections-prabowo-alleges-fraud-says-he-will-reject-official-vote-tally

Ingbar, J. (2004). *From Organizational Ethics: Where Values and Culture Meet*. Institute for Global Ethics.

Internet Organised Crime Threat Assessment (IOCTA). (2016). Report. Author.

Internet Watch Foundation. (2017). *Annual Report 2016*. IWF.

Interpol. (2015). *Crimes against Children*. Fact Sheet, Interpol. Retrieved from https://www.interpol.int/en/News-and-Events/News/2015/Online-child-sex-abuser-forums-targeted-in-INTERPOL-coordinated-operation

Işık, M. (2018). *İletişim ve Etik*. Eğitim Publications.

Ivancin, M. (2012). Business Ethics and the Return on Reputation. *Journal of Mass Media Ethics*, *27*(2), 150–152. do i:10.1080/08900523.2012.684593

Jackson, T. (2007). *Prosperity without growth: foundations for the economy of tomorrow*. Routledge.

Jaelani, A. (2017). Halal tourism industry in Indonesia: Potential and prospects. *International Review of Management and Marketing*, *7*(3), 25–34.

Jakubowska, H.(2017). Framing the Winter Olympic Games: A Content Analysisof Polish Newspapers Coverage of Female and Male Athletes. *Polish Sociological Review*, *1*(197), 67-81.

Jamal, T. (2019). *Justice and Ethics in Tourism*. Routledge Press. doi:10.4324/9781315162942

James, C. S., & Joseph, C. S. (2001). The transformational – transactional leadership model in practice. *Leadership and Organization Development Journal*, *22*(8), 383–393. doi:10.1108/01437730110410107

Jay, M., Kalet, A., Ark, T., Mc Macken, M., Messito, M. J., Richter, R., Schlair, S., Sherman, S., Zabar, S., & Gillespie, C. (2009). Physicians' Attitudes About Obesity And Their Associations With Competency And Specialty: A Cross-Sectional Study. *BMC Health Serv Res.*, *9*(106), 1-11.

Johnson, C. E. (2018). *Organizational ethics: A practical approach*. Sage Publications.

Johnson, R. E., Lanaj, K., & Barnes, C. M. (2014). The good and bad of being fair: Effects of procedural and interpersonal justice behaviors on regulatory resources. *The Journal of Applied Psychology*, *99*(4), 99635–99650. doi:10.1037/a0035647 PMID:24446913

Jon, R. (2018). *China's ByteDance leapfrogs Uber to becomes world's most valuable startup.* Retrieved from https://techcrunch.com/2018/10/26/chinas-bytedance-leapfrogs-uber-to-becomesworlds-most-valuable-startup/

Jones, C., Parker, M., & Ten Bos, R. (2005). *For business ethics.* Routledge. doi:10.4324/9780203458457

Jones, T. M. (1991). Ethical Decision Making by Individuals in Organizations: An Issue-Contingent Model. *Academy of Management Review, 16*(2), 366–395. doi:10.5465/amr.1991.4278958

Jose, A., & Thibodeaux, M. S. (1999). Institutionalization of ethics: The perspective of managers. *Journal of Business Ethics, 22*(2), 133–143. doi:10.1023/A:1006027423495

Jung, A. R. (2017). The influence of perceived ad relevance on social media advertising: An empirical examination of a mediating role of privacy concern. *Computers in Human Behavior, 70*, 303–309. doi:10.1016/j.chb.2017.01.008

Kalafatoğlu, Y. (2011). *Sosyal Medya ve E-Pazarlama İlişkisi.* Academic Press.

Kalan, I., & Yeşil, Y. (2010). Obezite ile ilişkili kronik hastalıklar. *Diyabet ve obezite, 78.*

Kaler, J., & Chryssides, G. (1996). *Essentials of business ethics.* McGraw-Hill.

Kamoche, K. (2007). 15 Strategy, knowledge. *Human Resource Management: Ethics and Employment: Ethics and Employment*, 252.

Kang, J. Y. M., & Kim, J. (2017). Online customer relationship marketing tactics through social media and perceived customer retention orientation of the green retailer. *Journal of Fashion Marketing and Management, 21*(3), 298–316. doi:10.1108/JFMM-08-2016-0071

Kant, I. (2001). *Lectures on ethics* (Vol. 2). Cambridge University Press.

Kapferer, J. N. (1992). *Strategic Brand Management.* Kogan Page.

Kaplan, A. M., & Haenlein, M. (2010). Users of the world, unite! The challenges and opportunities of Social Media. *Business Horizons, 53*(1), 59–68. doi:10.1016/j.bushor.2009.09.003

Kapoor, K. K., Tamilmani, K., Rana, N. P., Patil, P., Dwivedi, Y. K., & Nerur, S. (2018). Advances in social media research: Past, present and future. *Information Systems Frontiers, 20*(3), 531–558. doi:10.100710796-017-9810-y

Kaptein, M. (2008). Developing a measure of unethical behavior in the workplace: A stakeholder perspective. *Journal of Management, 34*(5), 978–1008. doi:10.1177/0149206308318614

Kapu, H. (2009). *Akademik bir disiplin olarak iş ahlakı. İşletmelerde iş etiği.* İstanbul Ticaret Odası Yayınları.

Karabay, M. E. (2010). *A Phenomenon of Strategic Competition in Financial Markets in the Transition to Knowledge Economy: An Application on the Banking Sector* (Unpublished PhD thesis). Marmara University Banking and Insurance Institute Banking Department.

Karabay, M. (2015). *Ethical and Ethical Leadership in Business.* Beta Publications.

Kari-Oca 1 Declaration. (n.d.). Retrieved from http://nafaforestry.org/ff/download/kari_oca.pdf

Kari-Oca 2 Declaration. (n.d.). Retrieved from https://www.ienearth.org/kari-oca-2-declaration/

Kars, N. (2010). *Haberin Tarihi, Kuramları, Söylemi ve Haberin Tarihi, Kuramları, Söylemi ve Radyo-Televizyon Haberciliği.* Derin Publications.

Kasar, K. S., & Akyol, A. (2019). Hemşirelik Öğrencilerinin Sağlıklı Yaşam Biçimi Davranışlarının Obezite Önyargı Düzeyine Etkisi. *Hemşirelikte Eğitim Ve Araştırma Dergisi, 16*(2), 79–86.

Kautilya. (1972). *The Kautilya Arthashastra*. New Delhi: MLBD Publishers.

Kayacan, M. (2005). *Anonim şirketlerin sosyal sorumlulukları ve etik değerler. In İSMMMO I. Uluslararası Muhasebe Denetimi Sempozyumu*. Nisan.

Keçecioğlu, T. (2006). *Stratejik insan kaynakları yönetimi*. Sistem yayıncılık.

Keeley, B., & Little, C. (2017). *The State of the Worlds Children 2017: Children in a Digital World*. UNICEF. Retrieved from: https://files.eric.ed.gov/fulltext/ED590013.pdf

Kelion, L. (2019). *Amazon sued over Alexa child recordings*. Retrieved from: https://www.bbc.com/news/technology-48623914

Keller, A. C., Smith, K. T., & Smith, L. M. (2007). Do gender, educational level, religiosity, and work experience affect the ethical decision-making of US accountants? *Critical Perspectives on Accounting, 18*(3), 299–314. doi:10.1016/j.cpa.2006.01.006

Keser, A. (2002). Değişen Yönleriyle Personel Yönetimi: İnsan Kaynakları Yönetimi. *ISGUC The Journal of Industrial Relations and Human Resources, 4*(1).

Khaliq, A. (2009). Leadership and Work Motivation from the Cross Cultural Perspective. *International Journal of Commerce and Management, 29*(1), 72–84.

Khan, G. F., & Vong, S. (2014). Virality over YouTube: An empirical analysis. *Internet Research*.

Khan, M. L. (2017). Social media engagement: What motivates user participation and consumption on YouTube? *Computers in Human Behavior, 66*, 236–247. doi:10.1016/j.chb.2016.09.024

Kheel, M. (2008). *Nature Ethics: An Ecofeminist Perspective*. Rowman and Littlefield.

Khushf, G. (1998). The scope of organizational ethics. *HEC Forum, 10*(2), 127–135. doi:10.1023/A:1008890902591 PMID:11645346

Kim, M. J., Lee, C. K., & Preis, M. W. (2016). Seniors' loyalty to social network sites: Effects of social capital and attachment. *International Journal of Information Management, 36*(6), 1020–1032. doi:10.1016/j.ijinfomgt.2016.04.019

Kim, Y., Wang, Y., & Oh, J. (2016). Digital media use and social engagement: How social media and smartphone use influence social activities of college students. *Cyberpsychology, Behavior, and Social Networking, 19*(4), 264–269. doi:10.1089/cyber.2015.0408 PMID:26991638

Kingsley Trezise, E. (1996). An introduction to business ethics for human resource management teaching and research. *Personnel Review, 25*(6), 85–89. doi:10.1108/00483489610148554

Kinran, S. B. (2006). *İş Ahlakı ve Etik Değerlerin Türk İşletmeciliğindeki Yeri*. Yayınlanmış Yüksek Lisans Tezi, İstanbul Üniversitesi SBE.

Kirel, Ç. (2000). *Örgütlerde Etik Davranışlar*. Yönetimi ve Bir Uygulama Çalışması, Anadolu Üniversitesi Yayınları.

Kline, W. (2009). Business as an ethical standard. *Journal of Private Enterprise, 24*(2), 35.

Knowledge, V. I. (2019). *The TikTok Strategy: Using AI Platforms to Take Over the World*. Academic Press.

Koh, H. C., & El'fred, H. Y. (2004). Organisational ethics and employee satisfaction and commitment. *Management Decision.*

Kolçak, M. (2012). *Meslek etiği.* Turkey Murathan Publications.

Kolodinsky, R. W. (2006). Wisdom, Ethics and Human Resources Management. *Human resource management ethics*, 47-69.

Kopperi, M. (1999). Business ethics in global economy. *EJBO. Electronic Journal of Business Ethics and Organization Studies.*

Korab-Karpowicz, J. W. (2018). Political Realism in International Relations. In *The Stanford Encyclopaedia of Philosophy.* <https://plato.stanford.edu/archives/sum2018/entries/realism-intl-relations/>

Köseoglu, M. A., Sehitoglu, Y., Ross, G., & Parnell, J. A. (2016). The Evolution of Business Ethics Research in the Realm of Tourism and Hospitality: A Bibliometric Analysis. *International Journal of Contemporary Hospitality Management, 28*(8), 1598–1621. doi:10.1108/IJCHM-04-2015-0188

Kovac, P., & Jukic, T. (2017). Declarations and Reality of Europeanized Public Administration in Eastern Europe: Journals Content Analysis in Slovenia and Croatia. *Transylvanian Review of Administrative Sciences, 50*(3), 127–145. doi:10.24193/tras.2017.0008

Kozak, M. (2012). *General Tourism Information.* Anadolu Üniversitesi Press.

Krafft, M., Arden, C. M., & Verhoef, P. C. (2017). Permission Marketing and Privacy Concerns— Why Do Customers (Not) Grant Permissions? *Journal of Interactive Marketing, 39*, 39–54. doi:10.1016/j.intmar.2017.03.001

Krakoff, S. A., & Rosser, E. (Eds.). (2012). *Tribes, Land, and the Environment.* Ashgate Publishing Company.

Kramer, A., & Chung, C. (2011, July). Dimensions of self-expression in Facebook status updates. In *Proceedings of the International AAAI Conference on Web and Social Media* (*Vol. 5*, No. 1). AAAI.

Kuçükoğlu, M. T. (2012). Ethical Values and Institutionalization of Ethics. *Journal of Law and Economics Studies, 4*(1), 177–185.

Küçükyılmaz M. (2013, Feb. 16). Sosyal Medyada Henüz Reşit Olamadık. *Star Gazetesi.*

Kumar, V. D., & Prabha, M. S. (2019). Getting glued to TikTok®. Undermining the psychology behind widespread inclination toward dub-mashed videos. *Arch Ment Health, 20*(2), 76–77. doi:10.4103/AMH.AMH_7_19

Kunkel, D., Wilcox, B. L., Cantor, J., Palmer, E., Linn, S., & Dowrick, P. (2004). Report of the APA task force on advertising and children. Washington, DC: American Psychological Association.

Kurtuluş, B. (2005). Business Ethics: Past and Present. *Journal of Social Politics Conferences*, (50), 737-759.

Kuznets, S. (n.d.). *National Income, 1929-1932.* Retrieved from https://fraser.stlouisfed.org/title/971

Laczniak, G. R., & Murphy, P. E. (2006). Normative perspectives for ethical and socially responsible marketing. *Journal of Macromarketing, 26*(2), 154–177. doi:10.1177/0276146706290924

Laczniak, G., & Murphy, P. (1993). *Ethical Marketing Decisions: the Higher Road.* Allen and Bacon.

Lafer, G. (2005). The critical failure of workplace ethics. *The ethics of human resources and labor relations*, 273-298.

Larmer, R. A. (1992). Whistleblowing and employee loyalty. *Journal of Business Ethics, 11*(2), 125–128. doi:10.1007/BF00872319

Lawler, E., & Boudreau, J. W. (2009). *Achieving excellence in human resources management: An assessment of human resource functions*. Stanford University Press. doi:10.1515/9780804771238

Lazar, S. (2020). War. In *The Stanford Encyclopaedia of Philosophy*. https://plato.stanford.edu/archives/spr2020/entries/war/

Lazar, J. (2001). *İletişim Bilimi* (C. Anık, Trans.). Vadi Publications.

Lee, I. (2018). Social media analytics for enterprises: Typology, methods, and processes. *Business Horizons*, *61*(2), 199–210. doi:10.1016/j.bushor.2017.11.002

Lee, S., & Jung, H. (2016). The effects of Corporate Social Responsibility on profitability The moderating roles of differentiation and outside investment. *Management Decision*, *54*(6), 1383–1406. doi:10.1108/MD-07-2015-0268

Legge, K. (2000). The ethical context of HRM: the ethical organisation in the boundaryless world. *Ethical issues in contemporary human resource management*, 23-40.

Lehtonen, T. (2018). *Sustainable investment and degrowth. In Strongly Sustainable Societies: Organising Human Activities on a Hot and Full Earth*. Routledge.

Leo'n-Soriano, R., Mun˜oz-Torres, M. J., & Chalmeta-Rosalen, R. (2010). Methodology for sustainability strategic planning and management. *Industrial Management & Data Systems*, *110*(2), 249–268. doi:10.1108/02635571011020331

Leopold, A. (1949). *A Sand County Almanac*. Oxford University Press.

Leszczynska, A. (2010). Manager's attitude toward environment. *Industrial Management & Data Systems*, *110*(8), 1234–1250. doi:10.1108/02635571011077852

Levitt, T. (1960). Marketing Myopia. *Harvard Business Review*, 45–56. PMID:15252891

Lewis, S. C., & Westlund, O. (2014). Big Data and Journalism: Epistemology, expertise, economics, and ethics. *Digital Journalism*, *3*(3), 447–466. doi:10.1080/21670811.2014.976418

Leymann, H. (1996). The content and development of mobbing at work. *European Journal of Work and Organizational Psychology*, *5*(2), 165–184. doi:10.1080/13594329608414853

Lickorish, L. J., & Kershaw, A. G. (1958). *The Travel Trade*. Practical Press.

Li, E. Y., McLeod, R. Jr, & Rogers, J. C. (2001). Marketing Information Systems in Fortune 500 Companies: A Longitudinal Analysis of 1980, 1990 and 2000. *Information & Management*, *38*(5), 307–322. doi:10.1016/S0378-7206(00)00073-2

Li, K., Wang, X., Li, K., & Che, J. (2016). Information privacy disclosure on social network sitess: An empirical investigation from social exchange perspective. *Nankai Business Review International*, *7*(3), 282–300. doi:10.1108/NBRI-02-2015-0005

Lin, D., Hanscom, L., Murthy, A., Galli, A., Evans, M., Neill, E., Mancini, M., Martindill, J., Medouar, F.-Z., Huang, S., & Wackernagel, M. (2018). Ecological Footprint Accounting for Countries: Updates and Results of the National Footprint Accounts, 2012–2018. *Resources*, *7*(3), 58–79. doi:10.3390/resources7030058

Livemint.com. (2019). Retrieved from https://www.livemint.com/industry/media/the-chinese-tech-guru-behind-tiktok-s-success-11574883537442.html

Livingstone, S., Carr, J., & Byrne, J. (2015). *One in three: Internet governance and children's rights*. Global Commission on Internet Governance Paper Series.

Lorenz, K. (1966). *Evolution and modification of behavior*. Methuen.

Lozano, J. (2007). Organizational ethics. In Ethics in the economy (Vol. 165, No. 186, pp. 165-186). Peter Lang Publishing Group in association with GSE Research.

Lull, J. (2001). Medya İletişim Kültür. Ankara, Turkey: Vadi Publication.

Lupton, D. (2014). The commodification of patient opinion: The digital patient experience economy in the age of big data. *Sociology of Health & Disease.*, *36*(6), 856–569. doi:10.1111/1467-9566.12109 PMID:24443847

Lupton, D., & Williamson, B. (2017). The datafied child: The dataveillance of children and implications for their rights. *New Media & Society*, *19*(5), 780–794. doi:10.1177/1461444816686328

Lussier, R. N., & Hendon, J. R. (2017). *Human resource management: Functions, applications, and skill development.* Sage publications.

Lussier, R. N., & Hendon, J. R. (2019). *Fundamentals of Human Resource Management: Functions, Applications, and Skill Development.* SAGE Publications, Incorporated.

Lutz, C., & Strathoff, P. (2014). Privacy Concerns and Online Behavior – Not so Paradoxical After All?Viewing the Privacy Paradox through different theoretical lenses. *SSRN Electronic Journal*, 81-99. doi:10.2139srn.2425132

Ma, Z. (2010). *The Status of Contemporary Business Ethics Research: Present and Future.* Academic Press.

MachiavelliN. (2006). *The Prince.* Retrieved from http://www.gutenberg.org/files/1232/1232-h/1232-h.htm

MacLachlan, A. (2006). An ethic of plurality: Reconciling politics and morality in Hannah Arendt. In History and Judgement. Vienna: IWM Junior Visiting Fellows' Conferences.

Malhotra, N. K., Kim, S. S., & Agarwal, J. (2004). *Internet Users' Information Privacy Concerns (IUIPC): The Construct, the Scale, and a Causal Model.* Academic Press.

Malhotra, N. K., Kim, S. S., & Agarwal, J. (2004). Internet users' information privacy concerns (IUIPC): The construct, the scale, and a causal model. *Information Systems Research*, *15*(4), 336–355. doi:10.1287/isre.1040.0032

Malhotra, N. K., & Miller, G. L. (1998). An integrated model for ethical decisions in marketing research. *Journal of Business Ethics*, *17*(3), 263–280. doi:10.1023/A:1005711311426

Manduchi, A., & Picard, R. (2009). Circulations, Revenues, and Profits in a Newspaper Market with Fixed Advertising Costs. *Journal of Media Economics*, *22*(4), 211–238. doi:10.1080/08997760903375902

Mangold, W. G., & Faulds, D. J. (2009). Social media: The new hybrid element of the promotion mix. *Business Horizons*, *52*(4), 357–365. doi:10.1016/j.bushor.2009.03.002

Marangoz, M. (2018). İnternette Pazarlama. *BETA Bulletin of Experimental Treatments for AIDS.*

March, J. G., & Simon, H. A. (1958). *Organizations.* Wiley.

Marcoux, A. M. (2006). The concept of business in business ethics. *Journal of Private Enterprise, 21*(2), 50-67.

Markham, A. N. (2016). *Ethnography in the Digital Internet Era From fields to flows, descriptions to interventions.* http://jeremydean.org/blog/wp-content/uploads/2016/02/ethnographyinthedigitalera2016finaldraft.pdf

Martin, K. D., & Murphy, P. E. (2017). The role of data privacy in marketing. *Journal of the Academy of Marketing Science*, *45*(2), 135–155. doi:10.100711747-016-0495-4

Martin, M., & Owen, T. (2015). *Introduction. In Routledge Handbook of Human Security.* Routledge.

Marwick, A., & Hargittai, E. (2019). Nothing to hide, nothing to lose? Incentives and disincentives to sharing information with institutions online. *Information Communication and Society, 22*(12), 1697–1713. doi:10.1080/1369118X.2018.1450432

Mascheroni, G., & Cuman, A. (2014). *Net Children Go Mobile: Final report.* Deliverables D6.4 and D5.2. Milano: EDucatt. Retrieved from http://netchildrengomobile.eu/reports/

Maslow, A. H. (1943). A theory of human motivation. *Psychological Review, 50*(4), 370–396. doi:10.1037/h0054346

Mason, P., & Mowforth, M. (1996). Codes of Conduct in Tourism. *Progress in tourism and hospitality research, 2*(2), 151-167.

Mason, R. O. (1986). Four Ethical Issues of the Information Age. Management Information Systems Research Center, 5-12.

Matelski, M. J. (1995). *TV Haberciliğinde Etik* (Ö. Düzgören, Trans.). Yapı Kredi Publications.

Maughan, T. (2017). The Promise ind the Hype of 'personalized Medicine'. *The New Bioethics: a Multidisciplinary Journal of Biotechnology and the Body, 23*(1), 13–20. doi:10.1080/20502877.2017.1314886 PMID:28517988

Mauro, N., Natale, S. M., & Libertella, A. F. (1999). Personal values, business ethics and strategic development. *Cross Cultural Management, 6*(2), 22–28. doi:10.1108/13527609910796942

Mavnacıoğlu, K. (2009). İnternette Kullanıcıların Oluşturduğu ve Dağıttığı İçeriklerin Etik Açıdan İncelenmesi: Sosyal Medya Örnekleri. *Fırat Üniversitesi İletişim Fakültesi Medya ve Etik Semp*, 7-9.

Mayer, D. (2001). Community, Business Ethics, and Global Capitalism. *American Business Law Journal, 38*(2), 215–260. doi:10.1111/j.1744-1714.2001.tb00289.x

Mayfield, A. (2008). *What is social media.* Academic Press.

Mcdonald, G. M. (2009). An anthology of codes of ethics. *European Business Review, 21*(4), 344–372. doi:10.1108/09555340910970445

McNamara, C. (1999). *Complete Guide to Ethics Management: A Ethics Toolkit for Managers.* Academic Press.

Meadows, D. H. (1972). *The limits to growth.* Universe Books.

Megone, C., & Robinson, S. J. (Eds.). (2002). *Case histories in business ethics.* Routledge. doi:10.4324/9780203361207

Mehrabad, M. S., & Brojeny, M. F. (2007). The development of an expert system for effective selection and appointment of the jobs applicants in human resource management. *Computers & Industrial Engineering, 53*(2), 306–312. doi:10.1016/j.cie.2007.06.023

Meral, K. Z., & Meral, Y. (2021). The Role of Social Media in Arab Spring. *e-Journal of New Media, 5*(1), 26-34.

Meral, Y., & Özbay, D. E. (2020). Electronic Trading, Electronic Advertising, and Social Media Literacy: Using Local Turkish Influencers in Social Media for International Trade Products Marketing. In N. Taskiran (Ed.), *Handbook of Research on Multidisciplinary Approaches to Literacy in the Digital Age* (pp. 224–261). IGI Global. doi:10.4018/978-1-7998-1534-1.ch012

Miceli, M. P., Near, J. P., & Dworkin, T. M. (2008). *Whistle-blowing in organizations.* Psychology Press. doi:10.4324/9780203809495

Mies, M., & Shiva, V. (1993). *Ecofeminism.* Zed Books. doi:10.5040/9781350219786

Milberg, S. J., Burke, S. J., Smith, H. J., & Kallman, E. A. (1995). Values, Personal Information, Privacy and Regulatory Approaches. *Communications of the ACM, 38*(12), 65–74. doi:10.1145/219663.219683

Milgram, S. (1963). Behavioral study of obedience. *Journal of Abnormal and Social Psychology, 67*(4), 371–378. doi:10.1037/h0040525 PMID:14049516

Milgram, S. (1974). *Obedience to authority: An experimental view*. Harpercollins.

Miltgen, C. L., & Smith, H. J. (2015). Makale adı: Exploring information privacy regulation, risks, trust, and behavior. *Information & Management, 52*(6), 741–759. doi:10.1016/j.im.2015.06.006

Miron, D., Petcu, M., & Sobolevschi, I. M. (2011). Corporate Social Responsibility and the sustainable competitive advantage. *Amfiteatru Economic, 13*, 29, 162–179.

Misirlis, N., & Vlachopoulou, M. (2018). Social media metrics and analytics in marketing–S3M: A mapping literature review. *International Journal of Information Management, 38*(1), 270–276. doi:10.1016/j.ijinfomgt.2017.10.005

Mitchell, T. R., Daniels, D., Hopper, H., George-Falvy, J., & Ferris, G. R. (1996). Perceived correlates of illegal behavior in organizations. *Journal of Business Ethics, 15*(4), 439–455. doi:10.1007/BF00380364

Monin, P., Noorderhaven, N., Vaara, E., & Kroon, D. (2013). Giving sense to and making sense of norms of justice in post-merger integration. *Academy of Management Journal, 56*(1), 256–284. doi:10.5465/amj.2010.0727

Moore, E. S., & Rideout, V. J. (2007). The online marketing of food to children: Is it just fun and games? *Journal of Public Policy & Marketing, 26*(2), 202–220. doi:10.1509/jppm.26.2.202

Morris, D. (2004). Defining a moral problem in business ethics. *Journal of Business Ethics, 49*(4), 347–357. doi:10.1023/B:BUSI.0000020866.05705.d2

Mucuk, İ. (2014). *Pazarlama İlkeleri*. Türkmen.

Muenjohn, N. (2008). *Leadership Theories and Concepts: Its Past and Present*. Paper presented SouthEast Asia Regional Conference: Globalization, Innovation, and Leadership, Kuala Lumpur.

Murphy, P. B. (2002). *Marketing ethics at the millennium: Review, reflections, and recommendations. In The Blackwell Guide to Business*. Blackwell.

Murphy, P. E., Laczniak, G. R., & Wood, G. (2007). An ethical basis for relationship marketing: A virtue ethics perspective. *European Journal of Marketing, 41*(1/2), 37–57. doi:10.1108/03090560710718102

Mustafa, B. (2011). Relationship between ethical climate, organizational trust and individual performance. *Atatürk University Journal of Economic and Administrative Sciences, 25*(1), 171–192.

Mutlu, E. (1991). *Televizyonu Anlamak*. Gündoğan Publications.

Myerson, R. B. (1997). *Game Theory: Analysis of Conflict*. Harvard University Press.

Næss, A. (1973). The Shallow and the Deep Long-Range Ecology Movement. *Inquiry, 16*(1), 95–100. doi:10.1080/00201747308601682

Næss, A. (1989). *Ecology, Community and Life Style*. Cambridge University Press. doi:10.1017/CBO9780511525599

Nahm, T. (2013). 1 Transhumanismus: Die Letzte Grosse Erzählung. In Reader zum Transhumanismus. Berlin: BoD.

Nairn, A., & Fine, C. (2008). Who's messing with my mind? The implications of dual-process models for the ethics of advertising to children. *International Journal of Advertising, 27*(3), 447–470. doi:10.2501/S0265048708080062

Nakano, C. (2007). The significance and limitations of corporate governance from the perspective of business ethics: Towards the creation of an ethical organizational culture. *Asian Business & Management*, *6*(2), 163–178. doi:10.1057/palgrave.abm.9200216

Naked prince news ethics. (2012a). Retrieved from https://www.telegraph.co.uk/news/politics/9712859/Leveson-Naked-Harry-and-Kate-pictures-show-internet-is-an-ethical-vacuum.html

Naked prince news ethics. (2012b). Retrieved from https://www.chicagotribune.com/lifestyles/ct-xpm-2012-08-23-sns-rt-us-britain-prince-nakedbre87l08c-20120822-story.html

Nalçaoğlu, H. (2005). *Medya ve Toplum İlişkisini Anlamak Üzere Bir Çerçeve. In Habercinin El Kitabı Dizisi: Medya ve Toplum.* Turkey IPS İletişim Vakfı Yayınları.

Nash, R. (1989). *The Rights of Nature: A History of Environmental Ethics.* The University of Wisconsin Press.

Near, J. P., & Miceli, M. P. (1985). Organizational dissidence: The case of whistle-blowing. *Journal of Business Ethics*, *4*(1), 1–16. doi:10.1007/BF00382668

Nebenzhal, I. D., & Jaffe, E. D. (1998). Ethical dimensions of advertising executions. *Journal of Business Ethics*, *17*(7), 805–815. doi:10.1023/A:1005850812845

Neiman, S. (2015). *Evil in Modern Thought: An Alternative History of Philosophy.* Princeton University Press.

Nemati, H. (2008). *Information Security and Ethics: Concepts, Methodologies, Tools, and Applications.* Information Science Reference. doi:10.4018/978-1-59904-937-3

Nixon, R. (2011). *Slow Violence and the Environmentalism of the Poor.* Harvard University Press. doi:10.4159/harvard.9780674061194

Norton, B. G. (1991). *Toward Unity among Environmentalists.* Oxford University Press.

Nunan, D., & Domenico, M. D. (2012). Market research and the ethics of big data. *International Journal of Market Research*, *55*(4), 505–520. doi:10.2501/IJMR-2013-015

Nurrachmi, R. (2019). *The Potential Industry of Islamic Tourism in ASEAN Countries.* Academic Press.

Nussbaum, M. (2000). *Women and Human Development: A Study in Human Capabilities.* Cambridge University Press. doi:10.1017/CBO9780511841286

Nussbaum, M. (2011). *Creating Capabilities: The Human Development Approach.* Belknap Press. doi:10.4159/harvard.9780674061200

O'Higgins, E., & Kelleher, B. (2005). Comparative perspectives on the ethical orientations of human resources, marketing and finance functional managers. *Journal of Business Ethics*, *56*(3), 275–288. doi:10.100710551-004-3898-y

O'Neil, D. (2001). Analysis of Internet Users' Level of Online Privacy Concerns. *Social Science Computer Review*, *19*(1), 17–31. doi:10.1177/089443930101900103

O'Reilly, T. (2007). What Is Web 2.0: Design Patterns and Business Models for the Next Generation of Software. *Munich Personal RePEc Archive*, 17-37. https://mpra.ub.uni-muenchen.de/4578/

Obesity and the Economics of Prevention Fit not Fat. (2010). OECD Publishing.

ODIHR. (2009). *Hate Crime Laws A Practical Guide.* Retrieved from http://www.osce.org/odihr /36426?download=true

Oğuz, N. Y. (1997). *Tıp Etiğinde Temel İlkeler.* Etik Bunun Neresinde, Funda Matbaacılık.

Okhovat, H. (2010, September). A Study on Religious Tourism Industry Management, Case Study: Islamic Republic of Iran. *International Journal of Academic Research, 2*(5).

Omar, B., & Dequan, W. (2020). *Watch, share or create: The influence of personality traitsand user motivation on TikTok mobile video usage*. Academic Press.

Omar, C. M. C. (2019). Exploring the Validation of Islamic Hospitality Management in Tourism Destination: Malaysian Perspectives. *Journal of Hospitality and Networks, 1*(1), 34–43.

Ortmeier, P. J., & Meese, E. III. (2009). *Leadership, Ethics and Policing: Challenges for the 21st Century* (2nd ed.). Pearson.

Ott, J. S. (1989). *The organizational culture perspective*. Dorsey Press.

Ouchi, W. G., & Wilkins, A. L. (1985). Organizational culture. *Annual Review of Sociology, 11*(1), 457–483. doi:10.1146/annurev.so.11.080185.002325

Ozcomert, G. H., Ozcelikay, G., Sar, S., & Asil, E. (2000). The Assessment of Benefıcence And Non-Malefıcıence prıncıples From The Perspectıve of Pharmaceutıcal Ethıcs. *Turkiye Klinikleri J Med Ethics, 8*(2), 101–104.

Özel. Kurguları, & Algıları. (n.d.). *Bildiriler Kitabı*, 397. Pew Research Center. Retrieved from: http://gradelevelreading.net/wp-content/uploads/2018/10/Role-of-Media-in-Supporting-Parent-Success-Pre-Readings-Combined.pdf

Özgener, Ş. (2004). *İş Ahlakının Temelleri*. Yönetsel Bir Yaklaşım, Nobel Yayın Dağıtım.

Özkalp, E., & Kırel, Ç. (2011). Örgütsel Davranış, Ekin Basım-Yayım, Ekim 2011. Bursa.

Page, S. (2011). *Tourism Management*. Routledge Press. doi:10.4324/9780080879369

Pamukçu, H., & Sarıışık, M. (2017). Helal Turizm Kavramı ve Gelişimi Üzerine Genel Bir Değerlendirme. *Uluslararası İktisadi ve İdari Bilimler Dergisi, 3*(1), 82–98.

Papacharissi, Z. (2002). The presentation of self in virtual life: Characteristics of personal home pages. *Journalism & Mass Communication Quarterly, 79*(3), 643–660. doi:10.1177/107769900207900307

Parker, M. (2003). Business, ethics and business ethics: Critical theory and negative dialectics. *Studying management critically*, 197-219.

Parker, M. (Ed.). (1998). *Ethics & organizations*. Sage. doi:10.4135/9781446280171

Patient, D. L., & Skarlicki, D. P. (2010). Increasing interpersonal and informational justice when communicating negative news: The role of the manager's empathic concern and moral development. *Journal of Management, 36*(2), 555–578. doi:10.1177/0149206308328509

Patrick, M. S. (2018). *TikTok video app has become a petri dish for youth culture*. Retrieved September 20, 2018, from https://www.japantimes.co.jp/news/2018/08/25/national/media-national/tiktok-video-appbecome-petri-dish-youth-culture/#.XIda5ohKjid

Patterson, P., & Wilkins, L. (1994). *Media ethics: issues and cases*. Brown & Benchmark, cop.

Payne, D., Raiborn, C., & Askvik, J. (1997). A global code of business ethics. *Journal of Business Ethics, 16*(16), 1727–1735. doi:10.1023/A:1005877726316

Peng, J., Zhang, G., Zhang, S., Dai, X., & Li, J. (2014). Effects of online advertising on automobile sales. *Management Decision, 52*(5), 834–851. doi:10.1108/MD-10-2012-0753

Pernice, I. (2019). Protecting the global digital information ecosystem: A practical initiative. *Internet Policy Review, 8*. https://policyreview.info/articles/news/protecting-global-digital-information-ecosystem-practical-initiative/1386

Perran, A. K. A. N. (2007). Uygulama açısından iş etiği kuralları ve evrensel turizm etiği ilkeleri. Anatolia. *Turizm Araştırmaları Dergisi, 18*(1), 7–20.

Persson, I., & Savulescu, J. (2014). *Unfit for the future*. Oxford University Press.

Peters, K., Chen, Y., Kaplan, A. M., Ognibeni, B., & Pauwels, K. (2013). Social media metrics—A framework and guidelines for managing social media. *Journal of Interactive Marketing, 27*(4), 281–298. doi:10.1016/j.intmar.2013.09.007

Peters, T. J., & Waterman, R. R. H. (1982). *Search of Excellence: Lessons from America's Best Run Companies* (Vol. 1). Harper & Row.

Petty, G. C., & Hill, R. B. (2005). Work ethic characteristics: Perceived work ethics of supervisors and workers. *Journal of Industrial Teacher Education, 42*(2), 5–20.

Pew Research Center. (2019). *Public's 2019 Priorities: Economy, Health Care, Education and Security All Near Top of List*. https://www.people-press.org/2019/01/24/publics-2019-priorities-economy-health-care-education-and-security-all-near-top-of-list/

Phelan, S. M., Burgess, D. J., Yeazel, M. W., Hellerstedt, W. L., Griffin, J. M., & van Ryn, M. (2015). Impact of weight bias and stigma on quality of care and outcomes for patients with obesity. *Obesity Reviews, 16*(4), 319–326.

Phillip, M. (2014). *Public ethics and political Judgment*. Committee on Standards in Public Life.

Pieper, J. (1991). *Philosophie*. Kontemplation, Weisheit.

Pinnington, A., Macklin, R., & Campbell, T. (Eds.). (2007). *Human resource management: Ethics and employment*. Oxford University Press.

Plumwood, V. (1993). *Feminism and the Mastery of Nature*. Routledge.

Politics. (1996). In *Collins English Dictionary* (7th ed.). Harper Collins Publishers.

Poon, M., & Tarrant, M. (2009). Obesity: attitudes of undergraduate student nurse and registerednurses. *Journal of Clinical Nursing, 18*, 2355-2365.

Porter, A. (2017). Bioethics and Transhumanism. *The Journal of Medicine and Philosophy, 42*(3), 237–260. doi:10.1093/jmp/jhx001 PMID:28499043

Porter, G. (2005). A "career" work ethic versus just a job. *Journal of European Industrial Training, 29*(4), 336–352. doi:10.1108/03090590510597160

Porter, M. E., & Kramer, M. R. (2006). Strategy & society. The link between competitive advantage and Corporate Social Responsibility. *Harvard Business Review*, (December), 5, 78–94. PMID:17183795

Porter, S. (1997). The patient and power: Sociological perspectives on the consequences of holistic care. *Health & Social Care in the Community, 51*(1), 17–20. doi:10.1111/j.1365-2524.1997.tb00091.x

Postman, N. (2010). *Televizyon Öldüren Eğlence* (O. Akınhay, Trans.). Ayrıntı Publications.

Pratley, P. (1995). *The essence of business ethics* (Vol. 111). Prentice Hall.

Preece, J., Nonnecke, B., & Andrews, D. (2004). The top five reasons for lurking: Improving community experiences for everyone. *Computers in Human Behavior, 20*(2), 201–223. doi:10.1016/j.chb.2003.10.015

Price, A. (2007). *Human resource management in a business context*. Cengage Learning EMEA.

Priori Data. (2020). Retrieved from https://prioridata.com/publishers/tiktok-pte-ltd-1322881000/profile?publisher_id=1322881000&from=1&free=true&paid=true&free_plus=true&paid_plus=true&downloads=true&metrics_country=GG&time_period=last_cal_month

Priori Data. (2020a). *Leading Android apps in the Google play store worldwide in November 2019, by number of downloads (in millions)*. Statista. Retrieved from www.statista.com/ statistics/693944/leading-android-apps-world wide-by-downloads/

Priori Data. (2020b). *Leading iphone apps in the Apple app store worldwide in November 2019, by number (in millions)*. Statista. Retrieved from www.statista.com/statistics/695791/ top-iphone-apps-worldwide-by-number-ofdownloads/)

PRNewswire.com. (2018). Retrieved from https://www.prnewswire.com/news-releases/tiktok-captures-the-worlds-most-creative-videos-in-year-in-rewind-compilation-300767154.html

Prothero, A. (1990). Green Consumerism and the Societal Marketing Concept: Marketing Strategies for the 1990s. *Journal of Marketing Management*, *6*(2), 91. doi:10.1080/0267257X.1990.9964119

Puhl, R. M., & Brownell, K. D. (2006). Confronting and coping with weight stigma: An investigation of overweight and obese adults. *Obesity (Silver Spring, Md.)*, *14*(10), 1802–1815.

Puhl, R. M., & Heuer, C. A. (2009). The stigma of obesity: A review and update. *Obesity (Silver Spring, Md.)*, *17*(5), 941–964.

Puhl, R. M., & King, K. M. (2013). Weight discrimination and bullying. *Best Practice & Research. Clinical Endocrinology & Metabolism*, *27*, 117–127.

Puhl, R., Wharton, C., & Heuer, C. (2009). Weight bias among dietetics students: Implications for treatment practices. *Journal of the American Dietetic Association*, *109*, 438–444.

Quan-Haase, A., & Young, A. L. (2010). Uses and gratifications of social media: A comparison of Facebook and instant messaging. *Bulletin of Science, Technology & Society*, *30*(5), 350–361. doi:10.1177/0270467610380009

Rafik Essa Beekun. (1997). *Islamic Business Ethics*. The International Institute of Islamic Thought.

Ramonet, I. (2000). *Medyanın Zorbalığı* (A. Derman, Trans.). Om Yayınevi.

Randall, D. M., & Gibson, A. M. (1990). Methodology in business ethics research: A review and critical assessment. *Journal of Business Ethics*, *9*(6), 457–471. doi:10.1007/BF00382838

Ratković-Njegova, B., & Šiđanin, I. (2014). Public broadcasting crisis as management crisis: A case study of radio television of Vojvodina. *Journal for East European Management Studies*, *19*(3), 348–367. doi:10.5771/0949-6181-2014-3-348

Regan, T. (1983). *The Case for Animal Rights*. University of California Press.

Reidenbach, R. E., & Robin, D. P. (1991). A Conceptual Model of Corporate A Conceptual Model of Corporate. *Journal of Business Ethics*, 273–284. doi:10.1007/BF00382966

Reiss, M. C., & Mitra, K. (1998). The effects of individual difference factors on the acceptability of ethical and unethical workplace behaviors. *Journal of Business Ethics*, *17*(14), 1581–1593. doi:10.1023/A:1005742408725

Remisová, A., Lasáková, A., & Krzykala-Schaefer, R. (2013). Corporate Social Responsibility in European countries: The keystones of the concept and intercultural connotations. *Journal for East European Management Studies*, *18*(4), 512–543. doi:10.5771/0949-6181-2013-4-512

Ren, Y., Harper, F. M., Drenner, S., Terveen, L., Kiesler, S., Riedl, J., & Kraut, R. E. (2012). Building member attachment in online communities: Applying theories of group identity and interpersonal bonds. *Management Information Systems Quarterly*, *36*(3), 841–864. doi:10.2307/41703483

Rest, J., Turiel, E., & Kohlberg, L. (1969). Level of moral development as a determinant of preference and comprehension of moral judgments made by others. *Journal of Personality, 37*(2), 225-252.

Reyes, I., Wijesekera, P., Reardon, J., On, A. E. B., Razaghpanah, A., Vallina-Rodriguez, N., & Egelman, S. (2018). "Won't somebody think of the children?" examining COPPA compliance at scale. *Proceedings on Privacy Enhancing Technologies*, *2018*(3), 63–83. doi:10.1515/popets-2018-0021

Reynolds, P. (2000). 'Profit and Principles: Business Ethics in Hotel Management Companies in Asia'. *Australian Journal of Hospitality Management*, *7*(1), 1–13.

Ricoeur, P. (1973). Ethics and culture. *Philosophy Today*, *17*(2), 153–165. doi:10.5840/philtoday197317235

Riechmann, J. (2000). *Un mundo vulnerable. Ensayos sobre ecología, ética y tecnociencia*. Los Libros de la Catarata.

Rigel, N., & Batuş, G. (2005). *21. Yüzyıl İletişim Çağını Aydınlatan Kuramcılar*. Su Publications.

Robehmed, N., & Berg, M. (2018). Highest-paid YouTube stars 2018: Markiplier, Jake Paul, PewDiePie and more. *Forbes*. Retrieved from https://www. forbes. com/sites/natalierobehmed/2018/12/03/highest-paid-youtube-stars-2018-markiplierjake-paul-pewdiepie-and-more

Robertson, D. C., & Schlegelmilch, B. B. (1993). Corporate institutionalization of ethics in the United States and Great Britain. *Journal of Business Ethics*, *12*(4), 301–312. doi:10.1007/BF01666534

Rojek-Adamek, P. (2018). From Relational Space to Engagement—Designers' Ethics and Responsibility in the Light of Empirical Research. *Polish Sociological Review, 1*(201), 95-107.

Rokhman, W. (2010). The effect of Islamic work ethics on work outcomes. *EJBO. Electronic Journal of Business Ethics and Organization Studies*.

Rollin, B. E. (1980). *Animal Rights and Human Morality*. Prometheus Books.

Rollin, B. E. (2016). *A New Basis for Animal Ethics: Telos and Common Sense*. University of Missouri Press.

Rollin, B. E., & Kesel, M. L. (1990). *The Experimental Animal in Biomedical Research*. Boca Raton: CRC Press.

Rose, A. (2007). Ethics and human resource management. *Newyork. McGrow-Hill*, *37*, 38.

Rossiter, J. R., & Percy, L. (1979). *Visual communication in advertising*. Columbia University, Graduate School of Business.

Royakkers, L., Timmer, J., Kool, L., & van Est, R. (2018). Societal and ethical issues of digitalization. *Ethics and Information Technology*, *20*(2), 127–142. doi:10.100710676-018-9452-x

Royal College of Nursing. (2003, Apr. 22). *Definingnursing*. Retrieved Jan. 23, 2012, Retrieved from https://www.rcn.org.uk/__data/assets/pdf_file/0008/78569/001998.pdf

Rozendaal, E., Buijzen, M., & Valkenburg, P. (2011). Children's understanding of advertisers' persuasive tactics. *International Journal of Advertising*, *30*(2), 329–350. doi:10.2501/IJA-30-2-329-350

Ruh, H. (2004). *Ethik im Management. Ethik und Erfolg verbunden sich*. Zuerich: Orell Fuessli Verlag AG.

Ruiz-Lozano, M., De-los-Ríos-Berjillos, A., & Millán-Lara, S. (2018). Spanish Hotel Chains Alignment with the Global Code of Ethics for Tourism. *Journal of Cleaner Production*, *199*, 205–213. doi:10.1016/j.jclepro.2018.07.133

Runes, D. D. (1964). Dictionary of Philosophy. Littlefields, Adams and Co.

Rusk, J.-D. (2014). Trust and Decision Making in the Privacy Paradox. *Proceedings of the Southern Association for Information Systems Conference*, 1-6.

Sabin, J. A., Marini, M., & Nosek, B. A. (2012). Implicit and explicit anti-fat bias among a large sample of medical doctors by BMI, race/ethnicity and gender. *PLoS One*, *7*(11).

Saeed, M., Ahmed, Z. U., & Mukhtar, S. M. (2001). International marketing ethics from an Islamic perspective: A value-maximi- zation approach. *Journal of Business Ethics*, *32*(2), 127–142. doi:10.1023/A:1010718817155

Safko, L. (2010). *The social media bible: tactics, tools, and strategies for business success*. John Wiley & Sons.

Samsum, N. (2017). Discussions on the changing meaning and current state of the work. *Journal of Open Education Applications and Research*, *3*(3), 160–210.

Samuels, C., Brown, Q., Leoschut, L., Jantjies, J., & Burton, P. (2013). *Connected dot com: Young people's navigation of online risks. Social media, ICT's and online safety*. Centre for Justice and Crime Prevention and UNICEF.

Sandberg, H. (2011). Tiger talk and candy king: Marketing of unhealthy food and beverages to Swedish children. *Communications*, *36*(2), 217–244. doi:10.1515/comm.2011.011

Sandler, R. (2007). *Character and Environment: A Virtue-Oriented Approach to Environmental Ethics*. Columbia University Press.

Saul, G. K. (1981). Business ethics: Where are we going? *Academy of Management Review*, *6*(2), 269–276. doi:10.5465/amr.1981.4287830

Sayli, H., & Kızıldağ, D. (2007). Yönetsel Etik ve Yönetsel Etiğin Oluşmasında İnsan Kaynakları Yönetiminin Rolünü Belirlemeye Yönelik Bir Analiz. *Afyon Kocatepe Üniversitesi Sosyal Bilimler Dergisi*, *9*(1), 231–251.

Schaeffer, J-M. (2007). *La fin de l'exception humaine*. Paris: Editions Gallimard.

Schein, E. H. (1985). Defining organizational culture. *Classics of Organization Theory*, *3*(1), 490-502.

Schein, E. H. (1990). Organizational culture. American Psychological Association.

Schlegelmilch, B. B. (1998). *Marketing ethics: An international perspective*. Cengage Learning EMEA.

Schlosberg, D. (2007). *Defining Environmental Justice: Theories, Movements, and Nature*. Oxford University Press. doi:10.1093/acprof:oso/9780199286294.001.0001

Schmidtz, D. (2011). Respect for Everything. *Ethics, Policy & Environment*, *14*(2), 127–138. doi:10.1080/21550085.2011.578359

Schreck, T., & Keim, D. (2012). Visual analysis of social media data. *Computer*, *46*(5), 68–75. doi:10.1109/MC.2012.430

Schuh, S. C., Zheng, M. X., Xin, K. R., & Fernandez, J. A. (2017). The interpersonal benefits of leader mindfulness: A serial mediation model linking leader mindfulness, leader procedural justice enactment, and employee exhaustion and performance. *Journal of Business Ethics*, 1-19.

Schuler, R. S. (2000). The internationalization of human resource management. *Journal of International Management*, *6*(3), 239–260. doi:10.1016/S1075-4253(00)00025-9

Schumann, P. L. (2001). A moral principles framework for human resource management ethics. *Human Resource Management Review*, *11*(1-2), 93–111. doi:10.1016/S1053-4822(00)00042-5

Schwartz, M. (2007). The "business ethics" of management theory. *Journal of Management History*, *13*(1), 43–54. doi:10.1108/17511340710715160

Schweitzer, A. (1923). *The Philosophy of Civilization*. A. & C. Black Ltd.

Schwoerer, C. E., May, D. R., & Rosen, B. (1995). Organizational characteristics and HRM policies on rights: Exploring the patterns of connections. *Journal of Business Ethics*, *14*(7), 531–549. doi:10.1007/BF00871982

Scott, E. (2016). Ethics and human resource management. In *Practicing professional ethics in economics and public policy* (pp. 215–221). Springer. doi:10.1007/978-94-017-7306-5_11

Scrambler, G., & Britten, N. (2001). System, lifeworld and doctor-patient interaction. In G. Scrambler (Ed.), *Habermas, critical theory and health* (pp. 45–67). Routledge.

Sebag, V. (2007). *Droit et bioéthique*. De Boeck & Lancier.

Seidman, G. (2013). The need to belong: Desire for interpersonal attachments as a fundamental human motivation. *Personality and Individual Differences*, *54*(3), 402–407. doi:10.1016/j.paid.2012.10.009

Selvi, M. S., & Şahin, S. (2012). Sustainable Tourism: Local Government Perspective The Example of Western Blacksea Region. *International Journal of Social and Economic Sciences*, *2*(2), 23–36.

Şen, E. (2014). *Kişisel Verilerin Kaydedilmesi*. Retrieved from http://www.haber7.com/yazarlar/prof-dr-ersan-sen/1235168-kisisel-verilerin-kaydedilmesi

Sen, A. (2009). *The idea of justice*. Penguin Books Limited.

Sen, A. (2015). *Birth of discourse. In Routledge Handbook of Human Security*. Routledge.

Sen, A. K. (1984). Well-Being, Agency and Freedom: The Dewey Lectures 1984. *The Journal of Philosophy*, *82*(4), 169–221.

Sen, A. K. (2013). The Ends and Means of Sustainability. *Journal of Human Development and Capabilities*, *14*(1), 6–20. doi:10.1080/19452829.2012.747492

Sentürk, F. K., Bayirli, M., & Güçlü, C. (2016). The Effects of Islamic Work Ethics and the Ethics of Love of Money over the Perception of Smuggling:The Example of the Alanya Tradesmen. *Turkish Journal of Business Ethics*, *9*(2), 208–220.

Seyfi, S., & Hall, C. M. (2019). Deciphering Islamic theocracy and tourism: Conceptualization, context, and complexities. *International Journal of Tourism Research*, *21*(6), 735–746. doi:10.1002/jtr.2300

Shao, G. (2009). Understanding the appeal of user-generated media: A uses and gratification perspective. *Internet Research*, *19*(1), 7–25. doi:10.1108/10662240910927795

Shaw, W. (1999). *Contemporary ethics: Taking account of utilitarianism*. Academic Press.

Shaw, W. H., & Barry, V. (2015). *Moral issues in business*. Cengage Learning.

Sheehan, M. (2004). Workplace mobbing: A proactive response. In *Workplace Mobbing Conference* (pp. 1-11). Academic Press.

Sheridan, J. E. (1992). Organizational culture and employee retention. *Academy of Management Journal*, *35*(5), 1036–1056.

Sheth, J. N., Sisodia, R. S., & Sharma, A. (2000). The Antecedents and Consequences of Customer-Centric Marketing. *Journal of the Academy of Marketing Science*, *28*(1), 55–66. doi:10.1177/0092070300281006

Sidgwick, H. (1877). Bentham and Benthamism in politics and ethics. *Fortnightly, 21*(125), 627–652.

Sikorski, C., Luppa, M., Glaesmer, H., Brähler, E., König, H. H., & Riedel-Heller, S. G. (2013). Attitudes of health care professionals towards female obese patients. *Obesity Facts*, *6*(6), 512–522.

Simmons, J. (2008). Ethics and morality in human resource management. *Social Responsibility Journal*, *4*(1/2), 8–23. doi:10.1108/17471110810856794

Singer, J. B. (2013). The Ethical Implications of an Elite Press. *Journal of Mass Media Ethics*, *28*(3), 203–216. doi:10.1080/08900523.2013.802163

Singer, P. (1975). *Animal Liberation*. New York: New York Review Press.

Singer, P. (1990). *Animal Liberation: A New Ethics for Our Treatment of Animals*. Avon Books.

Singer, P. (2011). *Practical ethics*. Cambridge University Press. doi:10.1017/CBO9780511975950

Singer, P. (2016). *Ethics in the Real World: 82 Brief Essays on Things That Matter*. Princeton University Press.

Singhapakdi, A., Karande, K., Rao, C. P., & Vitell, S. J. (2001). How important are ethics and social responsibility?-A multinational study of marketing professionals. *European Journal of Marketing*, *35*(1/2), 133–153. doi:10.1108/03090560110363382

Singhapakdi, A., Vitell, S. J., & Kraft, K. L. (1996). Moral intensity and ethical decision-making of marketing professionals. *Journal of Business Research*, *36*(3), 245–255. doi:10.1016/0148-2963(95)00155-7

Sintonen, T. M., & Takala, T. (2002). Racism and ethics in the globalized business world. *International Journal of Social Economics*, *29*(11), 849–860. doi:10.1108/03068290210446267

Slavtcheva-Petkova, V., Nash, V. J., & Bulger, M. (2015). Evidence on the Extent of Harms Experienced by Children as a Result of Online Risks: Implications for policy and research. *Information Communication and Society*, *18*(1), 48–62. doi:10.1080/1369118X.2014.934387

Small, M. W. (1993). Ethics in business and administration: An international and historical perspective. *Journal of Business Ethics*, *12*(4), 293–300. doi:10.1007/BF01666533

Smilansky, J. (2002). *Yeni İnsan Kaynakları, Çev*. Derya Atakan, Epsilon Yayıncılık.

Smith, A. F. (1978). Lawrence Kohlberg's cognitive stage theory of the development of moral judgment. *New Directions for Student Services*, 53-67.

Smith, M. J. (1998). *E-merging strategies of identity: The rhetorical construction of self in personal web sites* (Doctoral dissertation). Ohio University.

Smith, A. (1987). *The Essential Adam Smith*. WW Norton & Company.

Smith, A., Voß, J.-P., & Grin, J. (2010). Innovation Studies and Sustainability Transitions: The Allure of the Multi-level Perspective and Its Challenges. *Research Policy*, *39*(4), 435–448. doi:10.1016/j.respol.2010.01.023

Smith, C., Baveja, R., Grieb, T., & Mashouri, G. A. (2017). Toward a science of translational science. *Journal of Clinical and Translational Science*, *1*(4), 253–255. doi:10.1017/cts.2017.14 PMID:29657860

Smith, J., Milberg, S. J., & Burke, S. J. (1996). Information Privacy: Measuring Individuals' Concerns about Organizational Practices. *Management Information Systems Research Center*, *20*(2), 167–196. doi:10.2307/249477

Smith, M., & Duffy, R. (2004). *The Ethics of Tourism Development*. Routledge Press.

Smith, N. C. (2001). Ethical guidelines for marketing practice: A reply to Gaski and some observations on the role of normative marketing ethics. *Journal of Business Ethics*, *32*(1), 3–18. doi:10.1023/A:1010685524206

Smock, A. D., Ellison, N. B., Lampe, C., & Wohn, D. Y. (2011). Facebook as a toolkit: A uses and gratification approach to unbundling feature use. *Computers in Human Behavior*, *27*(6), 2322–2329. doi:10.1016/j.chb.2011.07.011

Smolentceva, N. (2019). *TikTok: World's most successful video app faces security concerns*. Retrieved from https://www.dw.com/en/tiktok-worlds-most-successful-video-app-faces-security-concerns/a-48063869

Snoeyenbos, M., & Hummel, J. (1979). Business ethics. *Teaching Philosophy*, *3*(2), 238–243. doi:10.5840/teachphil19793244

Solis, B. (2008). *The essential guide to social media.* http://www. scribd. com/doc/3283966/The-Essential-Guide-to-Social-Media

Somers, M. J. (2001). Ethical Codes of Conduct and Organizational Context: A Study of the Relationship Between Codes of Conduct, Employee Behavior and Organizational Values. *Journal of Business Ethics*, *30*(2), 185–195. doi:10.1023/A:1006457810654

Sparks, C. (2013). Global media studies: Its development and dilemmas. *Media Culture & Society*, *35*(1), 121–131. doi:10.1177/0163443712464566

Sparks, J. R., & Pan, Y. (2010). Ethical judgments in business ethics research: Definition, and research agenda. *Journal of Business Ethics*, *91*(3), 405–418. doi:10.100710551-009-0092-2

Sparrow, R., & Sparrow, L. (2006). In the hands of machines? The future of aged care. *Minds and Machines*, *16*(2), 141–161. doi:10.100711023-006-9030-6

Spash, C. L. (2002). *Greenhouse economics: Value and ethics* (Vol. 1). Taylor & Francis. doi:10.4324/9780203209059

Spence, L. (2000). What ethics in the employment interview. *Ethical issues in contemporary human resource management*, 43-58.

Spence, E., & Van Heekeren, B. (2004). *Advertising Ethics*. Pearson.

Sprott, D. E. (2008). The Policy, Consumer, and Ethical Dimensions of Covert Marketing. *Journal of Public Policy & Marketing*, *27*, 1, 4–6.

Stallworth, L. E., McPherson, T., & Rute, L. (2001). Discrimination in the workplace: How mediation can help.... *Dispute Resolution Journal*, *56*(1), 35.

Standard, D. (2012). *Global Muslim Lifestyle Travel Market 2012: Landscape and Consumer Need Study for Airlines, Destinastions, Hotels & Resorts Executive Summary*. https://www.dinarstandard.com/travel-study/

Stanford Libraries. (2019). *Guiding Principles on Ethics in Digital Health Produced During a Seminar at Stanford Libraries*. https://library.stanford.edu/node/156021

Stankevičienė, J., & Čepulytė, J. (2014). Sustainable value creation: Coherence of Corporate Social Responsibility and performance of socially responsible investment funds. *Economic Research-Ekonomska Istraživanja*, *27*(1), 882–898. doi:10.1080/1331677X.2014.976058

Stark, A. (1993). What's the matter with business ethics? *Harvard Business Review*, *71*(3), 38–40. PMID:10126154

Statista.com. (2019). Retrieved from https://www.statista.com/chart/16939/tiktok-app-downloads/

Statista.com. (2019). Retrieved from https://www.statista.com/statistics/248074/most-popular-us-social-networking-apps-ranked-by-audience/

Statistica.com. (2020a). *Most viewed YouTube channel owners of all time as of January 2020, by views (in billions).* Retrieved from: https://www.statista.com/statistics/373753/most-viewed-youtubers-all-time/

Statistica.com. (2020b). *Spending on advertising to children worldwide from 2012 to 2021, by format (in billion U.S. dollars).* Retrieved from: https://www.statista.com/statistics/750865/kids-advertising-spending-worldwide/

Staub, E. (2003). Notes on cultures of violence, cultures of caring and peace, and fulfillment of basic human needs. *Political Psychology, 24*(1), 1–19. doi:10.1111/0162-895X.00314

Sternberg, E. (2000). *Just business: Business ethics in action.* Academic Press.

Stoilova, M., Livingstone, S., & Nandagiri, R. (2019). *Children's data and privacy online: growing up in a digital age: research findings.* Academic Press.

Stone, C. D. (1972). Should Trees Have Standing? Toward Legal Rights for Natural Objects. *Southern California Law Review, 45*, 450–501.

Suchááek, J., Seìa, P., Friedrich, V., & Koutsk˘, J. (2014). Media portrayals of regions in the Czech Republic: Selected issues. *E+M. Ekonomie a Management, 17*(4), 125–140. doi:10.15240/tul/001/2014-4-010

Suhonen, R., Stolt, M., Virtanen, H., & Leino-Kilpi, H. (2011). Organizational ethics: A literature review. *Nursing Ethics, 18*(3), 285–303. doi:10.1177/0969733011401123 PMID:21558106

Sung, Y., Lee, J. A., Kim, E., & Choi, S. M. (2016). Why we post selfies: Understanding motivations for posting pictures of oneself. *Personality and Individual Differences, 97*, 260–265. doi:10.1016/j.paid.2016.03.032

Svenaeus, F. (2019). A Defense of Phenomenological Account of Health and Disease. *The Journal of Medicine and Philosophy, 44*(4), 459–478. doi:10.1093/jmp/jhz013 PMID:31356662

Svensson, G., & Wood, G. (2003). The dynamics of business ethics: A function of time and culture–cases and models. *Management Decision, 41*(4), 350–361. doi:10.1108/00251740310468195

Svensson, G., Wood, G., & Callaghan, M. (2010). A corporate model of sustainable business practices: An ethical perspective. *Journal of World Business, 45*(4), 336–345. doi:10.1016/j.jwb.2009.08.005

Szalkowski, A., & Jankowicz, D. (1999). The ethical problems of personnel management in a transition economy. *International Journal of Social Economics, 26*(12), 1418–1427. doi:10.1108/03068299910248513

Tahir, T. B. M., & Husin, M. H. B. (2017). Online social media and risks: An exploration into existing children practice. In *2017 International Conference on Electrical Engineering and Informatics (ICELTICs)* (pp. 195-200). IEEE. 10.1109/ICELTICS.2017.8253250

Tailor, A., & Ogden, J. (2009). Avoiding the term 'obesity': An experimental study of the impact of doctors' language on patients' beliefs. *Patient Education and Counseling, 76*, 260–264.

Tainter, J. A. (2006). Social complexity and sustainability. *Ecological Complexity, 3*(2), 91–103. doi:10.1016/j.ecocom.2005.07.004

Takahashi, M., Fujimoto, M., & Yamasaki, N. (2003, November). The active lurker: influence of an in-house online community on its outside environment. In *Proceedings of the 2003 international ACM SIGGROUP conference on Supporting group work* (pp. 1-10). 10.1145/958160.958162

Tanneberger, A., & Ciupitu-Plath, C. (2018). Nurses' weight bias in caring for obese patients: Do weight controllability beliefs ınfluence the provision of care to obese patients? *Clinical Nursing Research, 27*(4), 414–432.

Tanrısever, C., Pamukçu, H., & Batman, O. (2016). New Tourism Trends in the World and Their Adaptations to Turkey. *Gümüshane University Electronic Journal of the Institute of Social Science/Gümüshane Üniversitesi Sosyal Bilimler Enstitüsü Elektronik Dergisi, 7*(16).

Taş, H. Y., Demirdöğmez, M., & Küçükoğlu, M. (2017). Possible effects of Generation Z, our future, on working life. *OPUS International Journal of Society Research, 7*(13), 1031–1048.

Taylor, P. W. (1981). The Ethics of Respect for Nature. *Environmental Ethics, 3*(3), 197–218. doi:10.5840/enviroethics19813321

Taylor, P. W. (1986). *Respect for Nature: A Theory of Environmental Ethics.* Princeton University Press.

Tekin, Ö. A. (2014). *İslami Turizm.* Türkiye'deki ve Dünyadaki Genel Durumu Üzerine Bir.

Tekin, Ö. A. (2014). İslami Turizm: Dünyadaki ve Türkiye'deki Genel Durum Üzerine Bir İnceleme. *Journal of International Social Research, 7*(29).

Teo, T. S. (2002). Attitudes toward online shopping and the Internet. *Behaviour & Information Technology, 21*(4), 259–271. doi:10.1080/0144929021000018342

The World Commission on Environment and Development. (1987). *Our Common Future.* Oxford University Press.

Third, A., Bellerose, D., Dawkins, U., Keltie, E., & Pihl, K. (2014). *Children's rights in the digital age: A download from children around the world.* Academic Press.

Thiroux, J. P., & Krasemann, K. W. (1980). *Ethics: Theory and practice.* Glencoe Publishing Company.

Thompson, D. F. (2018). Political Ethics. In H. Lafollette (Ed.), *International Encyclopaedia of Ethics.* Wiley Blackwell.

TikTok.com. (2020). Retrieved from https://www.tiktok.com/about?lang=en

Togelius, J. (2011). A procedural critique of deontological reasoning. *DiGRA '11: Proceedings of the 2011 DiGRA International Conference: Think Design Play, 6.* http://www.digra.org/digital-library/publications/a-procedural-critique-of-deontological-reasoning/

Tomažič, T. (2016). The Growth of Covert Advertising in the Context of Local Self-Government. *Lex localis - Journal of Local Self-Government, 14*(4), 739–758.

Tomažič, T., & Jurišić, J. (2009). Covert advertising in the context of media ethics on the example of Slovenian press. *Medijski dijalozi, 2*(4), 57-67.

Tomažič, T., Boras, D., Jurišić, J., & Lesjak, D. (2014). Covert Advertising in the Context of Inadmissible Presentation of Information. *Industrial Management & Data Systems, 114*(1), 107–122. doi:10.1108/IMDS-04-2013-0204

Tomkiewicz, J. (2017). The Social Consequences of Income Distribution—Are Increasing Company Profits a Threat to Economic and Social Stability? *Polish Sociological Review, 2*(198), 203-218.

Torlak, Ö., Özdemir, Ş., & Erdemir, E. (2008). *İş Ahlakı Raporu.* İGİAD.

Treise, D., Weigold, M. F., Conna, J., & Garrison, H. (1994). Ethics in advertising: Ideological correlates of consumer perceptions. *Journal of Advertising, 23*(3), 59–69. doi:10.1080/00913367.1994.10673450

Trevino, L. K., & Nelson, K. A. (2016). *Managing business ethics: Straight talk about how to do it right*. John Wiley & Sons.

Trevino, L. K., & Weaver, G. R. (1994). Business ETHICS/BUSINESS ethics: One field or two? *Business Ethics Quarterly*, *4*(2), 113–128. doi:10.2307/3857484

Türk Dil Kurumu (TDK). (n.d.). Retrieved from http://tdk.gov.tr/?option=com_gts

Türkiye Bilişim Vakfı. (2009). Retrieved from http:// www.tbv.org.tr/default.asp?ID=30)

Türkiye İstatistik Kurumu. (2017). Retrieved from https://www.tuik.gov.tr/basinOdasi/haberler/2017_31_20170607.pdf

Türkkan, L. (2015). *Örgüt Kültürünün Örgütsel Bağlılığa Etkileri*. Yüksek Lisans Tezi, İstanbul Üniversitesi, Sosyal Bilimler Enstitüsü, Hastane ve Sağlık Kuruluşlarında Yönetim Bilim Dalı.

Turkle, S. (1995). *Life on the Screen: Identity in the Age of the Internet*. Simon and Schuster.

Tutan, A. A. (2006). İşletmelerde iş etiğinin değerlendirilmesi ve bir kamu kurumunda uygulanması, yayınlanmış yüksek lisans tezi. Selçuk Üniversitesi Fen Bilimleri Enstitüsü, Konya.

Tutar, H., & Yılmaz, M. K. (2003). *Genel İletişim Kavramlar ve Modeller* (5th ed.). Nobel Publications.

Tuten, T. L., & Bosnjak, M. (2001). Understanding differences in web usage: The role of need for cognition and the five factor model of personality. *Social Behavior and Personality*, *29*(4), 391–398. doi:10.2224bp.2001.29.4.391

Tuten, T. L., & Solomon, M. R. (2017). Social media marketing. *Sage (Atlanta, Ga.)*.

Tylka, T. L., Annunziato, R. A., Burgard, D., Daníelsdóttir, S., Shuman, E., Davis, C., & Calogero, R. M. (2014). The weight-inclusive versus weight-normative approach to health: Evaluating the evidence for prioritizing well-being over weight loss. *Journal of Obesity*, 18.

U.S. Department of State. (2013). Trafficking in Persons Report. U.S. Department of State.

UCS. (n.d.). *Union of Concerned Scientists Report*. Retrieved from https://www.ucsusa.org/resources/each-countrys-share-co2-emissions

ulut, M. (Ed.). (2013). *Gençlik ve sosyal medya araştırma raporu*. TC Gençlik ve Spor Bakanlığı.

UN Women. (2014). *World Survey on The Role of Women in Development 2014*: *Gender Equality and Sustainable Development*. Retrieved from https://www.unwomen.org/en/digital-library/publications/2014/10/world-survey-2014#view

UN. (1987). *Our Common Future*. Retrieved from https://sustainabledevelopment.un.org/content/documents/5987our-common-future.pdf

UN. (2007). *United Nations Declaration on The Rights of Indigenous Peoples*. Retrieved from https://www.un.org/development/desa/indigenouspeoples/declaration-on-the-rights-of-indigenous-peoples.html

UN. (2015). *Transforming Our World: The 2030 Agenda for Sustainable Development*. Retrieved from https://sustainabledevelopment.un.org/post2015/transformingourworld

UN. (2019). *World Population Prospects 2019*. Retrieved from https://population.un.org/wpp/

UNDP. (2014). *Environmental Justice. Comparative Experiences in Legal Empowerment*. Retrieved from https://www.undp.org/content/undp/en/home/librarypage/democratic-governance/access_to_justiceandruleoflaw/environmental-justice---comparative-experiences.html

UNDP. (2019). *Human Development Data (1990-2018)*. Retrieved from http://hdr.undp.org/en/data

UNESCO. (n.d.). *The Rio Declaration on Environment and Development*. Retrieved from http://www.unesco.org/education/pdf/RIO_E.PDF

UNICEF. (2017). *Risks and opportunities related to children's online practices: Ghana country report*. Author.

United Nations Children's Fund. (2015). *UNICEF Global Programme to Build Capacity to Tackle Online Sexual Exploitation*. Final Report to the Government of the United Kingdom of Great Britain and Northern Ireland, Reporting Period: 24 May 2015–30 March 2016, UNICEF Child Protection Section, Programme Division, New York.

United Nations Commission on Human Security. (2003). *Human Security Now*. United Nations.

United Nations Development Programme. (1994). *Human Development Report*. Oxford University Press.

United Nations. (1948). *United Nations' Universal Declaration of Human Rights of 1948 Articles 18 and 19 of the Universal Declaration of Human Rights in 2.2.2.*

Untila Kaplan, O. (2019). *Media literacy in a changing process in Turkey. 11th International Conference on Electronics, Computers and Artificial Intelligence (ECAI)*, Pitesti, Romania. 10.1109/ECAI46879.2019.9042089

UNWTO. (2019). *World Tourism Organization*. https://www.e-unwto.org/doi/pdf/10.18111/9789284421152

Ural, T. (2003). İşletme ve Pazarlama Etiği. Baskı, Ankara: Detay Yayıncılık.

Üsdiken, B., & Wasti, S. A. (2002). Türkiye'de akademik bir inceleme alanı olarak personel veya" insan kaynakları" yönetimi, 1972-1999. *Amme Idaresi Dergisi*, *35*(3), 1–37.

Usta, E., & Akyolcu, N. (2014). Cerrahi hemşirelerinin fazla kilolu/obez hasta bakımına ilişkin bilgi ve uygulamalarının değerlendirilmesi. *Florence Nightingale Hemşirelik Dergisi*, *22*, 1–7.

Uzun, R. (2007). *İletişim Etiği: Sorunlar ve Sorumluluklar*. Gazi Üniversitesi İletişim Fakültesi Basımevi.

Valentine, S., Godkin, L., Page, K., & Rittenburg, T. (2009). Gender and ethics: Ethical judgments, ethical intentions, and altruism among healthcare professionals. *Gender in Management*, *24*(2), 112–130. doi:10.1108/17542410910938808

Valkenburg, P. M., & Cantor, J. (2001). The development of a child into a consumer. *Journal of Applied Developmental Psychology*, *22*(1), 61–72. doi:10.1016/S0193-3973(00)00066-6

Valkenburg, P. M., & Peter, J. (2007). Who visits online dating sites? Exploring some characteristics of online daters. *Cyberpsychology & Behavior*, *10*(6), 849–852. doi:10.1089/cpb.2007.9941 PMID:18085977

Vallance, E., & Elizabeth, V. (1995). *Business ethics at work*. Cambridge University Press. doi:10.1017/CBO9781139166461

Van Dijck, J. (2009). Users like you? Theorizing agency in user-generated content. *Media Culture & Society*, *31*(1), 41–58. doi:10.1177/0163443708098245

Van Dijck, J. (2012). Facebook as a tool for producing sociality and connectivity. *Television & New Media*, *13*(2), 160–176. doi:10.1177/1527476411415291

Van Reijmersdal, E. A., Jansz, J., Peters, O., & Van Noort, G. (2010). The effects of interactive brand placements in online games on children's cognitive, affective, and conative brand responses. *Computers in Human Behavior*, *26*(6), 1787–1794. doi:10.1016/j.chb.2010.07.006

Van Reijmersdal, E. A., Rozendaal, E., Smink, N., Van Noort, G., & Buijzen, M. (2017). Processes and effects of targeted online advertising among children. *International Journal of Advertising, 36*(3), 396–414. doi:10.1080/02650487.2016.1196904

Van Reijmersdal, E. A., Tutaj, K., & Boerman, S. C. (2013). The effects of brand placement disclosures on scepticism and brand memory. *Communications, 38*(2), 127–146. doi:10.1515/commun-2013-0008

Van Vuuren, L. J., & Eiselen, R. J. (2006). A role for HR in corporate ethics? South African practitioners' perspectives. *SA Journal of Human Resource Management, 4*(3), 22–28. doi:10.4102ajhrm.v4i3.95

VanMeter, R. A., Grisaffe, D. B., & Chonko, L. B. (2015). Of "likes" and "pins": The effects of consumers' attachment to social media. *Journal of Interactive Marketing, 32*, 70–88. doi:10.1016/j.intmar.2015.09.001

Velasquez, M. G., & Velazquez, M. (2002). *Business ethics: Concepts and cases* (Vol. 111). Prentice Hall.

Venkat, A., & Larkin, G. L. (2014). Ethical Dilemmas Posed in the Care of Obese Patients in the Emergency Department. *Online Journal of Health Ethics / the University of Mississippi Medical Center, 10*(1), 5.

Venture, K. (2003). Pazarlama Araştırmaları Kapsamında Yaşanan Teknoloji-Tabanlı Değişim. *Ege Üniversitesi İktisadi ve İdari Bilimler Fakültesi*, 78-89.

Vermillion, L., Lassar, W., & Winsor, R. (2002). The Hunt-Vitell general theory of marketing ethics: Can it enhance our understanding of principal-agent relationships in channels of distribution? *Journal of Business Ethics, 41*(3), 267–285. doi:10.1023/A:1021284922440

Verstraeten, J. (1998). From business ethics to the vocation of business leaders to humanize the world of business. *Business Ethics (Oxford, England), 7*(2), 111–124. doi:10.1111/1467-8608.00096

Vickers, M. R. (2005). Business ethics and the HR role: Past, present, and future. *Human Resource Planning, 28*(1), 26–33.

Villiard, H., & Moreno, M. A. (2012). Fitness on facebook: Advertisements generated in response to profile content. *Cyberpsychology, Behavior, and Social Networking, 15*(10), 564–568. doi:10.1089/cyber.2011.0642 PMID:22963337

Vitell, S., & Hunt, S. (2006). The General Theory of Marketing Ethics: A Revision and Three Questions. *Journal of Macromarketing, 26*(2), 1–11.

Vogel, D. (1991). Business ethics: New perspectives on old problems. *California Management Review, 33*(4), 101–117. doi:10.2307/41166675

Waiguny, M. K., Nelson, M. R., & Terlutter, R. (2014). The relationship of persuasion knowledge, identification of commercial intent and persuasion outcomes in advergames—The role of media context and presence. *Journal of Consumer Policy, 37*(2), 257–277. doi:10.100710603-013-9227-z

Walker, K. (2000). "It's difficult to hide it": The presentation of self on Internet home pages. *Qualitative Sociology, 23*(1), 99–120. doi:10.1023/A:1005407717409

Walle, A. H. (1995). Business Ethics and Tourism: From Micro to Macro Perspectives. *Tourism Management, 16*(4), 263–268. doi:10.1016/0261-5177(95)00015-G

Walsh, A. J. (2007). HRM and the ethics of commodified work in a market economy. *Human resource management: Ethics and employment*, 102.

Walzer, M. (1973). Political Action: The Problem of Dirty Hands. *Philosophy & Public Affairs, 2*(2), 160-180. http://www.jstor.org/stable/2265139

Wang, J. L., Jackson, L. A., Zhang, D. J., & Su, Z. Q. (2012). The relationships among big five personality factors, self-esteem, narcissism, and sensation-seeking to Chinese university student's uses of Social Networking Sites(SNSs). *Computers in Human Behavior, 28*(6), 2313–2319. doi:10.1016/j.chb.2012.07.001

Wang, T. L. (2014). The usage behaviors, motivations and gratifications of using User-Generated Media: The case study of Taiwan's YouTube. *Advances in Journalism and Communication, 2*(04), 137–150. doi:10.4236/ajc.2014.24015

Warren, K. J. (1990). The Power and the Promise of Ecological Feminism. *Environmental Ethics, 12*(2), 125–146. doi:10.5840/enviroethics199012221

Warren, R., & Tweedale, G. (2002). Business ethics and business history: Neglected dimensions in management education. *British Journal of Management, 13*(3), 209–219. doi:10.1111/1467-8551.00238

WCEL. (n.d.). *IUCN World Declaration on the Environment Rule of Law.* Retrieved from https://www.iucn.org/commissions/world-commission-environmental-law/wcel-resources/environmental-rule-law

Weaver, G. R., & Trevino, L. K. (2001). The role of human resources in ethics/compliance management: A fairness perspective. *Human Resource Management Review, 11*(1-2), 113–134. doi:10.1016/S1053-4822(00)00043-7

Weiss, J. W. (1994). Business ethics: A managerial, stakeholder approach (No. s 287). Belmont, CA: Wadsworth Publishing Company.

Weiss, J. W. (2003). Business Ethics, A Stakeholder and Issues Management Approach. Thomson-South Western.

Weiss, J. (2008). *Business ethics: A stakeholder and issues management approach.* Cengage Learning.

Wellman, C. (1975). *Morals and ethics.* Scott, Foresman.

Wells, C. (2015). *Subjects between ethics and politics* (Unpublished doctoral dissertation). University of Vanderbilt, Nashville, TN.

Wells, D., & Schminke, M. (2001). Ethical development and human resources training: An integrative framework. *Human Resource Management Review, 11*(1-2), 135–158. doi:10.1016/S1053-4822(00)00044-9

Westin, A. F. (1968). Privacy And Freedom. *Washington and Lee Law Review, 25*(1), 166–170.

White, G. W. (2001). Business ethics. *Journal of Business & Finance Librarianship, 6*(4), 49–49. doi:10.1300/J109v06n04_05

Whitehouse, M. (2017). Financial analysts and their role in financial communication and investor relations. In A. V. Laskin (Ed.), *Handbook of financial communication and investor relations* (pp. 117–126). Wiley. doi:10.1002/9781119240822.ch10

White, L. (1967). The Historical Roots of Our Ecological Crisis. *Science, 155*(3767), 1203–1207. doi:10.1126cience.155.3767.1203 PMID:17847526

Whiteside, D. B., & Barclay, L. J. (in press). When wanting to be fair isn't enough: The effects of depletion and self-appraisal gaps on fair behavior. *Journal of Management.*

Whiting, A., & Williams, D. (2013). Why people use social media: A uses and gratifications approach. *Qualitative Market Research, 16*(4), 362–369. doi:10.1108/QMR-06-2013-0041

Why money and muscle still rule in Indian politics: an interview with Milan Vaishnav. (2017, January 16). *Reuters.* Retrieved from https://in.reuters.com/article/milan-vaishnav-book-politics-india-idINKBN1500B

Wikipedia.com. (2020). Retrieved from https://tr.wikipedia.org/wiki/Trol

Wiley, C. (1998). Reexaminating perceived ethics issues and ethics roles among employment managers. *Journal of Business Ethics*, *17*(2), 147–161. doi:10.1023/A:1005795731002

Wilson, V. J., McCormack, B. G., & Ives, G. (2005). Understanding the workplace culture of a special care nursery. *Journal of Advanced Nursing*, *50*(1), 27–38. doi:10.1111/j.1365-2648.2004.03346.x PMID:15788063

Wind, J., & Mahajan, V. (2001). *Digital MarketingGlobal Strategies from the World's Leading Experts*. Wiley.

Winstanley, D., Woodall, J., & Heery, E. (1996a). The agenda for ethics in human resource management. *Business Ethics: A European Review*, *5*(4), 187-194.

Winstanley, D., & Hartog, M. (2002). Ethics and human resource management: Introduction. *Business Ethics (Oxford, England)*, *11*(3), 200–201. doi:10.1111/1467-8608.00277

Winstanley, D., & Woodall, J. (2000). The ethical dimension of human resource management. *Human Resource Management Journal*, *10*(2), 5–20. doi:10.1111/j.1748-8583.2000.tb00017.x

Winstanley, D., Woodall, J., & Heery, E. (1996b). Business ethics and human resource management: Themes and issues. *Personnel Review*, *25*(6), 5–12. doi:10.1108/00483489610148491

Wittgenstein, L. (1965). A lecture on ethics. *The Philosophical Review*, *74*(1), 3–12. doi:10.2307/2183526

Woodall, J., & Winstanley, D. (Eds.). (2000). *Ethical issues in contemporary human resource management*. Macmillan Business.

Wooten, K. C. (2001). Ethical dilemmas in human resource management: An application of a multidimensional framework, a unifying taxonomy, and applicable codes. *Human Resource Management Review*, *11*(1-2), 159–175. doi:10.1016/S1053-4822(00)00045-0

World Inequality Lab. (2018). *World Inequality Report*. Author.

Wott, C. B., & Carels, R. A. (2010). Overt weight stigma, psychological distress and weight loss treatment outcomes. *Journal of Health Psychology*, *15*(4), 608–614.

Wray-Bliss, E. (2009). Ethics. In The Oxford handbook of critical management studies. OUP.

Wright, P., Friestad, M., & Boush, D. (2005). The development of marketplace persuasion knowledge in children, adolescents and young adults. *Journal of Public Policy & Marketing*, *24*(2), 222–233. doi:10.1509/jppm.2005.24.2.222

Wu, K.-W., Huang, S. Y., Yen, D. C., & Popova, I. (2012). The effect of online privacy policy on consumer privacy concern and trust. *Computers in Human Behavior*, *28*(3), 889–897. doi:10.1016/j.chb.2011.12.008

Wuryasti, F. (2013). *Halal Tourism, a New Concept of Tourism Activities in Indonesia*. Available from: http://www.travel.detik.com/read/201 3/10/30/152010/2399509/1382/

Yamaguchi, S. (2004). Nursing culture of an operating theater in Italy. *Nursing & Health Sciences*, *6*(4), 261–269. doi:10.1111/j.1442-2018.2004.00199.x PMID:15507046

Yarcan, Ş. (2007). A Conceptual Evaluation of Code of Ethics for Professional Tourist Guides. *Anatolia: Turizm Araştırmaları Dergisi*, *18*(1), 33–44.

Yaşar, İ. H. (2020). *Sosyal Medya ve Siyaset*. Orion Academy.

Yaylagül, L. (2013). *Kitle İletişim Kuramları*. Dipnot Publications.

Yegen, C. (2018). Doğru Haber Alma Hakkı ve Sosyal Medya Dezenformasyonunu Doğruluk Payı ve Yalansavar ile Tartışmak. *Erciyes İletişim Dergisi, 5*(4), 101-121.

Yeh, O. (2019). TikTok surpasses one billion installs on the app store and google play. *Sensor Tower*. Retrieved from https://sensortower.com/blog/tiktok-downloads-one-billion

Yeter, D. (2009). İş Etiği Kavramı Ve İşletmeler Açısından Önemi. *İller Bankası Genel Md. UİB AB Ve Dış İl. MD. paribus. tr. googlepages. com/y_demir2. doc. Son Erişim Tarihi, 2*, 2010.

Yıldırım, G., & Kadıoğlu, S. (2007). Basic Concepts of Ethics and Medical Ethics, C.U. *Journal of the Faculty of Medicine, 29*(2), 7–12.

Yıldırım, G., & Kadıoğlu, S. (2007). Etik ve tıp etiği temel kavramları. *CÜ Tıp Fakültesi Dergisi, 29*(2), 7–12.

Yildiz, A. (2007). *KOBİ'lerde İş Ahlakı: Adapazarı Örneği. In Yayınlanmış Yüksek Lisans Tezi, Sakarya Üniversitesi SBE.* Sakarya.

Yilmaz, G. (2006). *Örgüt Kültürü ve İnsan Kaynakları Yönetimi.* Yayınlanmış Yüksek Lisans Tezi, Marmara Üniversitesi SBE.

Yoon, Y., Giirhan-Canli, Z., & Schwarz, N. (2006). The Effect of Corporate Social Responsibility (CSR) Activities on Companies with Bad Reputations. *Journal of Consumer Psychology, 16*(4), 377–390. doi:10.120715327663jcp1604_9

Young, K. S. (1999). Internet addiction: Symptoms, evaluation and treatment. *Innov. Clin. Pract., 17*(17), 351–352.

Zapata, C. P., Olsen, J. E., & Martins, L. L. (2013). Social exchange from the supervisor's perspective: Employee trustworthiness as a predictor of interpersonal and informational justice. *Organizational Behavior and Human Decision Processes, 121*(1), 1–12. doi:10.1016/j.obhdp.2012.11.001

Zentes, J. (2017). *Strategic Retail Management.* Springer Fachmedien. doi:10.1007/978-3-658-10183-1

Zey-Ferrell, M., & Ferrell, O. C. (1982). Role-set configuration and opportunity as predictors of unethical behavior in organizations. *Human Relations, 35*(7), 587–604. doi:10.1177/001872678203500707

Zhang, X., Wu, Y., & Liu, S. (2019). Exploring short-form video application addiction: Socio-technical and attachment perspectives. *Telematics and Informatics, 42*, 101243. doi:10.1016/j.tele.2019.101243

Zhao, G., Chen, Y., & Brockner, J. (2015). What influences managers' procedural fairness towards their subordinates? The role of subordinates' trustworthiness. *Journal of Experimental Social Psychology, 59*, 96–112. doi:10.1016/j.jesp.2015.04.002

Ziller, C. (2017). Equal Treatment Regulations and Ethnic Minority Social Trust. *European Sociological Review, 0*(4), 0, 1–13. doi:10.1093/esr/jcx059

Zulkharnain, A., & Jamal, S. A. (2012). Muslim Guest Perception of Value towards Syariah Concept Hotel. In Current Issues in Hospitality and Tourism, Research and Innovations (pp. 337-340). Taylor & Francis Group.

Zulküfli, W. S. W., Rahman, S. A., Awang, K. W., & Man, Y. B. C. (2011). Developing the Framework for Halal Friendly Tourism in Malaysia. *International Business Management, 5*(6), 295–302.

About the Contributors

Meliha Nurdan Taskiran was born in Istanbul, Turkiye. She worked for many universities, such as Marmara, Kocaeli, 19 Mayıs, till 2016. Now, she is an emeritus professor working for Istanbul Medipol University, School of Communication since 2017. She has articles and books on visual communication, visual semiotics, discourse analyses, film analyses, semiological studies on films, tv serials and advertisements. She likes reading books on literacy and visual rhetorics. She has been giving graduate and postgraduate level courses on Media Literacy, Advertising, Visual literacy and Culture. She is also interested in animal communication and edutainment. She is married with two children.

Fatih Pinarbasi is a research assistant at Istanbul Medipol University and a PhD candidate at Yildiz Technical University. His major are business administration and marketing. His research areas are digital marketing and data mining. In his professional career, he has worked for British Standards Institution for one year as a digital marketing expert. He has been working for Istanbul Medipol University for 3 years.

* * *

Zafer Adiguzel, in 2006, Kocaeli University, Faculty of Economics and Administrative Sciences, In 2010, graduated from Cardiff Metropolitan University with an MBA in Finance. In 2016 graduated from Gebze Technical University, Department of Management and Organization. Continues academic career as a faculty member at Istanbul Medipol University, Medipol Business School.

Zuhal Akmeşe completed her undergraduate education in the Department of Social Sciences Teaching - Afyon Kocatepe University in 2005, and also in the Department of Radio, Television and Cinema – Ege University Faculty of Communication in 2010, as the highest ranked student. After having been granted with the master's degree in the Department of Radio, Television and Cinema – Ege University Social Sciences Institute, she received her PhD in the department of Radio Television and Cinema – Istanbul University Social Sciences Institute. She published many researches with coauthors in the field of communication, along with the presentation given in international congresses & symposiums, and a number of movies she directed. Moreover, documentaries of Zuhal Akmeşe, which are named as "Tenur (2010), Beş Kuruş (2010), Son Kale (2011), Her Yer Hiç Kimse (2012) were presented in many national and international festivals. Maintaining to present her theoretical and practical studies in the fields of photography, documentary cinematography and communication, Zuhal Akmeşe has been an academic member in Dicle University, Communication Faculty Department of Radio, Television and Cinema since 2010.

Özlem Arda graduated from Istanbul University Faculty of Communication Department of Radio Television and Cinema in 2004 and also Faculty of Hasan Âli Yücel Education Faculty English Language Teaching Department in 2005. After master degree Özlem Arda has gained her doctorate degree. One of her documentaries named "Abkhazians in Anatolia – Soul People" (2004) has been awarded as the Best Documentary by Safranbolu International 5th Golden Saffron Documentary Film Festival (3rd Place) and also as the Best Documentary by Columbia Tri-Star 1st National Short Film Festival (1st Place). She has been working for Connectist as an international journal since 2010 as co-editor. Also she is an Associate Professor at Istanbul University Facult of Communication since 2010. She has been teaching documentary as Interactive Documentaries Film and Documentary Film courses for undergraduate and graduate students for years.

Josep M. Basart is graduated in Computer Science (1984) and in Philosophy (1998) from the Universitat Autònoma de Barcelona (UAB). He received his PhD in Computer Science (1988) by the UAB. Since 1990 he is associate professor at the Engineering School (UAB) where he has taught Engineering Ethics and several subjects in Applied Mathematics and Computer Science. He has published more than 30 papers in reviewed publications. He has collaborated in 12 funded research projects and has been reviewer for some international journals. His present areas of interest are Professional Ethics, Applied Ethics, and Social Implications of Information and Communication Technologies.

Elif Baykal is an Assoc. Prof. at Istanbul Meidpol University. She graduated from Bosphorus University in 2006. She has an MBA degree from Istanbul Medipol University and a PhD from Yıldız Technical University. She is teaching Management and Organisation. She has several works about Leadership, Family Firms, Positive Organizational Behavior and Workplace Spirituality.

Bal Krishna Chaturvedi is a former Cabinet Secretary and Head of Civil Service of India with more than four decades of experience in making public policy. He has during this period interacted with politicians both at national and global level and seen ethical issues in political life from very close quarters He was member of planning commission (now Niti Ayog), has authored two books and written extensively on issues of corruption, governance, energy and social sector. He was a member of board of UNICEF and UNDP.

Meryem Türkan Işik is Assistant Professor, Mersin University, Faculty of Nursing, Fundamental Nursing Department, Turkey.

Tansu Işıkay received her bachelor's degree from Yıldız Technical University, Faculty of Economics and Administrative Sciences, Department of Business Administration in 2015. She received her Master's degree from Marmara University Institute of Social Sciences Marketing Department in 2017. She continues her doctorate studies in Yıldız Technical University, Department of Business Management. She has been working as a Research Assistant in Business Administration at Doğuş University since 2019.

Miguel Kottow is a physician, ophthalmologist, bioethicists, Doctor of Medicine, MA Sociology, Full Professor Universidad de Chile. Author of numerous papers, book chapters and books.

Tommi Lehtonen is the Director of Responsibility and Ethics at the University of Vaasa, Finland. He specializes in ethics, social philosophy, and philosophy of science, and has extensive experience in multi- and interdisciplinary research both in Finland and abroad. His expertise areas include the role of attitudes in decision-making and choice behaviour, values and ethics of management, and sustainability in economic, social and cultural contexts. Lehtonen is the editor of Perspectives on Culture, Values, and Justice (Cambridge Scholars Publishing, 2015) and the author of After Secularization (Magnus Publications, 2012). He has also published widely in philosophical and social scientific journals.

Kevser Zeynep Meral is completing her Ph.D. in Advertisement and Public Relations at Bahçeşehir University, Istanbul. She has completed Boğaziçi University, Master Program of European Studies after graduating from Boğaziçi University Department of Philosophy with TÜBİTAK scholarship (The Scientific and Technological Research Council of Turkey) of Undergraduate Basic Science Scholarship granted to students among the first 2500 in the General University Entrance Examination). She is fluent in English (TOEFL 105/IBT), French (Dame De Sion French High School graduate) and German. She has two years experience in Digital Media Services, as Social Media Specialist.

Sağbetullah Meriç completed his Master's Degree in Tourism and Hotel Management in Düzce University in 2013. He was hired as an academic at Yüzüncü Yıl University in 2013. He completed his PhD at Atatürk University in 2017. He is still working as an Assistant Professor at Yüzüncü Yıl University. ORCID: https://orcid.org/0000-0003-1949-623X.

Bernard E. Rollin is Professor of Philosophy, Professor of Animal Sciences, and Professor of Biomedical Sciences at Colorado State University, where he is also a University Distinguished Professor. He earned his B.A. in Philosophy at City College of New York, and his Ph.D. in Philosophy at Columbia University.

Tina Tomažič was born in Maribor (Slovenia). She graduated from the Faculty of Economics and Business and earned, in the Marketing field an MSc in Economics and Business Sciences. She became a PhD at the Philosophical Faculty in the field of Information and Communication Science. She is working as an Assistant Professor at the Institute of Media Communications, at the Faculty of Electrical Engineering and Computer Science, University of Maribor in Slovenia. During her work she has participated actively in scientific conferences in the country and abroad, and has published numerous books and articles, which appeared in significant journals (such as Industrial Management & Data Systems, Journal of Criminal Investigation and Criminology, Lex localis, Philosophy, ethics, and humanities in medicine, and many more). She participated among others in a targeted research project on the impact of digital technologies Slovenian AV production in the light of Europe 2020. In the academic year 2007/08 she has performed at the Croatian Studies, University of Zagreb, Department of Communication in Croatia.

Katja Udir Mišič (Ph.D., University of Ljubljana, 2014) completed her undergraduate degree on design at the Academy of fine arts, University of Ljubljana and MA and PhD studies in Marketing Communication at the Faculty of Social Sciences, University of Ljubljana. She is a teaching assistant at the Faculty of Electrical Engineering and Computer Science at the University of Maribor, Institute of Media Communications since 2003. Her main research interests lie in the fields of City Marketing,

Graphic marketing, Visual communication and Graphic design. Prior her to employment at the University of Maribor she collaborated as a designer with various advertising agencies and architecture offices.

Gülay Yildirim is Associate Professor, Sivas Cumhuriyet Üniversty, Department of History and Ethics, Sivas-Turkey.

Neslihan Yilmaz completed his undergraduate education at Eskişehir Anadolu University, Department of Business Administration. He completed his master's degree in Konya Selçuk University, Department of Management and Organization Science. He worked in the education sector for 5 years. He continued his career as Quality Management Specialist at Ar-Gev Project Consultancy. After completing the necessary training processes, he worked as chief auditor. He worked as a brand, certificate and project specialist. In addition, "Management Development", Strategic Planning "," Project Management "," Business Building and Development "" OHSAS 18001 "," OHSAS 18001 Occupational Health and Safety Management Systems "," Applied Entrepreneurship Training "," Lead Auditor " He also worked as ISO 9001: 2000 Quality Management System Internal Auditor ", 9001 ISO 9001: 2000 Quality Management System Basic-Documentation Training", "Foreign Trade" certificates and Applied Entrepreneurship Trainer and gave "sales and marketing". KOSGEB Support and Development Agency supported more than 60 projects accepted in the Financial Support Program. Since 2015, she has been teaching Entrepreneurship, Accounting and Behavioral Sciences at Medipol University. Technology Transfer Office. Digital Transformation: Reflection of the Mind to Digitality - Information Literacy, Digital Information, Strategy is the first chapter writing work.

Index

E

F

G

H

I

J

K

M

N

O

P

R

S

www.ingramcontent.com/pod-product-compliance
Lightning Source LLC
LaVergne TN
LVHW062119300926
849161LV00010B/2494
* 9 7 8 1 7 9 9 8 4 1 1 7 3 *